# Pesticides - Updates on Toxicity, Efficacy and Risk Assessment

*Edited by Marcelo L. Larramendy and Sonia Soloneski*

Published in London, United Kingdom

Pesticides - Updates on Toxicity, Efficacy and Risk Assessment
http://dx.doi.org/10.5772/intechopen.98116
Edited by Marcelo L. Larramendy and Sonia Soloneski

Contributors
Basana Gowda, Sophia Subhadarsini Pradhana, Totan Adak, Govindharaj Guru-Pirasanna-Pandi, Naveenkumar B. Patil, Mahendiran Annamalai, Prakash Chandra Rath, Parvaneh Maghami, Aida Doroudian, Mahdieh Emadi, Reza Hosseinzadeh, Kobun Rovina, Nasir Md Nur 'Aqilah, Kana Husna Erna, Joseph Vonnie Merrylin, Adriana Bender Moreira de Lacerda, Patricia Arruda de Souza Alcarás, Maria Alves Cristina Alves Corazza, Bianca Simone Zeigelboim, Adrian Fuente, Joseph O. Lalah, Peter O. Otieno, Joanne A. Ogunah, Zedekiah Odira, Motunrayo Ganiyat Akande, Solomon Usman Abraham, Johnson Caleb Ogunnubi, Dele Omoyele Omoyele Adeniyi, Gerardo Cebrián-Torrejón, Ahissan Innocent Adou, Garrick Bibian, Odile Bordelais, Léa Farouil, Muriel Sylvestre, Sarra Gaspard, Marie-Noëlle Sylvestre, S. Marimuthu, P. Pavithran, G. Gowtham, Rahul R. Bakade, S. Sundaresha, Mehi Lal, Nayereh Hamedi, Miroslava Mihailova Hristova-Cherbadzhi, Alaa Eldin Bayoumi



First published in London, United Kingdom, 2022 by IntechOpen
IntechOpen is the global imprint of INTECHOPEN LIMITED, registered in England and Wales, registration number: 11086078, 5 Princes Gate Court, London, SW7 2QJ, United Kingdom

British Library Cataloguing-in-Publication Data
A catalogue record for this book is available from the British Library

Additional hard and PDF copies can be obtained from orders@intechopen.com

Pesticides - Updates on Toxicity, Efficacy and Risk Assessment
Edited by Marcelo L. Larramendy and Sonia Soloneski
p. cm.
Print ISBN 978-1-80356-038-0
Online ISBN 978-1-80356-039-7
eBook (PDF) ISBN 978-1-80356-040-3

For EU product safety concerns:
IN TECH d.o.o., Prolaz Marije Krucifikse Kozulić 3, 51000 Rijeka, Croatia,
info@intechopen.com or visit our website at intechopen.com.

# Meet the editors

Marcelo L. Larramendy, Ph.D., is a Professor of Molecular Cell Biology at the School of Natural Sciences and Museum, and head of the Molecular Cytogenetics and Genotoxicology Laboratory at the National University of La Plata, Argentina. He is a senior researcher at the National Scientific and Technological Research Council of Argentina. He has previously been a member of the Executive Committee of the Latin American Association of Environmental Mutagenesis, Teratogenesis and Carcinogenesis. He has over 460 publications to his name, including scientific publications, research communications and conferences worldwide. A recipient of several national and international awards, Prof. Larramendy is a regular lecturer at the international A. Hollaender courses organized by the IAEMS, and a former guest scientist at NIH (USA) and the University of Helsinki, (Finland). His particular area of expertise is genetic toxicology and he has acted as a referee for more than 30 international scientific journals. In 2015 he was a member of the International Panel of Experts at the International Agency for Research on Cancer (IARC, WHO, Lyon, France) for the evaluation of DDT, 2.4-D and Lindane.

Sonia Soloneski, Ph.D., is an Assistant Professor of Molecular Cell Biology at the School of Natural Sciences and Museum of the National University of La Plata, Argentina. She is a member of the National Scientific and Technological Research Council (CONICET) of Argentina, the Latin American Association of Environmental Mutagenesis, Teratogenesis and Carcinogenesis (ALAMCTA), the Argentinean Society of Biology (SAB) and the Society of Environmental Toxicology and Chemistry (SETAC). Her particular research interests are genetic toxicology, mutagenesis and ecotoxicology. She has authored more than 400 contributions in the field, including scientific publications in peer-reviewed journals and research communications. She has served as a review member for more than 35 international scientific journals. She has been a plenary speaker at scientific conferences and a member of scientific committees.

# Contents

# Preface

The widespread use of pesticides worldwide in agricultural production and for the control of disease vectors is an indicator of the cultivation status of different countries. With the growth in the global population and the concomitant intensification of agricultural production, pesticide overuse and contamination have also increased. The pesticide industry produces more than 800 active ingredients for the manufacture of over 40,000 commercial formulations used in the agricultural sector. Understanding of the unsafe usage of pesticides is increasing as the deleterious consequences for living species and the environment appear following occupational, accidental or incidental exposure to different types of pesticides. Environmental risk assessment associated with the production, storage and usage of pesticides has an important part to play in regulatory decisions. Existing and new compounds should be continuously evaluated to guarantee appropriate safety standards.

This book contains valuable information on diverse pesticides encountered in both anthropogenic and natural environments and covers the toxicity, efficacy and risk assessment of several compounds that can negatively influence the health of living species and ecosystems.

The book begins with a chapter on weed control strategies, describing innovative herbicide encapsulation methodologies with polymeric materials. Chapter 2 examines the use of several forms of chemical pesticide and their effects on sunflower cultivation and sunflower breeding. Strategies described include obtaining new genetic variability hybrids and sunflower varieties resistant to pests and chemicals. Chapter 3 describes strategies for the management of fungicidal resistance in potato pathogens, and discusses alternative cultural practices, the use of bio-agents and green chemicals, and the elimination of disease sources. Chapter 4 categorizes biomimetic and hemisynthetic pesticides according to their effect on one or more biological systems, including pesticides that target the nervous system, endocrine system, digestive system and various cellular structures in several insect groups. Chapter 5 considers pesticide toxicity in Tetranychidae mites (Chelicerata, Arachnida), and how sub-lethal concentrations of pesticides can significantly reduce the population growth, life expectancy and predation habits of Phytoseiid mites. The chapter also discusses the best strategies in integrated programmes for the biological control of pest populations. Chapter 6 reviews pesticides employed in the care of pets to control unwanted organisms such as ticks, fleas, and mites. The chapter highlights the advantages of applying pesticides as well as the harmful side effects that may be inflicted on animal companions. Chapter 7 describes the effect of simultaneous exposure to noise and pesticides on the hearing and vestibular systems of endemic disease control agents who worked on public health programmes. The results showed that some areas of the peripheral and central hearing system, as well as the peripheral vestibular system, are likely to be affected by exposure. Chapter 8 discusses the molecular effects of pesticides on haematological biomacromolecules, especially haemoglobin, as the target of organophosphates, carbamates, pyrethroids and neonicotinoids, among others. Chapter 9 examines the relationship between the use of banned pesticides and the incidence of health conditions that affect people's quality of life in health, social and/or labour contexts in developing countries. Chapter 10 describes the toxicity analysis of active ingredients found in open markets in North Central Nigeria, showing how consumers used common active ingredients that were not approved for agricultural use in the area. Chapter 11 presents aspects of the chemistry, manufacture, import and regulation of pesticides in Kenya, as well as their usage and negative impacts on the environment. The chapter describes various categories of chemicals such as organochlorine, organophosphate, carbamate, pyrethroid and neonicotinoid insecticides, as well as fungicides, herbicides and biopesticides routinely used in Oriental African countries. Chapter 12 focuses on the application of nanotechnology and current methodologies for the efficient and rapid detection of residual carbamate pesticides in fruits and vegetables. Sensitive and selective pesticide detection techniques to facilitate detection without the need for complex equipment are discussed. The final chapter discusses the major exposure routes of pesticides in water bodies, mainly from agricultural sectors, and their effect on the ecosystem. Decontamination techniques to eliminate pesticide contaminants associated with adverse effects on humans and other life forms are described.

The editors of Pesticides - Updates on Toxicity, Efficacy and Risk Assessment are enormously grateful to all the authors who have contributed to this book. The contributions made by specialists in this field of research are gratefully acknowledged. We hope that the information presented here will continue to meet the expectations and needs of all those interested in the different aspects of pesticides.

The publication of this book is of great importance to those scientists, pharmacologists, physicians and veterinarians, as well as engineers, teachers, graduate students and administrators of environmental programmes, who are employing different investigations to understand both basic and applied aspects of the use and misuse of pesticides.

**Sonia Soloneski, Ph.D. and Marcelo L. Larramendy, Ph.D.**
School of Natural Sciences and Museum,
National University of La Plata,
La Plata, Argentina

Chapter 1

# Polymeric Systems for the Delivery of Herbicides to Improve Weed Control Efficiency

*S. Marimuthu, P. Pavithran and G. Gowtham*


## Abstract

Weeds are unwanted plants, which interfere with the crop production. Weeds compete with crops for resources, causing severe yield loss. Chemical weed control through herbicides is a quite effective and reliable strategy to manage weeds. Herbicides constitute a major share of the global pesticide market. However, the applied herbicides undergo losses in the agroecosystem in different ways (chemical degradation, microbial decomposition, photo-degradation, leaching, run-off, and volatilization), thus lowering the herbicidal action coupled with contaminating ecosystem and groundwater. Encapsulation of herbicides is an innovative approach that addresses issues associated with the application of herbicides for controlling weeds. Encapsulation represents the embedding of an active ingredient in shell of polymeric material to achieve the controlled release of the active ingredient at the desired rate. The encapsulation of herbicides enhances stability, solubility, and bioactivity and alters the release pattern of herbicide resulting in improved weed control efficiency. Further, encapsulation lowers the application rate of herbicides, which in turn reduces the residue carryover of herbicide in soil and minimizes the environmental hazards. Therefore, encapsulated herbicide formulation has greater significance in the future weed management and will become ground-breaking technology in the chemical era of weed control.


**Keywords:** weeds, herbicides, polymers, encapsulation, weed control

## 1. Introduction

Weeds are as old as agriculture that influence crop growth and yield. Weeds compete with crops for resources such as space, water, nutrients, and light, which indirectly affect crop growth. Weeds inflicted tremendous yield loss besides deteriorating crop and grain quality, chocking water flow in an irrigation channel, sheltering crop pests, and causing ailments in living beings. The estimate on yield reduction due to weeds was one-third among agricultural pests [1]. The average loss in agricultural production due to weed infestation were 5, 10, and 25% in most developed, developing, and least developing nations, respectively [2]. Weeds cause a yield reduction of 10–80% depending on crops, type of weed flora infested, and magnitude of crop-weed competition [3]. Generally, yield loss due to weeds is 37% in developing countries, where either 90–95% or complete crop failure is common in certain locations [4]. There are 30,000 weed species infesting various crops on a global scale, out of which, 18,000 weed species are problematic causing severe yield losses in crop production. Estimates of 826 weed species were recorded in India, of which, 198 weed species were reported as serious weeds, while 80 weed species were classified as very serious weeds [5].

The degree of weed infestation is at an increasing rate in agricultural and non-cropped lands. The cost incurred for the adoption of weed control measures increases the cost of cultivation and reduces profit for the farmers. Manual weeding and mechanical weed management practices were the major options to manage weeds in agricultural fields. However, due to demand for human labor and increase in labor wages force the farmers to use herbicides to manage weeds. An annual average of 2 million tonnes of pesticides is consumed worldwide, where consumption of herbicides positioned first (47.5%) [6]. Herbicides are now widely used to manage weeds in modern agriculture. However, herbicides undergo various losses in soils after application *viz*. photo-degradation, volatilization, leaching, microbial degradation, run-off, etc., which in turn lowers the weed control potential of herbicides. Moreover, off-site transport of applied herbicides causes groundwater contamination. Nowadays, low volume herbicides are available, which show higher herbicidal activity and weed control efficiency over conventional herbicide formulations. Nevertheless, low volume herbicides are also reported with faster degradation potential in soil and increase the risk of environmental pollution. Therefore, encapsulation is an innovative and promising approach for developing controlled release formulations. Encapsulated formulation minimizes the herbicide loss in agroecosystem and improves weed control efficiency at a lower dosage. Encapsulation regulates the release and availability of active ingredients in the targeted site. Encapsulation involves the loading of active ingredients in the secondary material. The secondary materials are generally polymeric systems that regulate the release of active ingredients into the environment through diffusion-mediated process. Thus, encapsulation enhances the herbicidal activity and achieves higher weed control efficiency at a lower application rate. With the background, the chapter discusses the encapsulation of herbicides in the polymeric system and the characteristics of formulations and their scope in weed management.

## 2. Weeds and their characteristics

The term "weed" refers to "a plant out of its place or a plant growing where it is not desired at that time" [7]. The definition implies that *Echinochloa* sp is a weed in rice fields, similarly, pigeon pea is also considered a weed in greengram fields. Rice var. Jaya is a weed in IR 8 rice fields. Weeds are notorious and unwanted plants, which affect crop production. Weeds are categorized into annuals, biennials, and perennials based on their ontogeny [8]. Annual weeds complete their life cycle in a season with abundant seed production, while biennials survive for two seasons, completing the vegetative phase in the first season and reproductive phase in the succeeding season. Perennial weeds live for more than two years and propagate through both seeds and underground storage parts such as tubers, rhizomes, stolon, etc. Seed propagation is the sole mechanism for dispersal in annual and biennial weeds, whereas perennial weeds are largely propagated through vegetative propagules (**Table 1**).

| Weed | Scientific name | Family | Ontogeny | Mode of propagation |
|---|---|---|---|---|
| Country mallow | *Abutilon indicum* | Malvaceae | Annual | Seeds |
| Indian copper leaf | *Acalypha indica* | Euphorbiaceae | Annual | Seeds |
| Bristly starbur | *Acanthospermum hispidum* | Asteraceae | Perennial | Achenes |
| Khaki weed | *Alternanthera echinata* | Amaranthaceae | Biennial | Seeds |
| Alligator weed | *Alternanthera philoxeroides* | Amaranthaceae | Perennial | Seeds |
| Slender amaranth | *Amaranthus viridis* | Amaranthaceae | Annual | Seeds |
| Blistering ammannia | *Ammannia baccifera* | Lythaceae | Annual | Seeds |
| Mexican prickly poppy | *Argemone mexicana* | Papavaraceae | Annual | Seeds |
| Cape ash | *Bergia capensis* | Elatinaceae | Annual | Seeds |
| Purple chloris | *Chloris barbata* | Poaceae | Perennial | Seeds |
| Field bindweed | *Convolvulus arvensis* | Convolvulaceae | Perennial | Seeds and roots |
| Spreading dayflower | *Cyanotis axillaris* | Commonlinaceae | Annual | Seeds and bits of stem |
| Bermuda grass | *Cynodon dactylon* | Poaceae | Perennial | Seeds and stolon |
| Purple nutsedge | *Cyperus rotundus* | Cyperaceae | Perennial | Seeds and tubers |
| Umbrella sedge | *Cyperus difformis* | Cyperaceae | Annual | Seeds |
| Flat sedge | *Cyperus iria* | Cyperaceae | Annual | Seeds |
| Creeping wood sorrel | *Oxalis corniculata* | Oxaidaecae | Perennial | Seeds and tuberous roots |
| Johnson grass | *Sorghum halpense* | Poaceae | Perennial | Seeds and rhizome |

**Table 1.**
*Characteristic description of some important weeds.*

Weeds adapt well to a diverse ecosystem, which makes weeds more competitive than crops. Weeds produce abundant seeds in a single season and enrich the weed seed bank. A field with a seed bank of 5500 seeds $m^{-2}$ will increase the seed count of 1,98,500 Nos. $m^{-2}$ in two years, when there is no adoption of weed control measures [9]. Weed seeds are lighter in weight and smaller. Weed *Phalaris minor* weighs a test weight of 1.5–2.1 g [10] compared to the test weight of wheat (40 g). Some weeds produce seeds without fertilization, i.e., apomixis (e.g., *Taraxacum* spp.) [9]. Further, weed seeds germinate earlier and establish rapidly before the establishment of crops. Certain weed species exhibit rapid seedling growth and attain earlier maturity. Carrot grass (*Parthenium* sp) enters the flowering phase four weeks after emergence [11]. Weeds produce flowers and set seeds well in advance before the harvest of a crop. Weeds exhibit environmental plasticity to withstand vagaries of climatic conditions (drought, heat, cold) and edaphic situations through better adaptive and survival mechanisms. Parthenium weed exhibits faster growth at elevated levels of carbon dioxide and temperature. Weeds are mostly self-sown plants, which do not require optimum climatic and soil conditions for establishment. Moreover, weeds are opportunistic plants, which colonize everywhere if it is not controlled properly (**Table 2**).

Some weeds produce seeds morphologically mimicking crop seeds thus escaping from physical separation. The maturity of wild mustard *Argemone mexicana* coincides

| Weed species | Seed production per plant [12–14] |
|---|---|
| Redroot pigweed | 1,17,400 |
| Common lambsquaters | 72,450 |
| Common purslane | 52,300 |
| Shepherd's purse | 38,500 |
| Carrot weed | 30,000 |
| Common ragweed | 3380 |
| Jungle rice | 460–740 |
| Wild oat | 250 |

**Table 2.**
*Seed production potential of weeds.*

with the harvest of mustard crop and produces seeds resembling mustard seeds [15]. Certain annual weeds produce more than one flush in a single season, which increases the weed seed bank in the soil. Carrot grass completes four to five generations in a year under ideal environmental conditions [16]. Weeds produce a huge number of seeds; however, not all seeds germinate at a time. Weed seeds have the ability to resist decaying and exhibit various modes of dormancy. Velvetleaf (*Abutilon theophrasti*) and Fieldbind weed (*Convolvulus arvensis*) showed dormancy due to hard seed coat [17, 18]. Weed seeds have more longevity and remain viable for many years due to the phenomenon of dormancy. Field sowthistle (*Sonchus* sp.) produces viable seeds even when plants are cut at the flowering stage (**Table 3**) [19].

Weeds compete with crops efficiently for foraging nutrients from the soil with better-structured mechanisms. Weeds extract and accumulate more nutrients than crops, which make crops starve for nutrients. Crop nutrient contents, especially nitrogen, are closely correlated with crop yield potential, while an intense competition of weeds for nitrogen reduces the crop yield significantly. Weeds exhaust a huge amount of nutrients in soil in each season, thereby making soil progressively deficient in soil nutrients, thus affecting the crop growth and yield (**Table 4**).

Weeds such as *Digitaria sanguinalis* (696), *Echinochloa colona* (674), *Cynodon dactylon* (813), *Tephrosia purpurea* (1108), and *Tridax procumbens* (1402) have higher transpiration coefficient than crops such as sorghum (394) and maize (352) [27]. Vegetative propagules of weeds (roots, stolons, rhizomes etc.) penetrate deeper soil strata and grow vigorously with larger food reserves supporting weeds to survive under stress conditions. Seeds of fieldbind weed present at a soil depth of 6 cm have

| Weed species | Seed viability | References |
|---|---|---|
| *Parthenium* sp. | 8–10 years | [16] |
| *Convolvulus arvensis* | 20 years or more | [20] |
| *Chenopodium album* | 1700 year | [21] |
| *Nelumbo nucifera* | More than 3000 years | [22] |
| *Stellaria media* | More than 20 years | [23] |

**Table 3.**
*Longevity of weed seeds.*

| Weed species | Nutrient content (%) [24–26] | | |
|---|---|---|---|
| | N | P | K |
| *Amaranthus viridis* | 3.16 | 0.06 | 4.51 |
| *Chenopodium album* | 2.59 | 0.37 | 4.34 |
| *Achyranthus aspera* | 2.21 | 1.63 | 1.32 |
| *Cyperus rotundus* | 2.17 | 0.26 | 2.73 |
| *Ipomea carnea* | 1.90 | 0.75 | 2.50 |
| *Cynodon dactylon* | 1.72 | 0.25 | 1.75 |
| *Parthenium hysterophorus* | 2.54 | 0.44 | 1.23 |

**Table 4.**
*Nutrient composition of weeds.*

the ability to germinate normally [28]. Similarly, weeds of carpetweed (*Trianthema portulacastrum*) have the potential to germinate from a soil depth of 9 cm [29]. Roots of sowthistle located in the soil depth of 50 cm produce shoots to reach above-ground [19]. Similarly, perennial weeds have regenerative ability while many weeds possess adaptive mechanism (disagreeable odor, bitter taste, spines, etc.), which repel animals from grazing (evasiveness). Animals, birds, winds, water, etc. disseminate weed seeds [30]. Field sowthistle disperses weeds to a distance of 100 m through wind [31]. Yellow mistletoe (*Loranthus europaeus*) is mostly dispersed through birds such as Mistle Thrush (*Turdus viscivorus*) and Eurasian jay (*Garrulus glandarius*) [32]. Most of the weeds exhibit $C_4$-type photosynthesis conferring the advantage to mitigate the impact of moisture stress during crop growth and utilize low levels of $CO_2$ in the crop microclimate for photosynthesis.

## 3. Impacts of weed infestation on crop production

Weeds are the major biotic threat, which affect yield and crop quality by exerting direct (allelopathy) and indirect (competition) influence on crops. Moreover, weeds serve as a reservoir of various crop insects and diseases. It also reduces the working efficiency of labor and agricultural machinery and increases the cost of cultivation.

The degree of competition of weeds on crops depends on weed flora infested, level and duration of weed infestation, competing ability of crops, and climatic factors that influence crop and weed growth. The yield reduction of crops due to weed infestation is directly correlated with the degree of weed competition. The increase of one kilogram of weed biomass reduces the crop biomass by one kilogram [33]. Weeds affect crop growth directly by releasing allelochemicals. Weed infestations cause 100% yield loss in crops if the weed remains uncontrolled. Weeds are responsible for 33% (one-third) of yield losses in crops among the agricultural pests in India [1]. The yield reduction due to weed infestation in various crops is presented in **Table 5**.

Similarly, the estimated yield loss of grain crops in Australia was 2.52 billion USD due to weed infestation [43]. India suffers an economic loss of USD 11 billion annually due to weeds. In addition, higher monetary losses due to weeds were documented in rice (USD 4420 million) followed by wheat and soybean (USD 3376 and 1559 million, respectively). Annual yield loss of 3 million tons in China due to weed infestation in grain crops [44].

| Crops | Per cent yield loss | Reference |
|---|---|---|
| Direct Seeded rice | 21.4 | [34] |
| Transplanted rice | 13.8 | [34] |
| Wheat | 10–60 | [35] |
| Maize | 25.3 | [34] |
| Sorghum | 25.1 | [34] |
| Pearlmillet | 27.6 | [34] |
| Fingermilet | 50 | [35] |
| Greengram | 30.8 | [34] |
| Chickpea | 10–50 | [35] |
| Redgram | 16–65 | [35] |
| Lentil | 30–35 | [35] |
| Horsegram | 30 | [35] |
| Soybean | 31.4 | [34] |
| Groundnut | 35.8 | [34] |
| Niger | 20–30 | [35] |
| Cotton | 40–60 | [35] |
| Jute | 30–70 | [35] |
| Sugarcane | 25–50 | [35] |
| Sugarbeet | 70 | [36] |
| Tomato | 92–95 | [37] |
| Okra | 40–80 | [38] |
| Cabbage | 45–80 | [39] |
| Cauliflower | 61 | [40] |
| Carrot | 90 | [41] |
| Radish | 86 | [42] |
| Peas | 25–30 | [40] |

**Table 5.**
*Yield reduction due to weed infestation.*

Weed infestation reduces crop and grain quality [45]. Certain weed species set seeds coinciding with crop maturity and few weeds produce seeds, which resemble crop seeds. Therefore, weed seeds have a chance to form admixture with crop seeds during the harvest thud affecting grain quality. Mustard seeds get contaminated with seeds of Mexican prickly poppy (*Argemone mexicana*) during the harvest. Weed infestation affects the quality of leafy and other vegetable crops. Commercially available wheat grain for household purposes was found to be contaminated with seeds of *Phalaris minor* @ 2–3 g/kg of grain [46]. Similarly, leaves of Loranthus (*Dendrophthoe falcate*) were plucked unwittingly impairing tea quality.

Weeds act as collateral hosts for various crop insects, diseases, and nematodes. Weeds act as a reservoir for various pests providing food and habitat that in turn affect crops. Weeds acting as hosts for pests and diseases are listed in **Table 6**.

| Weeds | Crop insects/disease | Crop | Reference |
|---|---|---|---|
| *Brachiaria mutica*, *Digitaria marginata*, *Dinebra retroflexa*, *Echinochloa crusgalli*, *Leersia hexandra* | Blast disease (*Pyricularia grisea*) | Rice | [47] |
| *Mikania cordata*, *Bidens biternata*, *Emilia sonchifolia*, *Polygonum chinense* and *Lantana camara* | Tea mosquito bug (*Helopeltis theivora*) | Tea | [48] |
| *Anagallis arvensis*, *Convolvulus arvensis* and *Chenopodium album* | Alternaria blight | Mustard | [49] |
| *Elytrigia spp*, *Agropyron spp*, *Festuca spp*, *Dactylis spp*, *Phleum spp* and *Lolium spp* | Stem rust | Wheat | [50] |

**Table 6.**
*Weeds act as shelter for insect pest and diseases.*

Weeds interfere with the movement of laborers while carrying out various farm operations *viz*. weeding, fertilizer application, spraying of chemicals, etc. Weeds also cause physical discomfort such as itching, allergy symptoms in human beings, and reduce efficiency during field operations. Parthenium weed causes human-related ailments such as asthma, skin rashes, eczema, swelling and itching of mouth and nose, etc. [51]. Fields infested with weeds such as *Argemone mexicana* and *Amaranthus spinosus* possess thorns and spines, respectively, which in turn restrict the movement of farm laborers causing hindrance to carrying out field operations.

Weed-free environment is prerequisite for attaining the maximum possible yield. Therefore, weed management practices raise the cost of cultivation and reduce the profit for farmers. The average cost of weed control is ₹4000 $ha^{-1}$ for winter season crops, while it is ₹6000 $ha^{-1}$ for crops that are grown during the rainy season [52]. Similarly, grain growers in Australia spent $113 per hectare for weed control [43].

## 4. Weed management strategies

Weed infestations are dynamic in nature. The adoption of high-input agricultural practices, use of high-yielding dwarf varieties and hybrids, and adoption of monoculture cause weed shift and composition of weeds. Moreover, invasion of alien weeds and consequences of climate change also determine the weed composition and weed dominance in field conditions. Therefore, ideal weed management strategies are crucial for establishing favorable environment for crops.

Weed management methods that commonly adopted in agriculture are prevention, cultural methods, mechanical methods, chemical weed management, and biological method. Weed management on a farm become successful when adoption of various weed management techniques as an integrated approach.

Cultural method encompasses crop management practices ranging from field preparation to crop harvest. Cultural method provides a favorable crop environment for crops to establish well to compete with weeds. Cultural method minimizes the yield reduction and maintains the purity of harvested grain. Similarly, cultural methods prevent the enrichment of weed seed bank. Cultural methods are cost-effective, feasible to adopt, and ecologically sound in nature; however, these are labor-intensive methods.

Mechanical and physical methods involve physical removal of weeds before sowing or planting crops or during the crop period. The method intends to either kill weeds or make them less favorable for weed seed germination and establishment.

It includes tillage operations, manual weeding, hoeing, sickling, digging, dredging, chaining, mowing, cutting, stale seedbed, flooding burning, flaming, and mulching. This method is highly effective in controlling perennial weeds and reducing annual weed infestation in cropped lands. It saves time and labor for weeding. However, weeds found closely to crop are not removed through physical methods. Mechanical method warrants optimum soil moisture for weeding operations.

The use of chemicals was the third era of agronomical practices, which created a major impact in agriculture by substituting labor and mechanical energy [53]. The word "herbicide" is derived from Latin "herba" and "caedere" meaning "plant" and "to kill," respectively. It implies that herbicides are used to kill the plant. Chemical weed control is the only strategy in areas of labor scarcity and, where mechanical and manual weeding is not feasible [54]. Herbicides are greatly differed in chemical structure, mobility in plants, mechanism of action, polarity, solubility, selectivity, etc. The pre-emergence herbicides control weeds that are emerged from soil. Selective herbicides with reference to crops are useful to eliminate mimicry weeds. Herbicides are effective in controlling problematic and perennial weeds. Chemical weed control is the cost-effective and reliable option compared to other weed management strategies. However, chemical weed control has certain limitations *viz*. herbicide drift, groundwater contamination, residual effect on succeeding crop, and risk of developing herbicide-resistant biotypes.

Biological weed management involves the use of living organisms such as disease-causing organisms, insects, animals, fish, and competitive plants to suppress the growth of weeds. Biological control does not eradicate weeds completely but it will minimize weed population. Biological control measures are effective against introduced weeds [55, 56]. The remarkable examples of the success of biological weed control were the eradication of Prickly pear (*Opuntia* spp.) in Australia and Lantana in Hawaii [57, 58].

Among different weed management strategies, chemical weed management is quite efficient, convenient, and economical to control weeds. There are different herbicides that are commercially available in the market to manage weeds. However, there are many factors that govern when, where, and how a particular herbicide is used for managing weeds.

## 5. Herbicides in weed management

Herbicides are a crucial component in chemical weed management. Due to labor shortage and hike in labor wages, farmers are forced to use herbicides in their fields. Herbicides are extensively used at a large scale to control weeds both in cropped and non-cropped areas. The application of herbicide has made remarkable transformation in agricultural production. Herbicides replace the manual and mechanical weed control in modern-day agriculture [59]. Hay [60] described the progress of herbicide evolution in agriculture.

Chemicals such as oil wastes, rock salts, copper salts, crushed arsenical ores, and sulphuric acid were used initially in the 1920s for eradicating weeds infested railway tracks, roads, and timber yards [61]. Pokorny synthesized 2,4-D herbicide in 1941 and found that 2,4-D was selectively toxic to broadleaved weeds. This work was the foundation for the development of herbicides. Herbicides occupy a major share (47.5%) in the pesticide market followed by insecticides and fungicides [6]. There are a variety of herbicides, which are commercially available for use *viz*. selective, nonselective, systemic, and contact herbicides. Herbicides are greatly varied in site of action and show selectivity for the control of weeds without affecting crops [62]. Plant factors include exposure of meristems to spray droplets of herbicides, leaf traits and root morphology affect the selectivity of herbicides. Plant characteristics of genetic make-up also influence the selectivity of herbicides. Herbicides kill target species alone without affecting nontarget species. Herbicide Resistance Action Committee (HRAC) [63] grouped herbicides based on mode of action are listed in **Table 7**.

Low-dose herbicides such as pyrazosulfuron-ethyl, sulfosulfuron, metsulfuron-methyl, Quizalofop-ethyl, bispyribac sodium, etc. are available in the market, which

| Mode of action | Herbicide family | Herbicide |
|---|---|---|
| Inhibition of Acetyl CoA Carboxylase (ACCase) | Cyclohexanediones (DIMs) | Alloxydim, Butroxydim, Clethodim, Sethoxydim, Tepraloxydim, Tralkoxydim |
| | Aryloxyphenoxy-propionates (FOPs) | Clodinafop-propargyl, Cyhalofop-butyl, Diclofop-methyl, Fenoxaprop-ethyl, Fluazifop-butyl, Isoxapyrifop, Metamifop, Quizalofop-ethyl |
| | Phenylpyrazoline | Pinoxaden |
| Inhibition of Acetolactate Synthase (ALS) | Pyrimidinyl benzoates | Bispyribac-sodium, Pyriminobac-methyl, Pyrithiobac-sodium |
| | Sulfonanilides | Pyrimisulfan, Triafamone |
| | Triazolopyrimidine—Type 1 | Cloransulam-methyl, Diclosulam, Florasulam, Flumetsulam, Metosulam |
| | Triazolopyrimidine—Type 2 | Penoxsulam, Pyroxsulam |
| | Sulfonylureas | Amidosulfuron, Azimsulfuron, Bensulfuron-methyl, Chlorimuron-ethyl, Chlorsulfuron, Ethoxysulfuron, Flazasulfuron, Flucetosulfuron, Halosulfuron-methyl, Imazosulfuron, Metsulfuron-methyl, Orthosulfamuron, Pyrazosulfuron-ethyl, Rimsulfuron, Sulfosulfuron, Triasulfuron, Tribenuron-methyl, Trifloxysulfuron-Na, Triflusulfuron-methyl |
| | Imidazolinones | Imazamox, Imazapic, Imazapyr, Imazethapyr |
| | Triazolinones | Flucarbazone-Na, Propoxycarbazone-Na, Thiencarbazone-methyl |
| Inhibition of Photosynthesis at PS II—Serine 264 Binders | Triazines | Atrazine, Cyanazine, Cyprazine, Desmetryne, Dimethametryn, Prometon, Prometryne, Simetryne, Simazine, Terbumeton, Terbuthylazine, Terbutryne, Trietazine |
| | Triazolinone | Amicarbazone |
| | Triazinones | Ethiozin, Hexazinone, Isomethiozin, Metamitron, Metribuzin |
| | Uracils | Bromacil, Isocil, Lenacil, Terbacil |
| | Phenylcarbamates | Chlorprocarb, Desmedipham, Phenisopham, Phenmedipham |
| | Pyridazinone | Chloridazon, Brompyrazon |
| | Ureas | Benzthiazuron, Chloroxuron, Difenoxuron, Diuron, Fenuron, Fluometuron, Isoproturon, Linuron, Monuron, Neburon, Tebuthiuron |
| | Amides | Pentanochlor, Propanil |

| Mode of action | Herbicide family | Herbicide |
|---|---|---|
| Inhbition ofPhotosynthesis at PS II—Histidine 215 Binders | Nitriles | Bromofenoxim, Bromoxynil, Ioxynil |
| | Phenyl-pyridazines | Pyridate |
| | Benzothiadiazinone | Bentazon |
| PS I Electron Diversion | Pyridiniums | Cyperquat, Diquat, Paraquat |
| Inhibition of Protoporphyrinogen Oxidase (PPO) | Diphenyl ethers | Acifluorfen, Chlornitrofen, Fluoronitrofen, Nitrofen, Oxyfluorfen |
| | Phenylpyrazoles | Pyraflufen-ethyl |
| | N-Phenyl-oxadiazolones | Oxadiargyl, Oxadiazon |
| | N-Phenyl-imides | Fluthiacet-methyl, Pentoxazone, Trifludimoxazin, Tiafenacil |
| Inhibition of Phytoene Desaturase | Phenyl ethers | Beflubutamid, Diflufenican, Picolinafen |
| | N-Phenyl heterocycles | Flurochloridone, Norflurazon |
| | Diphenyl heterocycles | Fluridone, Flurtamone |
| Inhibition of Hydroxyphenyl Pyruvate Dioxygenase (HPPD) | Triketones | Tembotrione, Tefuryltrione, Bicyclopyrone, Fenquinotrione |
| | Pyrazoles | Pyrasulfotole, Topramezone, Pyrazolynate, Pyrazoxyfen |
| | Isoxazoles | Isoxaflutole |
| Inhibition of Homogentisate Solanesyltransferase | Phenoxypyridazine | Cyclopyrimorate |
| Inhibition of Deoxy-D-Xyulose Phosphate Synthase | Isoxazolidinone | Clomazone, Bixlozone |
| Inhibition of Enolpyruvyl Shikimate Phosphate Synthase (ESPS) | Glycine | Glyphosate |
| Inhibition of Glutamine Synthetase | Phosphinic acids | Glufosinate-ammonium, Bialaphos or bilanafos |
| Inhibition of Dihydropteroate Synthase | Carbamate | Asulam |
| Inhibition of Microtubule Assembly | Dinitroanilines | Fluchloralin, Isopropalin, Nitralin, Oryzalin, Pendimethalin, Trifluralin |
| | Pyridines | Dithiopyr, Thiazopyr |
| | Phosphoroamidates | Butamifos, DMPA |
| | Benzoic acid | DCPA |
| | Benzamides | Propyzamide |
| | Carbamates | Carbetamide, Chlorpropham, Propham |
| Inhibition of Cellulose Synthesis | Triazolocarboxamide | Flupoxam |
| | Benzamides | Isoxaben |
| | Alkylazines | Triaziflam, Indaziflam |
| | Nitriles | Dichlobenil, Chlorthiamid |
| Uncouplers | Dinitrophenols | Dinosam, Dinoseb, DNOC |

| Mode of action | Herbicide family | Herbicide |
|---|---|---|
| Inhibition of Very Long-Chain Fatty Acid Synthesis | Azolyl-carboxamides | Cafenstrole, Fentrazamide, Ipfencarbazone |
| | α-Thioacetamides | Anilofos, Piperophos |
| | Isoxazolines | Pyroxasulfone, Fenoxasulfone |
| | Oxiranes | Indanofan, Tridiphane |
| | α-Chloroacetamides | Acetochlor, Alachlor, Butachlor, Dimethachlor, Metazachlor, Metolachlor, Pretilachlor, Propachlor, |
| | α-Oxyacetamides | Mefenacet, Flufenacet |
| | Thiocarbamates | Butylate, Cycloate, EPTC, Molinate, Thiobencarb, Tiocarbazil, Tri-allate |
| | Benzofurans | Benfuresate, Ethofumesate |
| Auxin Mimics | Pyridine-carboxylates | Picloram, Clopyralid, Aminopyralid |
| | Pyridyloxy-carboxylates | Triclopyr, Fluroxypyr |
| | Phenoxy-carboxylates | 2,4,5-T, 2,4-D, 2,4-DB, Fenoprop, MCPA, MCPB |
| | Benzoates | Dicamba, Chloramben, TBA |
| | Quinoline-carboxylates | Quinclorac, Quinmerac |
| | Pyrimidine-carboxylates | Aminocyclopyrachlor |
| | Phenyl carboxylates | Chlorfenac, Chlorfenprop |
| Auxin Transport Inhibitor | Aryl-carboxylates | Naptalam, Diflufenzopyr-sodium |
| Inhibition of Fatty Acid Thioesterase | Benzyl ether | Cinmethylin, Methiozolin |
| Inhibition of Solanesyl Diphosphate Synthase | Diphenyl ether | Aclonifen |
| Inhibition of Lycopene Cyclase | Triazole | Amitrole |

**Table 7.**
*HRAC classification of herbicides.*

control weeds efficiently at lower concentration. However, active molecules are lost through various degradation processes in agro-ecosystem and reduce the weed control efficiency. The offsite movement of herbicides also poses serious environmental hazards in certain circumstances.

## 6. Issues associated with chemical weed control

Each herbicide molecule is unique in its herbicidal activity. The nature of herbicide, soil, and climatic conditions influence the behavior and weed control efficiency of herbicides. Herbicides are subjected to various forms of degradation on reaching soils, which in turn reduce the weed control activity of herbicides. Herbicidal activity and persistence of herbicides in soil are determined by various factors *viz.* soil sorption coefficient, leaching potential, and volatilization behavior of herbicide molecule. Soil with high content of clay or organic matter facilitates more adsorption of herbicide, while dry soils have more unoccupied binding sites, promoting the binding of herbicide

| Process of degradation | Description | Factors affecting degradation | Examples |
|---|---|---|---|
| **Transport of active molecules (physical process)** | | | |
| Volatilization | Lost *via* evaporation from soil surface | Vapor pressure, temperature and wind velocity | Dinitroanilines, Thiocarbamates [64] |
| Adsorption | Interactions with soil | Organic matter, clay content, Soil moisture | Bipyridinium [65], Pendimethalin [66] |
| Leaching | Offsite transport of herbicide molecules into soil | Herbicide solubility, soil texture and rainfall | Bromacil, diuron [67], thifensulfuron-methyl [68], sulfentrazone [69] |
| Physical drift | Transport of spray droplets by wind | Wind velocity and droplet size | — |
| **Degradation of active molecules** | | | |
| Photo-decomposition | Degraded by sunlight | Chemical structure, duration and intensity of exposure to sunlight | Dinitramine, Nitralin, Fluchloralin [70], Paraquat [71] |
| Chemical degradation | Breakdown of active molecules into metabolite through different chemical process (hydrolysis, oxidation-reduction reaction, etc.) | Chemical nature | Sulfonylurea [72] |
| Microbial degradation | Degradation of active molecules through soil microbes. | Soil pH, moisture content, organic matter and temperature | Sulfonylurea [72], Oxyfluorfen [73], 2,4-D [74], Glyphosate [75] |

**Table 8.**
*Fate of herbicides in agro-ecosystem*

molecules thus affecting the herbicidal activity. Soil microbial population also influences the fate of applied herbicide in agro-ecosystem. The pre and postemergence herbicides experience different modes of loss in soils. In spite of loss of herbicidal activities, maintenance of herbicides above the threshold level is crucial to achieve the desired effect on weeds. The fate of herbicides in solid is summarized in **Table 8** [64, 76].

Direct application of herbicide in soil as pre-emergence or pre-plant incorporation poses a serious threat to the environment compared to other methods of herbicide applications. Leaching of herbicides especially ureas, sulfonylurea, and uracil herbicides contaminates groundwater. Herbicides with higher solubility, mobility, and sorption to soil particles are categorized with higher potential herbicides for groundwater contamination. Herbicides that persist in the soil impede the germination of succeeding crops through phytotoxicity effect. Persistence of herbicide in soil is listed in **Table 9**. Further, nontarget plant species are also affected due to spray drift and inappropriate application of herbicides.

Herbicide poses serious health hazards such as cancer, neurological disorders, and respiratory and reproductive related problems on the prolonged exposure to herbicide [78–82].

Herbicide-resistant weeds are superweeds, which evolve resistance against the use of single or multiple herbicides. The factors for the development of herbicide resistance among weeds are due to the repeated application of same herbicide or

| **Persistence in soil** [77] | | | |
|---|---|---|---|
| **Less than 1 month** | **1–3 months** | **3–6 months** | **More than 6 months** |
| 2,4-D, MCPA, Glyphosate | Butachlor, Alachlor, Halosulfuron, Pyrazosulfuron-ethyl, Metribuzin, Bispyribac-sodium, Fluzifop-butyl, Metsulfuron-methyl, Oxyfluorfen | Pendimethalin, Fluchloralin, Isoproturon, Imazethapyr, Oxadiazon, Linuron | Atrazine, Simazine, Paraquat, Diquat, Chlorsulfuron, Diuron, Bromacil, Imazapyr, Sulfentrazone, Trifluralin, Picloram |

**Table 9.**
*Persistence of herbicides in soil.*

herbicides with a similar mode of action [83]. There are 266 weed species, which developed resistance against herbicides. Further, infestation of herbicide-resistant weeds has been reported in 71 countries [84]. The control of superweeds requires alternate strategies other than herbicides, which incur additional cost for managing resistant weeds. Herbicide-resistant weeds also pose weed shift in specific regions.

Application of selective herbicides increases risk of infestation of nonselective weeds. Herbicides do not exert consistent weed control since interaction of herbicides with the environment is dynamic in nature. Herbicides also affect non-target weed species in certain regions posing the threat to biodiversity. Therefore, chemical weed control has several issues on the herbicide use efficiency besides posing threat to nontarget sites.

## 7. Herbicide encapsulation: an innovative approach

Conventional herbicide formulations are recommended at a higher dosage over the minimum threshold level to complement the herbicide losses encountered in agroecosystem to achieve higher weed control [85]. Further, a significant quantity of applied herbicides undergoes various degradation paths causing environmental pollution. Herbicide encapsulation is the smart delivery approach, which addresses and resolves the constraints of conventional chemical weed management. Encapsulation involves the entrapment of herbicides in polymeric systems to safeguard the active molecules from the environmental vulnerability and achieve controlled release of herbicides in the target environment. The active ingredients are encapsulated in the shell materials for improving weed control efficiency through prolonged release of active ingredients in the soil. Encapsulation promotes the stability of active ingredients and reduces the herbicide requirement significantly by minimizing the loss of herbicides into the environment [86–89]. Herbicide encapsulation is a versatile technology performed at nano and micro-scale by incorporating active ingredients into the suitable carrier [90]. The assembly of active ingredients and carrier material resulted in sustained release of active ingredients for a longer period at the desired rate. Similarly, encapsulated formulation reduces herbicide dosage coupled with slow-release results in reducing the residue buildup in soil and eliminating phytotoxicity [91]. Sulfentrazone, a pre-emergence herbicide was encapsulated using calcium alginate and calcium chloride as cross-linker [92]. The resultant formulation offered controlled release of sulfentrazone and minimized the leaching potential of herbicide. Similarly, encapsulation of atrazine with starch polymer impeded volatilization [93].

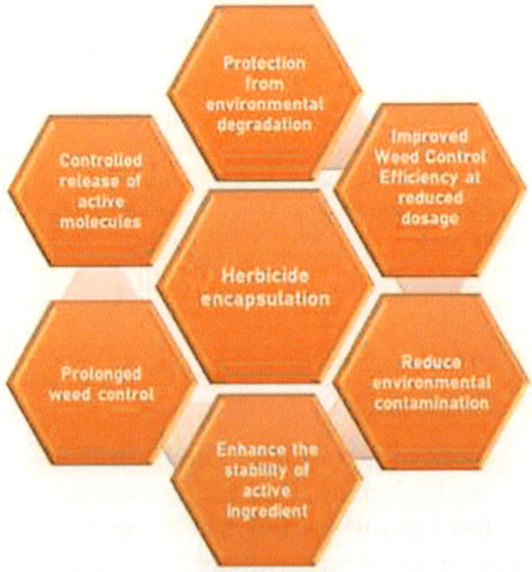


**Figure 1.**
*Advantages of herbicide encapsulation.*

Nano-encapsulated atrazine in poly epsilon-caprolactone carrier system exhibited targeted weed control at ten times lower dose of the recommended level of herbicide [94]. In addition, it reduced the soil mobility of atrazine in soils. Meanwhile, smart delivery of herbicide shows higher efficacy of weed control and exhausts the weed seed bank resulting in less emergence of weeds (**Figure 1**) [95].

## 8. Polymers for herbicide encapsulation

Generally, carriers are polymeric materials that are employed for the encapsulation of herbicides to develop a smart delivery system. There are numerous carrier materials (natural, synthetic, and semisynthetic polymers) available for herbicide encapsulation. However, synthetic polymer has less significance than natural and semisynthetic polymer since it is not degradable in nature and remains as a contaminant in the soil. In contrast, natural carrier materials are advantageous since they are eco-friendly, biocompatible, cost-effective, easily available, and biodegradable in nature [96, 97]. Alginates, chitosan, starch, pectin, lignin, Arabic gum, cyclodextrin, cellulose, and gelatin are the biopolymers employed for herbicide encapsulation [90, 98–101]. Commonly used synthetic polymers are polycaprolactone, polyurethane, polyvinyl alcohol, and polystyrene sulfonate [99, 102, 103]. Semisynthetic polymers are natural polymers with side-chain modification through the replacement of hydrogen from hydroxyl group of glucose repeating units with ethyl, methyl, carboxymethyl, and carboxyethyl moieties.

### 8.1 Natural polymers

Alginate is an anionic linear polysaccharide polymer that naturally exists in the cell walls of brown seaweed *viz. Ascophyllum nodosum*, *Laminaria hyperborea,* and *Macrocystis pyrifera* [104, 105]. Alginate has been explored for the controlled release of active compounds via ionotropic gelation method [92]. Leaching potential of sulfentrazone herbicide was reduced by developing sustained release of herbicide by exploiting alginate polymeric system. Tebuthiuron was encapsulated using alginate as carrier material to impede leachability of herbicides in agroecosystem [106]. Similarly, starch is a homopolysaccharide that is made up of two distinct molecules of amylose and amylopectin [90].

Starch is extensively found in cereal grains, roots, tubers, and fruits, which is also employed as a carrier for smart release of herbicide. Herbicides such as 2,4-D and 2,4,5-T were encapsulated with corn, wheat, potato, and cassava starches [107]. Encapsulation with wheat and potato starches exhibited slower release of herbicide because of higher amylose content and molecular weight of starch in wheat and potato starch. Sulfentrazone herbicide was encapsulated using starch *via* solvent evaporation method for season-long weed control by reducing leaching potential of herbicide [108]. Atrazine was encapsulated by utilizing starch as carrier and resultant formulation minimize the volatilization loss over the conventional formulation [93].

Chitosan is a nontoxic, biodegradable, and biocompatible polymer obtained through the deacetylation of chitin, which is usually found in the cell walls of fungi and bacteria. Chitosan is a cationic linear polysaccharide, which is highly efficient carrier system for agrochemicals [89, 109]. Paraquat-loaded chitosan/tripolyphosphate nanoparticles reduced the soil sorption of paraquat, thus improving the stability of herbicide [99]. Cellulose and its derivatives are explored as a carrier system for the smart delivery of active compounds. The formulation was developed by mixing chitosan and glyphosate at different molar ratios in water for the mart delivery of glyphosate, where chitosan polymer plays a dual role as eco-friendly adjuvant and polymeric carrier of glyphosate facilitating prolonged release of herbicide [110].

Cellulose is a polysaccharide that is biodegradable in nature and available in abundance at a lower cost. Alginate/cellulose-based delivery system containing imazethapyr offered the extended-release of active material [100].

Pectin is a polysaccharide and an anionic biopolymer, which is abundantly present in higher plants' cell walls. Pectin is a biodegradable, nontoxic, and easily available natural polymer. Pectin is composed of D-galacturonic acid units, which are linked by α-(1-4) glycosidic linkage [111]. Nowadays, pectin is also explored as a carrier system for the controlled release of the active ingredient due to its characteristics *viz.* more stability at acidic and high-temperature conditions, gelation property, non-toxicity, biocompatibility, and easily available at a cheaper cost [112]. Six percent pectin and two percent calcium chloride were found as optimum concentration for smart release of herbicides *via* ion gelation technique [113]. Metsulfuron-methyl loaded in pectin nanoparticles were found to be effective with higher herbicidal activity at a lower application rate as compared to the commercial herbicide [87].

Lignin is another important polymer that is obtained as a byproduct in pulp and paper industries. Lignin exhibits UV shielding property and antimicrobial activity, which attracted lignin to explore as a polymer for the delivery of herbicide. Moreover, lignin is relatively available in abundance at a lower cost [114, 115]. Dicamba herbicide was encapsulated in lignosulfonate carrier system for sustained release [116]. Lignin-polyethylene glycol-based chloridazon and metribuzin were synthesized for minimizing leachability of herbicides in light-textured soils [117].

Cyclodextrin is a cyclic oligosaccharide consisting of glucose units, derived through enzymatic conversion of starch. β-Cyclodextrin is a highly preferred molecule for encapsulation of active molecules since it is easily available at a lower cost [118]. Cyclodextrin has a unique structure, which enables it to form inclusion complex with hydrophobic active molecules. Terbuthylazine herbicide molecule was encapsulated using cyclodextrin, which showed improved solubility and bioavailability of herbicide molecule [119].

Guar gum is a neutral polysaccharide made up of the main chain of D-mannopyranose residues linked together by β-(1,4) glycosidic bonds and a secondary chain of D-galactopyranose residues linked together by α-(1,6) glycosidic bonds. Solubility of guar gum in cold water rises in proportion to the galactose/mannose molar ratio [120]. Herbicide formulation of guar gum-g-cl-polyacrylate/bentonite clay hydrogel composite was employed for pre-emergence application, while guar gum-g-cl-poly N-isopropylacrylamide nano hydrogel was used for the post-emergence application [121]. The encapsulation efficiencies of imazethapyr into guar gum-g-cl-polyacrylate/bentonite clay hydrogel composite ranged from 75.99 to 98.96% and guar gum-g-cl-poly N-isopropylacrylamide nano hydrogel ranged from 67.98 to 80.90%. The time to release 50 percent of the loaded imazethapyr ($t_{1/2}$) was between 0.06 and 4.8 days in CGNHG, while it was from 4.4 to 12.6 days in GG-HG system, Encapsulation of bioherbicides were also attempted using natural carrier materials such as Arabic gum, Persian gum/gelatin and gelatin [101].

### 8.2 Synthetic polymers

Polycaprolactone is biodegradable and hydrophobic polyester belonging to the aliphatic family. Polycaprolactone is utilized as a smart delivery vehicle for various active ingredients since it is biocompatible, cost-effective and possesses unique mechanical properties [122, 123]. Encapsulation of pretilachlor in polycaprolactone polymer enhanced the stability and herbicidal activity of herbicide [124]. Encapsulated atrazine and paraquat herbicides in poly-ε-carpolactone carrier system minimized the environmental impacts associated with the use of herbicides [125]. Similarly, poly-ε-caprolactone based atrazine nanocapsules reduced the soil mobility of herbicide and showed higher weed control efficiency at a lower application rate [94, 126, 127].

Polyurea is a product derived from the reaction of isocyanates and amines. Polyurea is used as shell material for herbicide encapsulation since it has high thermal stability and is available at a lower cost. Polyurea was utilized as a polymer for encapsulation of oxyfluorfen to reduce the phytotoxic effect on non-target plants [128]. Polyurea-based pretilachlor microcapsule formulation was synthesized through polymerization, which was found to be efficient in controlling weeds [129]. Polyurea-based pendimethalin encapsulated formulation reduced the usage of organic solvents during the manufacture of emulsifiable concentrate formulation eliminating the environmental pollution due to its application [130]. Pendimethalin was encapsulated using shell material made up of polyurethane urea to improve weed control efficiency [131].

Polyvinyl alcohol is a water-soluble polymer being widely explored for herbicide encapsulation. Glyphosate, a non-selective herbicide was encapsulated using polyvinyl alcohol and polystyrene sulfonate to minimize the herbicide loss in the environment [103]. Further, polyurethane polymeric systems are also utilized for the controlled delivery of active ingredients. Polyurethane is a synthetic polymer composed of urethane units, which is biocompatible and biodegradable in nature [132]. Trifluralin loaded in polyurethane network through interfacial polymerization protected the active ingredient from volatilization and photodegradation [133].

Polylactic acid is a biodegradable polymer derived from renewable sources such as corn, wheat, and rice. Polylactic acid is an aliphatic semicrystalline polyester which is hydrolyzable, eco-friendly, and biocompatible in nature [134, 135]. Microparticles of metazachlor herbicide were synthesized with low molecular weight polylactic acid for the controlled release of active molecules [136, 137]. Encapsulation of metolachlor herbicide was also attempted using a high molecular weight of polylactic acid for smart delivery of herbicide [138]. Similarly, Poly (lactic-co-glycolic acid) is a biopolymer composed of monomers of lactic and glycolic acids [139], exploited as carrier system for the smart delivery of atrazine herbicide to reduce the environmental impacts associated with application of herbicide [140].

### 8.3 Semi-synthetic polymers

Cellulose and its derivatives are exploited as a carrier system for the smart delivery of active compounds in agriculture. The two primary classes of cellulose derivatives are cellulose ethers and cellulose esters, which have varied levels of mechanical and physicochemical properties.

Ethyl cellulose is a derivative of cellulose in which the hydroxyl group of cellulose is substituted with the ethyl ester group [141]. Ethyl cellulose is a hydrophobic polymer utilized for improving the stability of the active ingredient to achieve higher use efficiency. 2,4-D herbicide was loaded in ethyl cellulose microspheres to achieve sustained release of herbicides [142]. Ethyl cellulose-loaded alachlor formulation reduced the soil mobility of herbicide which achieved prolonged weed control at a lower application rate [143]. Norfluazon based controlled release system using ethyl cellulose reduced the soil mobility of herbicide and protected the active ingredient from photodegradation [144, 145]. Solvent evaporation method was utilized to introduce atrazine, a broadleaf weed control herbicide, into ethyl cellulose-controlled release formulations [146], to sustain the release of herbicide.

Carboxymethyl cellulose is a cellulose derivative that is anionic in nature with high solubility in water. Carboxylmethyl cellulose readily forms gel in solutions of multivalent cations, such as aluminum or iron cations, to generate hydrogels. Controlled release formulations of acetochlor were synthesized with various modified forms of clay/carboxymethyl cellulose (**Figures 2–4** and **Table 10**) [147].

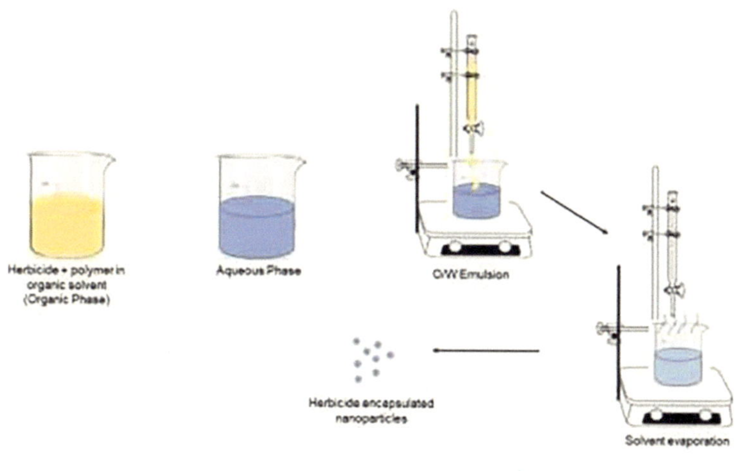


**Figure 2.**
*Solvent evaporation technique.*

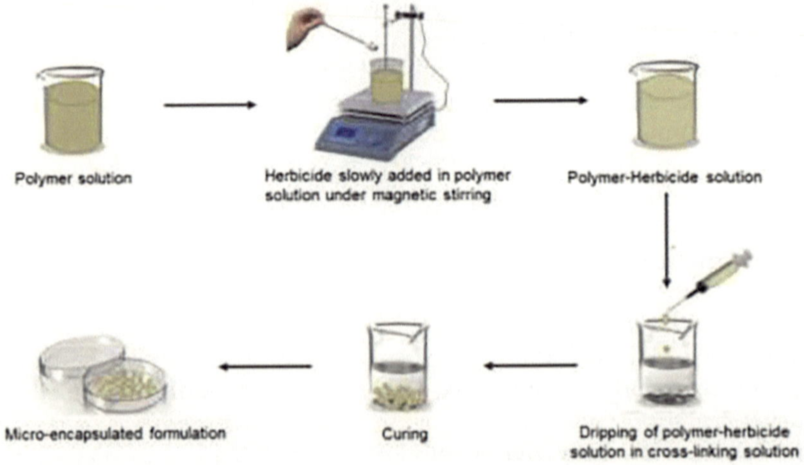


**Figure 3.**
*Ion gelation method.*

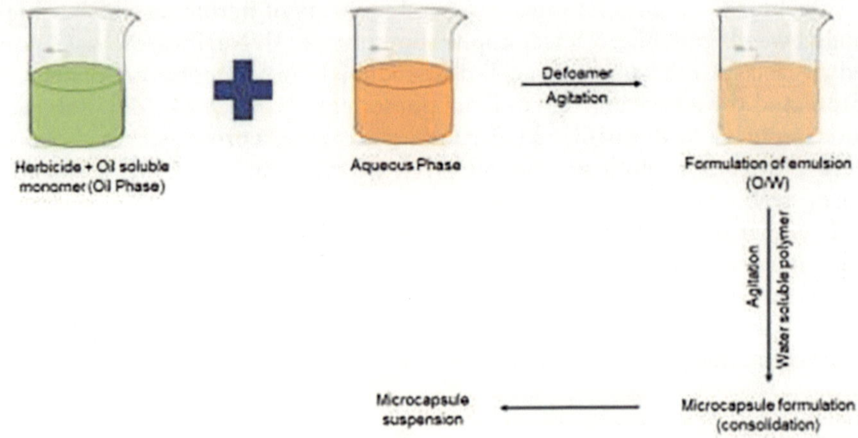


**Figure 4.**
*Preformed polymerization technique.*

| Herbicide | Polymer | Encapsulation technique | Characteristics of formulation | Authors |
|---|---|---|---|---|
| Trifluralin | Polyurethane | Interfacial polymerization | Microencapsulation protected herbicide from photo degradation and volatilization | [133] |
| Pretilachlor | Poly ε-caprolactone | Interfacial deposition of preformed polymer | Encapsulation efficiency was 99.5 ± 1.3%. Enhanced herbicidal activity with less environmental toxicity | [124] |
| 2,4-D | Ethyl cellulose | Emulsion solvent evaporation | Encapsulation efficiency of 7.7–27% with prolonged release of 2,4-D | [142] |

| Herbicide | Polymer | Encapsulation technique | Characteristics of formulation | Authors |
|---|---|---|---|---|
| Tebuthiuron | Sodium alginate | Ion gelation technique | Controlled release carrier system for tebuthiuron | [106] |
| Chloridazon and Metribuzin | Lignin and ethyl cellulose | | Reduced leaching and photo degradation of herbicides | [117] |
| Terbuthylazine | β-cyclodextrin | Kneading method | Improved herbicide solubility | [119] |
| Atrazine | Poly ε-caprolactone with chitosan as coating agent | Modified interfacial deposition of preformed polymer | Improved adhesive property of herbicide on foliage of target weeds | [98] |
| Paraquat | Chitosan/ tripolyphosphate | Ionic gelation | Encapsulation efficiency of polymeric system was 65% with stability of 60 days. Reduced Soil sorption of herbicide in soils | [99] |
| Tribenuron-methyl | Zein | Solvent evaporation | Encapsulation efficiency was 81± 3% with enhanced solubility, controlled release formulation improved weed control | [148] |
| Sulfentrazone | Sodium alginate | Ionotropic gelation | Minimized herbicide leaching into the soil | [92] |
| Metazachlor | Polylactic acid/ polyethylene glycol | Solvent evaporation | Controlled release system for the delivery of herbicides for prolonged weed control | [137] |
| Pendimethalin | Starch | Solvent evaporation | Slow release system depends on soil moisture availability and non-toxic to earthworms | [88] |
| Metsulfuron-methyl | Pectin | Emulsification | Encapsulation efficiency of 63 ± 2% with increased herbicidal activity at lower dose | [87] |
| Tebuthiuron | Sodium alginate | Ionotropic gelation | Reduced herbicide loss due to leaching | [149] |
| Imazapic and Imazapyr | Alginate/chitosan and Chitosan/ tripolyphosphate | Ionotropic gelation | Enhanced herbicidal activity and less toxic | [150] |
| Imazethapyr | Alginate and Alginate/cellulose | Ionotropic gelation | Extended release of Imazethapyr for 30 days of application | [100] |
| Atrazine | Poly ε-caprolactone | Interfacial deposition of preformed polymer | Improved post-emergence activity at lower dose (ten-fold lower than recommended levels) in controlling target weeds. Reduced soil mobility | [126, 127, 151] |
| Norflurazon | Ethyl cellulose | Solvent evaporation | Prolonged release and reduced soil mobility and offered protection from photo degradation | [144] |

| Herbicide | Polymer | Encapsulation technique | Characteristics of formulation | Authors |
|---|---|---|---|---|
| Atrazine | Poly ε-caprolactone | Interfacial deposition | Enhanced pre-emergence herbicidal activity at ten times of lower dose | [94] |
| Oxyfluorfen | Polyurea | Interfacial polymerization | Reduced phytotoxicity in rice | [128] |
| Metazachlor | Poly lactic acid | Solvent evaporation | Enhanced herbicidal activity on target plants | [138] |
| Atrazine | Poly (lactic-*co*-glycolic acid) | Modified precipitation method | Encapsulation efficiency of 50% | [140] |
| Norflurazon | Ethyl cellulose | Oil in water emulsion through solvent evaporation | Controlled release formulation (depends on active ingredient loaded, emulsifying and pore forming agent) | [145] |
| Alachlor | Ethyl cellulose | Oil in water solvent evaporation | Reduced herbicide loss due to leaching by 39% and minimized the risk of groundwater contamination Encapsulated formulation showed better efficacy for 30 days of application | [143] |
| Metazachlor | Poly (lactic acid) | Oil in water solvent evaporation | Encapsulation efficiency of 30%. Release rate of herbicide depends on particle size and loading efficiency | [136] |
| Savory essential oil (Bioherbicides) | Arabic gum, Persian gum/ gelatin and Persian gum | Complex coacervation | Better stability for 42 days Increment in herbicidal activity with encapsulation | [101] |
| Glyphosate | Metal nanoparticles such as iron oxide and silver nanoparticle and water soluble polymer such as polyvinyl alcohol and poly Styrene Sulfonate | Spray drying method | Higher encapsulation and weed control efficiency | [103] |
| Sulfentrazone | Starch | Solvent evaporation | Reduced horizontal and vertical leachability potential of herbicide Offered season long weed management in black gram | [108] |
| Metribuzin and Tribenuron | Poly(3-hydroxubutyrate) | High energy ball milling | Higher efficiency of weed control | [96] |
| Picloram | Chitosan and sodium ligno-sulfonate | Layer by layer technique | Altered the release of herbicide and improved photo stability of herbicides | [152] |

| Herbicide | Polymer | Encapsulation technique | Characteristics of formulation | Authors |
|---|---|---|---|---|
| Atrazine and alachlor | Starch | Solvent evaporation | Encapsulated formulation resulted IN reduced mobility and volatilization losses | [93] |
| Dicamba | Copper chitosan nanoparticles | Green chemical reduction method | Reduced leaching losses | [153] |
| Pendimethalin | Polyurethane urea | Interfacial polymerization | Encapsulation efficiency was 53.2–89.1% with enhanced stability | [131] |

**Table 10.**
*Brief overview of herbicide encapsulation.*

## 9. Release profile of encapsulated herbicides

Herbicide encapsulation protects the active compound from different losses *viz.* leaching, volatilization, adsorption, photodecomposition, etc. The loss of herbicides is controlled by altering the release rate of active ingredients from the polymeric systems. Therefore, herbicide encapsulation serves as a platform to design herbicide formulation with varying release patterns of active molecules. Encapsulated formulation modified the herbicide release profile. The encapsulation of herbicides minimizes the adverse consequence in soil environment due to use of herbicides. Similarly, encapsulation technique offers an extended period of weed control at a lower dosage.

The particle size of the formulation greatly influences the release rate of active ingredients into the environment [142, 154–156]. Polymer-solvent ratio, water diffusion rate, pH of the releasing medium, molecular weight of the polymer, nature of interaction between shell and core materials (active molecules), polymers, methodology, and preparation conditions also govern the release profile of active molecules [143, 153, 157–159].

Encapsulation of metribuzin and chloridazon in lignin-polyethylene glycol system coated with ethyl cellulose (20%) and dibutyl sebacate (2.25%) resulted in the controlled release formulation and time taken for the delivery of 50% corresponding herbicide were 16.94 and 65.39 h, respectively [117]. Release kinetics study of paraquat loaded in pectin/chitosan/tripolyphosphate nanoparticles revealed that polymeric system sustained release of paraquat compared to that of conventional formulation where a significant amount of paraquat was not released until 30 min of incubation [160]. Similarly, alginate/chitosan-based paraquat nano-formulation modified the release profile of paraquat, which achieved 100% herbicide release in eight hours of incubation. The release of paraquat was extended for two hours compared to that of free form of paraquat in water medium [161]. Conventional paraquat released 92% of active molecules after 350 min of incubation, while paraquat from chitosan/tripolyphosphate nanoparticles diffused only 72% during the same period [99]. The commercial formulation of imazethapyr released more than 76% of active herbicide molecules in less than one day, whereas the time taken for fifty percent release of the active molecule from alginate and alginate/cellulose beads were 11.30 and 43.73 days respectively [100]. Laboratory studies on the release profile of starch-encapsulated atrazine revealed that 70% of active ingredients were delivered in three days, while the remaining quantities of herbicides were released over 16 days of incubation. However, the maximum release was noticed after 15 days of application under field conditions as against the peak release of herbicide, which was observed in three days of incubation in

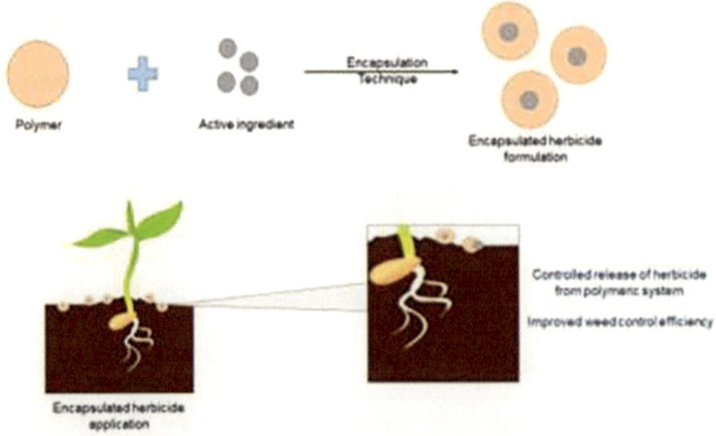


**Figure 5.**
*Overview of encapsulated herbicide formulation.*

*in vitro* study [162]. Moreover, multilayer encapsulation of active ingredients resulted in the reduction of burst release and extended the release period (**Figure 5**) [152].

## 10. Longevity of weed control by encapsulated herbicides

Encapsulated herbicide formulation delivers active material to the target environment in a sustained pattern thus protecting active molecules from environmental vulnerability and eventually resulted in an efficient and extended period of weed control. Ethyl cellulose-based microencapsulated alachlor formulation showed greater herbicidal activity even after 30 days of application achieving long-term weed control [143]. Efficacy of free and encapsulated metribuzin in poly (3-hydroxybutyrate) was tested against *Avena fatua* as target species [96]. The results revealed that encapsulated metribuzin offered prolonged weed control of 70 days against *Avena fatua*, whereas 40% of germinated weeds were observed at 42 days with application of conventional metribuzin formulation. Similarly, metazachlor and pendimethalin were encapsulated separately using terpolymer (L-Lactide/Glycolide/PEG) where the weed control was effective against target weed species for 2–3 months [163]. Encapsulated formulation of pendimethalin herbicide delivered herbicide sustainably during the period of forty days to achieve season-long weed control [88].

## 11. Residual effect of encapsulated herbicide formulation

Encapsulation of herbicides exhibits the same herbicidal activity at a lower dose as compared to its conventional formulation. Controlled release formulation reduced the amount of active ingredient applied to the environment thus reducing the residue buildup in agro-ecosystem. Herbicide encapsulation conferred the controlled release of active material to maintain the threshold level of herbicides for an extended period to control weeds [164]. Encapsulated herbicide formulation was not active in soil to affect the succeeding crop [165]. Residual effect of poly (ε-caprolactone) based atrazine formulation was validated on soybean plants [94]. The results showed that nano-formulation enhanced the short-term without causing a long-term residual effect.

## 12. Conclusion

Weeds are the crucial yield-limiting factor in crop production that affect crop growth and yield dither directly or indirectly. There are different weed management options are available; however, chemical weed control strategy is quite effective among them. Herbicide is an important component in weed management and registered a major share in the pesticide market to improve crop productivity. Application of herbicide poses several environmental consequences since herbicides are subjected to different degradation processes in agro-ecosystem. Encapsulation of herbicides is an innovative strategy and offers a controlled release system to address the issues of chemical weed control. Polymers are unique and explored their specific characteristics for the encapsulation of active ingredients. There are numerous polymers available to design smart release formulation. Herbicide encapsulation improves stability of active ingredient and safeguards the active molecules from environmental vulnerability. Further, it enhances the bioactivity, which helps to achieve prolonged weed control with higher efficiency. Encapsulated formulation achieves the same weed control efficiency at lower dosage as compared to the conventional formulation. Research evidence showed that there was no significant residue carryover due to application of encapsulated formulation. Therefore, the development of encapsulated herbicide formulation has greater scope in crop protection. Moreover, encapsulated formulation will make a greater revolution in the chemical era to manage weeds at a lower rate of application. However, costlier instruments are required for designing and characterization of encapsulated formulation, and regulatory evaluation of nano-formulation are few limitations for development and commercialization of smart herbicides.

## Conflict of interest

The authors declare no conflict of interest.

## Author details

S. Marimuthu[1*], P. Pavithran[2] and G. Gowtham[1]

1 Department of Nano Science and Technology, Tamil Nadu Agricultural University, Coimbatore, Tamil Nadu, India

2 Department of Agronomy, Tamil Nadu Agricultural University, Coimbatore, Tamil Nadu, India

*Address all correspondence to: sm20@tnau.ac.in

## References


[1] DWR. Vision 2050. Directorate of Weed research. Indian Council of Agricultural Research. 2015:1-32

[2] Oerke EC. Crop losses to pests. The Journal of Agricultural Science. 2006;**144**(1):31-43

[3] Rao A, Chauhan B. Weeds and weed management in India: A review. Indian Journal of Weed Science. 2015;**2015**:87-118

[4] ISWS. Proceedings of ISWS Biennial Conference on Weed Management for Enhancing Farmers' Income and Food Security. Goa, India: ICAR-CCARI; 2020. p. 270

[5] Choudhury P, Singh R. Role of herbicides in agriculture. Indian Farming. 2015;**65**:66-72

[6] De A, Bose R, Kumar A, Mozumdar S. Worldwide pesticide use. In: Targeted Delivery of Pesticides Using Biodegradable Polymeric Nanoparticles. Dordrecht: Springer; 2014. p. 5-6

[7] Buchholtz KP. Report of the terminology committee of the Weed Science Society of America. Weeds. 1967;**15**:388-389

[8] Talaka A, Rajab YS. Weed biology and ecology: A key to successful weed management and control. Journal of Agriculture Veterinary Science (IOSR-JAVS). 2013;**2**(3):11-14

[9] Naylor R. Weed seed biology. In: Thomas B, Murphy DJ, Murray BG, editors. Encyclopedia of Applied Plant Sciences. Amsterdam: Elsevier; 2003. p. 1500-1508

[10] Garcia-Baudin JM, Salto T, Aguirre R. Preliminary note on the germination of *Phalaris brachystachys* Link and *Phalaris minor* Retz. In: Proceedings of the 6th International Colloquium on Weed Ecology, Biology and Systematics. Montpellier: European Weed Research Society; 1980. p. 123-131

[11] Jayachandra. Parthenium weed in Mysore state and its control. Current Science. 1971;**5**:568-569

[12] Wilson RG. Biology of weed seed in the soil. In: Altieri MA, Liebman M, editors. Weed Management in Agroecosystems: Ecological Approaches. Boca Raton, FL: CRC Press; 1988. pp. 25-39

[13] Joshi S. Biocontrol of *Parthenium hysterophorus* L. Crop Protection. 1991;**10**(6):429-431

[14] Raju RA, Reddy MN. Control of jungle rice in the Godavari delta. Indian Farming. 1989;**39**(2):30-31

[15] Gupta RK. Foodborne infectious diseases. Food Safety in the 21st Century. 2017;**1**:13-28

[16] Gnanavel I, Natarajan SK. *Parthenium hysterophorus* L.: A major threat to natural and agro eco-systems in India. International Journal of Agriculture, Environment and Biotechnology. 2013;**6**(2):261-269

[17] Cardina J, Sparrow DH. Temporal changes in velvetleaf (*Abutilon theophrasti*) seed dormancy. Weed Science. 1997;**45**(1):61-66

[18] Qasem JR. Weed seed dormancy: The ecophysiology and survival strategies. In: Seed Dormancy and Germination. London, UK: IntechOpen; 2019. pp. 20-54

[19] Stevens OA. Perennial Sow Thistle: Growth and Reproduction. North Dakota: Agricultural Experiment Station; 1924

[20] Americanos PG, Labrada R, Caseley JC, Parker C. *Convolvulus arvensis* L. Estudio FAO. Producción y Protección Vegetal (FAO). Rome: FAO; 1996. p. 120

[21] Odum S. Germination of ancient seeds. Dansk Botanisk Arkiv. 1965;**24**:1-70

[22] Libby WF. Radiocarbon Dating. 2nd ed. Chicago; IL: University of Chicago Press; 1955

[23] Gulden RH, Shirtliffe SJ. Weed seed banks: Biology and management. Prairie Soils and Crops Journal. 2009;**2**:46-52

[24] Biradar DP, Shivakumar KS, Prakash SS, Pujar BT. Bionutrient potentiality of *Parthenium hysterophorus* and its utility as green manure in rice ecosystem. Karnataka Journal of Agricultural Sciences. 2006;**19**(2):256

[25] Reddy KS, Gopinath KA, Kumari VV, Ramesh K. Weed-nutrient interactions in agricultural systems. Indian Journal of Fertilizer. 2018;**14**:50-58

[26] Mahanta K, Jha DK. Nutritional status of vermicompost produced from weed biomass and rice straw as influenced by earthworm species and seasons. Indian Journal of Weed Science. 2009;**41**:211-215

[27] Kanitkar NV, Sirur SS, Gokhale DH. Dryfarming in India. 2nd ed. New Delhi: Indian Council of Agricultural Research; 1960

[28] Asgharipour MR. Effects of planting depth on germination and the emergence of field Bindweed (*Convolvulus arvensis* L.). Asian Journal of Agricultural Sciences. 2011;**3**(6):459-461

[29] Balyan RS, Bhan VM. Germination of horse purslane (*Trianthema portulacastrum*) in relation to temperature, storage conditions, and seeding depths. Weed Science. 1986;**34**(4):513-515

[30] Benvenuti S. Weed seed movement and dispersal strategies in the agricultural environment. Weed Biology and Management. 2007;**7**(3):141-157

[31] Hume L, Archibold OW. The influence of a weedy habitat on the seed bank of an adjacent cultivated field. Canadian Journal of Botany. 1986;**64**(9):1879-1883

[32] Cramp S, Brooks DJ. Handbook of the Birds of Europe, the Middle East and North Africa: The Birds of the Western Palearctic. Warblers: Oxford University Press; 1993

[33] Rao VS. Principles of Weed Science. 2nd ed. New York: Science Publishers; 2000. p. 526

[34] Gharde Y, Singh PK, Dubey RP, Gupta PK. Assessment of yield and economic losses in agriculture due to weeds in India. Crop Protection. 2018;**107**:12-18

[35] Rao AN, Wani SP, Ladha JK. Weed Management Research in India: An Analysis of the Past and Outlook for Future4. Jabalpur: Directorate of Weed Research; 2014. pp. 1-26

[36] Soltani N, Dille JA, Robinson DE, Sprague CL, Morishita DW, Lawrence NC, et al. Potential yield loss in sugar beet due to weed interference in the United States and Canada. Weed Technology. 2018;**32**(6):749-753

[37] Bakht T, Khan IA. Weed control in tomato (*Lycopersicon esculentum* Mill.) through mulching and herbicides. Pakisthan Journal of Botany. 2014;**46**(1):289-292

[38] Patel RH, Shroff J, Dutta S, Meisheri TG. Weed dynamics as

influenced by soil solarization: A review. Agricultural Reviews. 2005;**26**(4):295-300

[39] Akshatha V, Prameela KP, Narayanankutty C, Menon MV. Weed management in cabbage (*Brassica oleracea* var. *capitata* L.). Journal of Tropical Agriculture. 2019;**56**(2):187-190

[40] Singh M, Kaul A, Pandey V, Bimbraw AS. Weed management in vegetable crops to reduce the yield losses. International Journal of Current Microbiology and Applied Sciences. 2019;**8**(7):1241-1258

[41] Singh RK, Bisen N, Rani M. Flurochloridone-A promising herbicide for weed management in carrot. Indian Journal of Weed Science. 2017;**49**(4):409-410

[42] Singh S, Chhokar RS, Gopal R, Ladha JK, Gupta RK, Kumar V, et al. Integrated Weed Management: A Key to Success for Direct-seeded Rice in the Indo-Gangetic Plains. Los Banos, Philippines: International Rice Research Institute; 2009. pp. 261-278

[43] Llewellyn R, Ronning D, Ouzman J, Walker S, Mayfield A, Clarke M. Impact of weeds on Australian grain production: The cost of weeds to Australian grain growers and the adoption of weed management and tillage practices. In: Report for GRDC. Australia: CSIRO; 2016

[44] Mesterhazy A, Oah J, Popp J. Losses in the grain supply chain: Causes and solutions. Sustainability. 2020;**12**(6):2342

[45] Rao AN. The historical and future perspective of Weed Science research in India. In: Fifty Years of Weed Research in India. Jabalpur: Indian Society of Weed Science; 2018. p. 1-23

[46] Das TK. Weed Science: Basics and Applications. Jain Brothers: New Delhi; 2008. p. 901

[47] Nutter FF. The Role of Plant Disease Epidemiology in Developing Successful Integrated Disease Management Programs. In: Ciancio A, Mukerji KG, editors. General Concepts in Integrated Pest and Disease Management. Integrated Management of Plants Pests and Diseases. Dordrecht: Springer; 2007. p. 45-79

[48] TBI. Plant Protection Code. Kolkata, West Bengal: Tea Board of India (TBI); 2019. p. 58

[49] Rathore SS, Shekhawat K, Premi OP, Kandpal BK. Major Weeds of Rapeseed–Mustard in India. All India Coordinated Research Project on Rapeseed–Mustard. Sewar, Bharatpur, India: ICAR–Directorate of Rapeseed–Mustard Research; 2012. p. 36

[50] Skolotneva ES, Lekomtseva SN, Kosman E. The wheat stem rust pathogen in the central region of the Russian Federation. Plant Pathology. 2013;**62**(5):1003-1010

[51] Balaich JN. The effects of the natural product parthenin on the development of plasmodium in the anopheline mosquito [Thesis]. Maryland: Johns Hopkins University; 2015

[52] Yaduraju NT, Mishra JS. Smart weed management: A small step towards doubling farmers' income. Indian Journal of Weed Science. 2018;**50**(1):1-5

[53] Zimdahl LR. Introduction to Chemical Weed Control: Fundamentals of Weed Science. 5th ed. New York: Academic Press; 2018. pp. 391-416

[54] Rahman M. Herbicidal weed control: Benefits and risks. Advances in Plants & Agriculture Research. 2016;**4**(5):00153

[55] Telkar SG, Gurjar GN, Dey JK, Kant K, Solanki SP. Biological weed

control for sustainable agriculture. International Journal of Economic Plants. 2015;**2**(4):181-183

[56] Kevan P, Shipp L. Biological control and biotechnological amelioration in managed ecosystems. In. Moo-Young M, editor. Comprehensive Biotechnology. 2nd edition. Amsterdam: Elsevier; 2011. p. 757-761

[57] Evans HC. Biological control of weeds. In: Agricultural Applications. Berlin Heidelberg, Germany: Springer; 2002. pp. 135-152

[58] DeFelice MS. Prickly pear cactus, *Opuntia* spp.: A spine-tingling tale. Weed Technology. 2004;**18**(3):869-877

[59] Menne H, Kocher H. HRAC classification of herbicides and resistance development. Modern Crop Protection Compounds. 2007;**23**:5-26

[60] Hay JR. Gains to the grower from weed science. Weed Science. 1974;**22**(5):439-442

[61] Green MB, Hartley GS, West TF. Chemicals for crop improvement and pest management. 3rd edition. USA: Pergamon Press; 1987

[62] Blanco FM, Ramos YG, Scarso MF, Jorge LA. Determining the Selectivity of Herbicides and Assessing Their Effect on Plant Roots: A Case Study with Indaziflam and Glyphosate Herbicides. London: InTech; 2015. 275-97 p

[63] HRAC herbicide classification [Internet]. 2020. Available from: https://hracglobal.com/tools/hrac-moa-2020-revision-description-and-master-herbicide-list [Accessed February 11, 2022].

[64] Blasioli S, Braschi I, Gessa CE. The Fate of Herbicides in Soil. In: Andreas Kortekamp, editors. Herbicides and Environment. London: Intechopen; 2011. p. 175-193

[65] Tadeo-Jalife NR, Vasquez-Medrano R, Quevedo IR. Assessing the adsorption of bipyridinium herbicides on model soil granular media. Frontiers in Water. 2021;2

[66] Sharipov U, Kocarek M, Jursik M, Nikodem A, Boruvka L. Adsorption and degradation behavior of six herbicides in different agricultural soils. Environmental Earth Sciences. 2021;**80**(20):1-4

[67] El-Nahhal Y, Hamdona N. Adsorption, leaching and phytotoxicity of some herbicides as single and mixtures to some crops. Journal of the Association of Arab Universities for Basic and Applied Sciences. 2017;**22**:17-25

[68] Cessna AJ, Elliott JA, Bailey J. Leaching of three sulfonylurea herbicides during sprinkler irrigation. Journal of Environmental Quality. 2010;**39**(1):365-374

[69] Shaner DL. Field dissipation of sulfentrazone and pendimethalin in Colorado. Weed Technology. 2012;**26**(4):633-637

[70] Parochetti JV, Dec GW. Photodecomposition of eleven dinitroaniline herbicides. Weed Science. 1978;**26**(2):153-156

[71] Florencio MH, Pires E, Castro AL, Nunes MR, Borges C, Costa FM. Photodegradation of Diquat and Paraquat in aqueous solutions by titanium dioxide: Evolution of degradation reactions and characterisation of intermediates. Chemosphere. 2004;**55**(3):345-355

[72] Grey TL, McCullough PE. Sulfonylurea herbicides' fate in soil: Dissipation, mobility, and

other processes. Weed Technology. 2012;**26**(3):579-581

[73] Mohamed AT, El-Hussein AA, El Siddig MA, Osman AG. Degradation of oxyfluorfen herbicide by soil microorganisms, biodegradation of herbicides. Biotechnol. 2011;**10**:274-279

[74] Stibal M, Bælum J, Holben WE, Sorensen SR, Jensen A, Jacobsen CS. Microbial degradation of 2, 4-dichlorophenoxyacetic acid on the Greenland ice sheet. Applied and Environmental Microbiology. 2012;**78**(15):5070-5076

[75] Kanissery R, Gairhe B, Kadyampakeni D, Batuman O, Alferez F. Glyphosate: Its environmental persistence and impact on crop health and nutrition. Plants. 2019;**8**(11):499

[76] Gavrilescu M. Fate of pesticides in the environment and its bioremediation. Engineering in Life Sciences. 2005;**5**(6):497-526

[77] Janaki P, Sharma N, Chinnusamy C, Sakthivel N, Nithya C. Herbicide residues and their management strategies. Indian Journal of Weed Science. 2015;**47**(3):329-344

[78] Parron T, Requena M, Hernandez AF, Alarcon R. Association between environmental exposure to pesticides and neurodegenerative diseases. Toxicology and Applied Pharmacology. 2011;**256**(3):379-385

[79] Hernandez AF, Parron T, Alarcon R. Pesticides and asthma. Current Opinion in Allergy and Clinical Immunology. 2011;**11**(2):90-96

[80] Nerozzi C, Recuero S, Galeati G, Bucci D, Spinaci M, Yeste M. Effects of Roundup and its main component, glyphosate, upon mammalian sperm function and survival. Scientific Reports. 2020;**10**(1):1-9

[81] Portier CJ. A comprehensive analysis of the animal carcinogenicity data for glyphosate from chronic exposure rodent carcinogenicity studies. Environmental Health. 2020;**19**(1):1-7

[82] Jowa L, Howd R. Should atrazine and related chlorotriazines be considered carcinogenic for human health risk assessment? Journal of Environmental Science and Health, Part C. 2011;**29**(2): 91-144

[83] Duary B. Recent advances in herbicide resistance in weeds and its management. Indian Journal of Weed Sciences. 2008;**24**:124-135

[84] Heap I. The International Herbicide-Resistant Weed Database. 2022. Available from www.weedscience.org [Accessed: February 12, 2022].

[85] Schreiber MM, Shasha BS, Trimnell D, White MD. Controlled release herbicides. Monograph Series of the Weed Science Society of America. USA: Weed Science Society of America; 1987. p. 177-191

[86] Nuruzzaman MD, Rahman MM, Liu Y, Naidu R. Nanoencapsulation, nano-guard for pesticides: A new window for safe application. Journal of Agricultural and Food Chemistry. 2016;**64**(7):1447-1483

[87] Kumar S, Bhanjana G, Sharma A, Dilbaghi N, Sidhu MC, Kim KH. Development of nanoformulation approaches for the control of weeds. Science of the Total Environment. 2017;**586**:1272-1278

[88] Kumar P, Chinnamuthu CR. Assembly of nanoencapsulated pendimethalin herbicide using solvent

evaporation method for season long weed control under irrigated ecosystem. International Journal of Pure and Applied Bioscience. 2017;5:349-357

[89] Pirzada T, de Farias BV, Mathew R, Guenther RH, Byrd MV, Sit TL, et al. Recent advances in biodegradable matrices for active ingredient release in crop protection: Towards attaining sustainability in agriculture. Current Opinion in Colloid & Interface Science. 2020;**48**:121-136

[90] Campos EV, de Oliveira JL, Fraceto LF, Singh B. Polysaccharides as safer release systems for agrochemicals. Agronomy for Sustainable Development. 2015;**35**(1):47-66

[91] Mogul MG, Akin H, Hasirci N, Trantolo DJ, Gresser JD, Wise DL. Controlled release of biologically active agents for purposes of agricultural crop management. Resources, Conservation and Recycling. 1996;**16**(1-4):289-320

[92] Dourado Junior SM, Nunes ES, Marques RP, Rossino LS, Quites FJ, Siqueira JR, et al. Controlled release behavior of sulfentrazone herbicide encapsulated in Ca-ALG microparticles: Preparation, characterization, mathematical modeling and release tests in field trial weed control. Journal of Materials Science. 2017;**52**(16):9491-9507

[93] Wienhold BJ, Sadeghi AM, Gish TJ. Effect of starch encapsulation and temperature on volatilization of atrazine and alachlor. 1993;**22**:162-166

[94] Preisler AC, Pereira AE, Campos EV, Dalazen G, Fraceto LF, Oliveira HC. Atrazine nanoencapsulation improves pre-emergence herbicidal activity against *Bidens pilosa* without enhancing long-term residual effect on *Glycine max*. Pest Management Science. 2020;**76**(1):141-149

[95] Chinnamuthu CR, Kokiladevi E. Weed management through nanoherbicides. Application of Nanotechnology in Agriculture. 2007;**10**:978-971

[96] Volova T, Shumilova A, Zhila N, Sukovatyi A, Shishatskaya E, Thomas S. Efficacy of slow-release formulations of metribuzin and tribenuron methyl herbicides for controlling weeds of various species in wheat and barley stands. ACS Omega. 2020;**5**(39):25135-25147

[97] Azwa ZN, Yousif BF, Manalo AC, Karunasena W. A review on the degradability of polymeric composites based on natural fibres. Materials & Design. 2013;**47**:424-442

[98] Grillo R, Rosa AH, Fraceto LF. Poly (ε-caprolactone) nanocapsules carrying the herbicide atrazine: Effect of chitosan-coating agent on physico-chemical stability and herbicide release profile. International Journal of Environmental Science and Technology. 2014;**11**(6):1691-1700

[99] Grillo R, Pereira AE, Nishisaka CS, De Lima R, Oehlke K, Greiner R, et al. Chitosan/tripolyphosphate nanoparticles loaded with paraquat herbicide: An environmentally safer alternative for weed control. Journal of Hazardous Materials. 2014;**278**:163-171

[100] Nornberg AB, Gehrke VR, Mota HP, Camargo ER, Fajardo AR. Alginate-cellulose biopolymeric beads as efficient vehicles for encapsulation and slow-release of herbicide. Colloids and Surfaces A: Physicochemical and Engineering Aspects. 2019;**583**:123970

[101] Taban A, Saharkhiz MJ, Khorram M. Formulation and assessment of nano-encapsulated bioherbicides based on biopolymers and essential oil. Industrial Crops and Products. 2020;**149**:112348

[102] Sikder A, Pearce AK, Parkinson SJ, Napier R, O'Reilly RK. Recent trends in advanced polymer materials in agriculture related applications. ACS Applied Polymer Materials. 2021;**3**(3):1203-1217

[103] Viji N, Chinnamuthu CR. Encapsulation of glyphosate herbicide with metal nanoparticles and water soluble polymers for controlled release of herbicide. Agricultural Research Journal. 2019;**56**(2):231-235

[104] Smidsrod O, Skja G. Alginate as immobilization matrix for cells. Trends in Biotechnology. 1990;**8**:71-78

[105] Joshi PP, Van Cleave A, Held DW, Howe JA, Auad ML. Preparation of slow release encapsulated insecticide and fertilizer based on superabsorbent polysaccharide microbeads. Journal of Applied Polymer Science. 2020;**137**(39):49177

[106] Faria DM, Dourado SM, Nascimento JP, Nunes ED, Marques RP, Rossino LS, et al. Development and evaluation of a controlled release system of TBH herbicide using alginate microparticles. Materials Research. 2016;**20**:225-235

[107] Zhifeng Z, Cuiyun Y, Liping W. Effect of starch sources on the release rates of herbicides encapsulated. Wuhan University Journal of Natural Sciences. 2006;**11**(2):423-426

[108] Kannamreddy V, Chinnamuthu CR, Marimuthu S, Bharathis C. Synthesizing nanoencapsulated sulfentrazone herbicide and optimizing time and dose for season long weed management in irrigated blackgram (*Vigna mungo* L.). Legume Research: An International Journal. 2021:1-8

[109] Kashyap PL, Xiang X, Heiden P. Chitosan nanoparticle based delivery systems for sustainable agriculture. International Journal of Biological Macromolecules. 2015;**77**:36-51

[110] Rychter P. Chitosan/glyphosate formulation as a potential, environmental friendly herbicide with prolonged activity. Journal of Environmental Science and Health, Part B. 2019;**54**(8):681-692

[111] BeMiller JN. An introduction to pectins: Structure and properties. In: Chemistry and Functions of Pectin. Washington, DC: American Chemical Society; 1986. pp. 2-12

[112] Yadav P, Pandey P, Parashar S. Pectin as natural polymer: An overview. Research Journal of Pharmacy and Technology. 2017;**10**(4):1225

[113] Pavithran P, Marimuthu S, Chinnamuthu CR, Lakshmanan A, Bharathi C, Kadhiravan S. Synthesis and characterization of pectin beads for the smart delivery of agrochemicals. 2021;**33**(22):136-155

[114] Yearla SR, Padmasree K. Preparation and characterisation of lignin nanoparticles: Evaluation of their potential as antioxidants and UV protectants. Journal of Experimental Nanoscience. 2016;**11**(4):289-302

[115] Farooq M, Zou T, Riviere G, Sipponen MH, Osterberg M. Strong, ductile, and waterproof cellulose nanofibril composite films with colloidal lignin particles. Biomacromolecules. 2018;**20**(2):693-704

[116] Weiss R, Ghitti E, Sumetzberger-Hasinger M, Guebitz GM, Nyanhongo GS. Lignin-based pesticide delivery system. ACS Omega. 2020;**5**(8):4322-4329

[117] Flores-Cespedes F, Daza-Fernandez I, Villafranca-Sanchez M, Fernandez-Perez M, Morillo E,

Undabeytia T. Lignin and ethylcellulose in controlled release formulations to reduce leaching of chloridazon and metribuzin in light-textured soils. Journal of Hazardous Materials. 2018;**343**:227-234

[118] Del Valle EM. Cyclodextrins and their uses: A review. Process biochemistry. 2004;**39**(9):1033-1046

[119] Garrido EM, Rodrigues D, Milhazes N, Borges F, Garrido J. Molecular encapsulation of herbicide terbuthylazine in native and modified β-cyclodextrin. Journal of Chemistry. 2017:1-9

[120] Vaughn SF, Kenar JA, Felker FC, Berhow MA, Cermak SC, Evangelista RL, et al. Evaluation of alternatives to guar gum as tackifiers for hydromulch and as clumping agents for biodegradable cat litter. Industrial Crops and Products. 2013;**43**:798-801

[121] Kumar V, Singh A, Das TK, Sarkar DJ, Singh SB, Dhaka R, et al. Release behavior and bioefficacy of imazethapyr formulations based on biopolymeric hydrogels. Journal of Environmental Science and Health, Part B. 2017;**52**(6):402-409

[122] Guarino V, Gentile G, Sorrentino L, Ambrosio L. Polycaprolactone: Synthesis, properties, and applications. In. Mark HF, editor. Encyclopedia of Polymer Science and Technology. 4th edition. London: Wiley; 2002. p. 1-36

[123] Mckeen L. Renewable resource and biodegradable polymers. In. Mckeen, editor. The Effect of Sterilization on Plastics and Elastomers. Amsterdam: Elsevier; 2012. p. 305-317

[124] Diyanat M, Saeidian H, Baziar S, Mirjafary Z. Preparation and characterization of polycaprolactone nanocapsules containing pretilachlor as a herbicide nanocarrier. Environmental Science and Pollution Research. 2019;**26**(21):21579-21588

[125] Grillo R, Oliveira HC, Renata L, Fraceto LF. Polymeric nanoparticles as carrier systems for herbicides. Journal of Nanomedicine and Nanotechnology. 2015;**6**(4):119

[126] Oliveira HC, Stolf-Moreira R, Martinez CB, Grillo R, de Jesus MB, Fraceto LF. Nanoencapsulation enhances the post-emergence herbicidal activity of atrazine against mustard plants. PLoS One. 2015;**10**(7):e0132971

[127] Sousa GF, Gomes DG, Campos EV, Oliveira JL, Fraceto LF, Stolf-Moreira R, et al. Post-emergence herbicidal activity of nanoatrazine against susceptible weeds. Frontiers in Environmental Science. 2018;**6**:12

[128] Rao J, Chandrani AN, Powar A, Chandra S. Design and application of polyurea microcapsules containing herbicide (oxyfluorfen). Designed Monomers and Polymers. 2020;**23**(1):155-163

[129] Chen H, Liu X, Deng S, Wang H, Ou X, Huang L, et al. Pretilachlor releasable polyurea microcapsules suspension optimization and its practical herbicidal security investigations. Frontiers in Chemistry. 2020;**8**:826

[130] Zhang DX, Zhang XP, Luo J, Li BX, Wei Y, Liu F. Causation analysis and improvement strategy for reduced pendimethalin herbicidal activity in the field after encapsulation in polyurea. ACS Omega. 2018;**3**(1):706-716

[131] Yılmaz H, Enginar H, Cifci C. Microencapsulation of pendimethalin with polyurethane-urea and determination of its stability. Journal of Taibah University for Science. 2021;**15**(1):685-694

[132] Chiono V, Sartori S, Calzone S, Boffito M, Tonda-Turo C, Mattu C, Gentile P, Ciardelli G. Synthetic biodegradable medical polyurethanes. In. Zhang XC, editor. Science and Principles of Biodegradable and Bioresorbable Medical Polymers. Amsterdam: Elsevier; 2017. p. 189-216

[133] Daneshvari G, Yousefi AR, Mohammadi M, Banibairami S, Shariati P, Rahdar A, et al. Controlled-release formulations of trifluralin herbicide by interfacial polymerization as a tool for environmental hazards. 2021;**11**(6):13866-13877

[134] Jamshidian M, Tehrany EA, Imran M, Jacquot M, Desobry S. Poly-lactic acid: Production, applications, nanocomposites, and release studies. Comprehensive Reviews in Food Science and Food Safety. 2010;**9**(5):552-571

[135] Karkri M. Thermal conductivity of biocomposite materials. In. Sadasivuni KK, Ponnamma D, Kim J, Cabibihan JJ, AlMaadeed MA, editors. Biopolymer Composites in Electronics. Amsterdam: Elsevier; 2017. p. 129-153

[136] Stloukal P, Kucharczyk P, Sedlarik V, Bazant P, Koutny M. Low molecular weight poly (lactic acid) microparticles for controlled release of the herbicide metazachlor: Preparation, morphology, and release kinetics. Journal of Agricultural and Food Chemistry. 2012;**60**(16):4111-4119

[137] Kucharczyk P, Pavelkova A, Stloukal P, Koutny M, Sedlarik V. Preparation of Nanoparticles Based on Biodegradable Chain Linked PLA/PEG Polymer for Controlled Release of Herbicides Metazachlor. Nanocon: Tanger Ltd, Brno; 2013. pp. 448-452

[138] Salac J, Sopík T, Stloukal P, Janasova N, Jursik M, Koutny M. Slow release formulation of herbicide metazachlor based on high molecular weight poly (lactic acid) submicro and microparticles. International Journal of Environmental Science and Technology. 2019;**16**(10):6135-6144

[139] Makadia HK, Siegel SJ. Poly lactic-co-glycolic acid (PLGA) as biodegradable controlled drug delivery carrier. Polymers. 2011;**3**(3):1377-1397

[140] Schnoor B, Elhendawy A, Joseph S, Putman M, Chacon-Cerdas R, Flores-Mora D, et al. Engineering atrazine loaded poly (lactic-co-glycolic acid) nanoparticles to ameliorate environmental challenges. Journal of Agricultural and Food Chemistry. 2018;**66**(30):7889-7898

[141] Rekhi GS, Jambhekar SS. Ethylcellulose: A polymer review. Drug Development and Industrial Pharmacy. 1995;**21**(1):61-77

[142] Elbahri Z, Taverdet JL. Optimization of an herbicide release from ethylcellulose microspheres. Polymer Bulletin. 2005;**54**(4):353-363

[143] Sopena F, Cabrera A, Maqueda C, Morillo E. Ethylcellulose formulations for controlled release of the herbicide alachlor in a sandy soil. Journal of Agricultural and Food Chemistry. 2007;**55**(20):8200-8205

[144] Perez-Martínez JI, Morillo E, Maqueda C, Gines JM. Ethyl cellulose polymer microspheres for controlled release of norfluazon. Pest Management Science: Formerly Pesticide Science. 2001;**57**(8):688-694

[145] Sopena F, Cabrera A, Maqueda C, Morillo E. Controlled release of the herbicide norflurazon into water from ethylcellulose formulations. Journal of Agricultural and Food Chemistry. 2005;**53**(9):3540-3547

[146] Cea M, Cartes P, Palma G, Mora ML. Atrazine efficiency in an andisol as affected by clays and nanoclays in ethylcellulose controlled release formulations. Revista de la ciencia del suelo y nutrición vegetal. 2010;**10**(1):62-77

[147] Li J, Li Y, Dong H. Controlled release of herbicide acetochlor from clay/carboxylmethylcellulose gel formulations. Journal of Agricultural and Food Chemistry. 2008;**56**(4):1336-1342

[148] Heydari M, Yousefi AR, Nikfarjam N, Rahdar A, Kyzas GZ, Bilal M. Plant-based nanoparticles prepared from protein containing tribenuron-methyl: Fabrication, characterization, and application. Chemical and Biological Technologies in Agriculture. 2021;**8**(1):1-11

[149] Macedo DF, Dourado SM, Nunes ES, Marques RP, Moreto JA. Controlled release of TBH herbicide encapsulated on Ca-ALG microparticles: Leaching and phytointoxication plants. Planta Daninha. 2019;**37**: e019189934

[150] Maruyama CR, Guilger M, Pascoli M, Bileshy-José N, Abhilash PC, Fraceto LF, et al. Nanoparticles based on chitosan as carriers for the combined herbicides imazapic and imazapyr. Scientific Reports. 2016;**6**(1):1-5

[151] Sousa BT, Pereira S, do Espírito A, Fraceto LF, de Oliveira HC, Dalazen G. Effectiveness of nanoatrazine in post-emergent control of the tolerant weed *Digitaria insularis*. Journal of Plant Protection Research. 2020;**60**(2):185-192

[152] Wang X, Zhao J. Encapsulation of the herbicide picloram by using polyelectrolyte biopolymers as layer-by-layer materials. Journal of Agricultural and Food Chemistry. 2013;**61**(16):3789-3796

[153] Wuana RA, Nnamonu LA, Sha'Ato R, Itodo HU. Characterization of formulated copper chitosan nanoparticles for the controlled release studies of dicamba dimethylamine herbicide. ChemSearch Journal. 2019;**10**(2):74-87

[154] Carr ME, Wing RE, Doane WM. Clay as a carrier in starch encapsulated herbicides prepared by extrusion processing. Starch-Starke. 1994;**46**(1):9-13

[155] Mahmoudian M, Torbati S, AliMirzayi N, Nozad E, Kochameshki MG, Shokri A. Preparation and investigation of poly (methylmethacrylate) nano-capsules containing haloxyfop-R-methyl and their release behavior. Journal of Environmental Science and Health, Part B. 2020;**55**(4):301-309

[156] Yadav S, Yadav RB, Chuhan SS, Kumar R, Kumar V. Efficacy of different herbicides and its combination against the weed flora of transplanted rice: A review. International Journal of Current Microbiology and Applied Science. 2020;**9**(8):2057-2068

[157] Zhu Z, Yu C, Zhou X, Wang L. Starch modification for slow release of herbicide encapsulated with concentrated starch paste. Wuhan University Journal of Natural Sciences. 2007;**12**(3):515-521

[158] Fernandez-Perez M, Villafranca-Sanchez M, Flores-Cespedes F, Daza-Fernandez I. Lignin-polyethylene glycol matrices and ethylcellulose to encapsulate highly soluble herbicides. Journal of Applied Polymer Science. 2015;**132**(6):41422

[159] Rebitski EP, Darder M, Aranda P. Layered double hydroxide/sepiolite hybrid nanoarchitectures for the controlled release of herbicides.

Beilstein Journal of Nanotechnology. 2019;**10**(1):1679-1690

[160] Rashidipour M, Maleki A, Kordi S, Birjandi M, Pajouhi N, Mohammadi E, et al. Pectin/chitosan/tripolyphosphate nanoparticles: Efficient carriers for reducing soil sorption, cytotoxicity, and mutagenicity of paraquat and enhancing its herbicide activity. Journal of Agricultural and Food Chemistry. 2019;**67**(20):5736-5745

[161] dos Santos SM, Cocenza DS, Grillo R, de Melo NF, Tonello PS, de Oliveira LC, et al. Paraquat-loaded alginate/chitosan nanoparticles: Preparation, characterization and soil sorption studies. Journal of Hazardous Materials. 2011;**190**(1-3):366-374

[162] Mills MS, Thurman EM. Reduction of nonpoint source contamination of surface water and groundwater by starch encapsulation of herbicides. Environmental Science & Technology. 1994;**28**(1):73-79

[163] Lewicka K, Dobrzynski P, Rychter P. PLAGA-PEG-PLAGA terpolymer-based carriers of herbicides for potential application in environment-friendly. Controlled Release Systems of Agrochemicals Materials. 2020;**13**(12):2778

[164] Collins RL, Doglia S, Mazak RA, Samulski ET. Controlled release of herbicides-theory. Weed Science. 1973;**21**(1):1-5

[165] Pereira AE, Grillo R, Mello NF, Rosa AH, Fraceto LF. Application of poly (epsilon-caprolactone) nanoparticles containing atrazine herbicide as an alternative technique to control weeds and reduce damage to the environment. Journal of Hazardous Materials. 2014;**268**:207-215

Chapter 2

# Pesticides and Sunflower Breeding

*Miroslava Hristova-Cherbadzhi*


## Abstract

The amount and quality of yields depend on the successful protection of crops from diseases, pests, weeds, and abiotic factors. The sunflower is a plant in which most diseases and pests are overcome genetically. The chemical method is also used in the production of sunflower, but it is important to say that there are still no genetically modified (GM) sunflowers on the market. By applying the classical breeding methods, new sunflower elite inbreed B lines that are resistant to two classes of herbicides (imidazolinones and sulfonylureas) were received. The aim of this study is to summarize the knowledge of pesticides and their use, as well as the breeding methods and resistance to herbicides in the sunflower.


**Keywords:** pesticides, herbicides, sunflower, hybridization, new trait

## 1. Introduction

The amount and quality of yields depend on the successful protection of crops from diseases, pests, weeds, and abiotic factors. The first attempts to do so were in ancient times when diseased plants and plant pests were manually removed. Later on, additional measures were taken, such as the application of different crop rotation schemes. The first chemicals in agriculture were used only two centuries ago (for fungicides). Nowadays, there are attempts to apply biological methods of plant protection, for example, attracting wild birds.

Nonetheless, the most widely used method is still the chemical one, which is based on the application of certain chemicals—*pesticides* (from Latin, pestis—infection, and cedo—to kill), which affect the living cells of the adversary and result in either their death or the inhibition of their development. Its main tasks are:

1. to protect plants from diseases and pests;

2. to destroy pests and weeds;

3. to treat the diseased plants if possible.

The chemical ingredient that determines the biological action of different types of pesticides is commonly referred to as the *active substance*. When applying the chemical method, it is assumed that this substance will show its action only when it comes into contact with the pathogens (bacteria and fungi), harmful insects, *Acari*, nematodes, rodents, slugs, harmful birds, weeds, or whatever else is being combated. For this purpose, the used detergent is applied either directly on the adversary, on its food, or on the surface on which it moves (soil, water basins, and air). However, the implementation of the pesticides often happens when plants are the most vulnerable and, thus, can be fatally damaged. Anyhow, this overlap is close to unavoidable.

*Advantages:* The chemical method is popular in agricultural practice because it allows to effectively protect even the most threatened crops with minimal human labor costs. Moreover, it makes it possible to quickly cover large areas using modern machines such as ground-based field sprayers, aerosol generators, and agricultural aircraft. These reduce the cost of production and contribute to the improvement of quality while meeting the growing demands of the market. All in all, they guarantee multiple return on investments.

*Disadvantages:* When applied alone, the chemical method never leads to the complete elimination of the danger of a given adversary. Instead, a system of methods for plant protection must be applied—often a combination of different pesticides. On top of that, the improper use of chemicals creates conditions for the pollution of not only the treated agricultural products, but also their immediate surroundings—neighboring crops and water bodies, the soil, and the environment. These lead to the poisoning of wildlife (bees, beetles, ants, fish, etc.). The favorable ratio between beneficial and harmful insects is also violated and that leads to the massive multiplication of the harmful ones as well as the appearance of new pest species that previously were of no concern. In the case of weeds, at the expense of destroyed species, others that are not affected by the same chemical agent appear and take over. In the worst-case scenario, resistant populations of the adversaries are obtained, so measures to prevent and overcome this phenomenon must be taken.

Some of the *requirements* of a pesticide are as follows:

- its physical and chemical properties should not deteriorate when stored (up to 2 years after its production);
- during the preparation of the working solution not to hydrolyze and not to form secondary metabolic compounds that are phytotoxic;
- to not cause deterioration of biochemical composition and nutritional qualities of the resulting plant products.

Additionally, how well a detergent does its job depends on a number of factors, such as the type of chemical, the adversary being controlled, the used machines, the environmental conditions, and so on. Additionally, for the successful application of the chemical method of plant protection, it is necessary to know:

- the basic composition and physical properties of the applied detergent;
- the doses and concentrations at which the best results are obtained;
- the rules to be followed during their application in practice;
- the mechanism of their action on pests and on the protected plants;
- the extent to which these agents are harmful to natural biocenoses and agrophytocenoses and measures to prevent pollution;

- the impact of the given pesticide on humans.

The *grouping* of pesticides:

- according to the pest against which they are applied:
    - *Fungicides*: they are used to prevent and combat plant diseases (*Fungicides* against fungi; *Bactericides* against bacteria; *Viricides* against viruses and mycoplasmas).
    - *Zoocides*: they are used to control harmful plant pests (*Insecticides* against insects; *Acaricides* against herbivorous *Acari*; *Nematicides* against nematodes; *Rodenticides* against rodents; *Lymacides* against slugs; *Avicides* against birds).
    - *Herbicides*: they are used to control weeds in crops.
- according to their mode of action: systemic, penetrating, contact, gastric and gas;
- according to the origin of the active substance, which they contain: of mineral origin, of plant origin, and synthetic;
- according to the method of their application—powders (for dusting), soluble powders (stabilizers are added to them to keep the suspension stable), granular, liquids (solutions and emulsions), and gaseous detergents (ex., *Fumigants*);
- according to the time of their application—for spraying in winter, for application during the vegetation period, etc.;
- according to the mechanism of their action on pathogenic organisms—protective, lethal, and curative;
- according to the place of their application.

*Powder pesticides* (for dusting with agricultural aircraft): It is easy and simple to use them as no prior preparation is required, no water is used, and no working solutions are prepared. At the same time, however, they greatly pollute the work area as they are easily blown away. That also makes it so they can be applied only during windless hours (especially in dry weather). Furthermore, the chemicals are poorly retained on the plants and easy to wash away, which leads to a rapid reduction in their effectiveness and more pollution.

*Soluble powders*: Spraying is a widely used method of applying these. Well-trained workers, the presence of special containers and clothing, and increased inspection of sprayers for corrosion of metal parts are all a must because the working solution has to be prepared with great care. However, a smaller amount of detergent per unit area is used, which covers the plants well due to its improved retention, especially if wetting agents and adhesives are added to the solution. In addition, of great importance for the good absorption and distribution of the applied chemicals are the surface of the leaf blade (presence of waxy coating and hairiness), the angle of the leaves on the stem, their shape and size, as well as their immobility after the passing of the sprayer. As a rule, plants wet from rain or dew are never sprayed since, when the drops of the working solution fall on the wet plant parts, they are repelled, quickly diluted with the available water, and tend to flow down in large drops from the leaves.

*Granular pesticides*: The advantages of using them are that their application overlaps with sowing, that they allow for reduced contact of the worker with the detergent, and the possibility of using more toxic substances while limiting environmental pollution by the targeted application of the chemicals in, for example, rows and nests.

*Aerosol mist* covers large areas and penetrates well into the crowns of perennial species and of crops with a merged surface, but cannot be used even in the slightest wind, because it is easily carried outside the treated areas, which causes environmental pollution. Owing to this disadvantage, aerosols are used mainly in greenhouses and empty warehouses.

*Fumigants* are highly toxic gases, which quickly cover entire warehouses and penetrate everywhere. This requires fumigation to be carried out only by well-trained workers with appropriate work clothes and gas masks in compliance with all rules for safe work with chemicals.

The distinction between pesticides is not strict, as a number of detergents from one group can be applied against two or more groups of adversaries. For example, dinitro-ortho-cresol detergents have been used as insecticides for winter spraying, have both good fungicidal and acaricidal effects, have been used as contact herbicides in cereals and legumes, and have been good nematicides when imported into the soil. A number of fungicides (caratan, morastan, acrex, acricide, etc.) used to control powdery mildew were also good acaricides. Zinc phosphide has been used to control harmful rodents, but it has also been a successful remedy against some harmful insects (mole cricket, woodlice, etc.) and against soil-dwelling nematodes.

In any case, chemical products reduce agricultural harvest. Biotechnology could improve this by helping in developing insect-resistant genetically modified (GM) or herbicide-resistant GM crops. Although sustainable management calls for complete knowledge of the biology of the target adversary and its relationship with other components of the agroecosystem, the areas sown with GM crops increase every year. There are currently 32 GM crops worldwide, eight of which are grown in the European Union. Up-to-date data on the global area of transgenic crops and the resulting desired effect (new trait, enzyme, gene or factor) can be found at https://www.isaaa.org/gmapprovaldatabase/default.asp.

So, what is better: organic food, pesticides treatment, or GM crops?

*Organic food* is safer than regular food but is trickier to produce. It is important to have a proper rotation sequence with all crops, exact dates of planting and harvesting, and tillage practices. Besides, production can be high risk due to potential losses from diseases, insects, birds, and weeds. Biological (beneficial insects, pathogens, and host resistance) or mechanical (temperature, weather events, and trapping) control may be performed.

Pesticides treatment: In the strip-till and no-till technology, with the reduction of the soil treatments, the soil surface remains covered by the residues of the previous crops, and weed control can be difficult. In place of cultivation, a farmer can suppress weeds by managing a cover crop, mowing, crimping, or herbicide application. However, these may result in an increase in total farm expenses or even worse, environmental pollution, and oversaturation of the soil with detergents that adversely affect the development of the next crop.

*GM crops,* in general, need fewer field operations, such as tillage, which allows more residues to remain in the ground, sequestering more $CO_2$ in the soil and reducing greenhouse gas emissions. Owing to the likelihood of genetically modified organisms

**Figure 1.**
*A plant infested by Downy mildew.*

(GMOs) causing problems in humans and animals when consumed, spatial insulation is done to avoid unwanted cross-pollination, i.e., separation of fields with GM crops from confectionery cultures.

It is difficult to find a balance between all these, but using technologies for growing crops, with timely pest control, with the least possible chemical treatment and the best varieties selected for a given microclimate, high yields can be obtained. In fact, using both pesticides and biotech crops is the most sustainable option.

To further examine the topic, special attention will be given to pesticides used on sunflowers and on sunflower breeding itself.

The chemical method is often used in sunflower production (**Figure 1**), but it is important to say that there is still no GM sunflower on the market. All of the herbicide-tolerant traits of the sunflower were grown through traditional plant growing and not biotech means. While genetic engineers aim to produce GMO versions of many food crops, they probably will not succeed in manipulating the sunflower's genes any time soon for two reasons. First, it is difficult to genetically change the sunflower. Second, the sunflower has many wild species to which transgenes can switch, and if that occurs, the result will be mass multiplication of the infected wild forms and pollution of the environment. Consequently, pesticide treatment and sunflower resistance to the used chemicals is crucial.

## 2. Resistance to herbicides in the sunflower

Over the last decade, increasing amounts of sunflower fields have been treated with pesticides. For example, herbicides have been extensively used for weed control in North America since 1973. That is done in attempts to combat the growing weed problem caused by the weeds' competition for moisture, nutrients, and depending on species for light and space as well. One result of the mentioned competition is substantial yield losses in sunflower production ranging from 20 to 70%. Herbicide use is an effective solution when planting sunflowers in a no-till or minimum-till cropping system. In fact, the chemical method may be beneficial if ground cover is needed to prevent soil erosion from wind and water. Nonetheless, that is not always the case. One of the registered preplant herbicide glyphosate, used for nonselective annual perennial grass and broadleaf weed control, is still the subject of much debate, as several studies report its negative effects on the environment [1]. Anyhow, new methods for testing the outcomes of the environmental exposure to glyphosate in sunflower production are being proposed, so the given results may end up being different [2].

However, the only way chemical treatment is possible is if the grown sunflower species are tolerant to the used pesticides. Thus, herbicide-resistant crops are becoming increasingly common in agricultural production. Berville et al. treated seeds of $F_1$ hybrids with gamma rays (100 Gy, 200 Gy, 300 Gy, and 400 Gy) and 0.2% ethyl methyl sulfonate and obtained mutants tolerant to bifenox and glyphosate [3]. Furthermore, the knowledge of sunflower genetics and breeding has been greatly expanded since the time that Škorić defined his hybrid model [4–6].

*Resistance to herbicides* from the class of imidazolinones (IMI) and of sulfonylureas (SU) is becoming one of the most important sunflower traits. Its benefits can be observed in Spain where imidazolinone resistance (transferred by sunflower breeding) has resulted in a broad spectrum of weed control (over 40 broadleaf and 20 grass weed species) and is highly effective in the control of the parasite *broomrape* (*Orobanche cumana* Wallr.). This tolerance has potential to be applied in all regions of the world for controlling several broadleaf weeds and even may control the broomrape in areas of the world where this parasitic weed attacks sunflower [7]. Be that as it may, the broomrape produces an extremely large number of seeds, and it is likely that if this control measure is widely used, isolation with herbicide resistance will become an issue. Previously, sulfonylurea (*SU*) and imidazolinone (*IMI*) herbicides were widely used to control wide sunflowers in the fields of corn, soybean, and other crop rotations that later developed herbicide resistance [1].

According to Sala et al., there are two primary mechanisms of *herbicide tolerance* (*HT*) in sunflower [8]:

i. tolerance caused by mutations in target sites of the herbicide (*target-site tolerance*);

ii. tolerance caused by mutations in nontarget sites (*nontarget-site tolerance*).

Target-site tolerance involves a reduced sensitivity of target specific enzymes or proteins, and thus, this type of tolerance is mostly monogenic—as *IMI* and *SU* resistance [9]. Nontarget tolerance, on the other hand, involves several mechanisms, such as reduced uptake or translocation of the herbicide, increased rate of herbicide detoxification, decreased rate of herbicide activation, or sequestration of the herbicide away from the target site into the vacuole or the apoplast [10]. Target- and nontarget-site mechanisms can also be implemented together, such as in one of the current technologies of weed control, *Imisun* sunflowers [11, 12].

### 2.1 Development of IMI-resistant sunflower

Imidazolinone (Imazethapyr, Persuit) resistance in wild population of annual sunflower (*Helianthus annuus* L.) was first identified in Kansas in 1996, in a soybean field treated with the herbicide for 7 consecutive years [13].

The USDA-ARS (North Dakota State University (NDSU), Fargo, ND, USA) research group transferred this resistance into cultivated sunflower genotypes and released the public populations oil maintainer IMISUN-1 (Reg. no. GS-18, PI 607927), oil restorer IMISUN-2 (Reg. no. GS-19, PI 607928), confection maintainer IMISUN-3

(Reg. no. GS-20, PI 607929), and confection restorer IMISUN-4 (Reg. no. GS-21, PI 607930) [14]. Similar programs with the aim to incorporate *IMI* resistance from the wild *H. annuus* from Kansas into elite lines and developed *IMI*-resistant hybrids were run by Alonso in Spain [7], by Malidža et al. in Novi Sad, Serbia [15], and by several private companies in Argentina [16].

Notable results were achieved by Sala et al. [17], who obtained mutants resistant to imidazolinones by inducing mutations with a solution of ethyl methanesulfonate. The authors identified an *IMI*-resistant single partially dominant nuclear gene that they coded *CHLA-PLUS* and proved at the molecular level (with simple-sequence repeat (*SSR*) marker for the *AHASL1* gene) that while it is different from $Imr_1$, both of them are allelic variants of the locus *AHASL1*.

It has been shown experimentally that the gene CHLA-PLUS has a higher degree of *IMI* resistance than the gene $Imr_1$ $Imr_2$. Breeding centers wishing to use the *CHLA-PLUS* gene for breeding purposes have to sign a contract with the company BASF. At the same time, BASF provides a protocol for screening for resistance at the molecular level (*Clearfield®Protocol SF30*). This established trademark production system for sunflower provides growers with a new technology, which ensures broad-spectrum post-emergent grass and broadleaf weed control combined with high-performing sunflower hybrids from leading seed companies or public institutions.

However, in recent years, probably due to overdose or incomplete absorption of the herbicide by the plants and accumulation in the soil, there has been a problem with the next year's wheat crop (crop rotation). Because of that, *IMI*-resistant wheat breeding selection programs have been launched.

### 2.2 Development of resistant to sulfonylurea (tribenuron-methyl)

With sunflower breeding for *IMI* resistance, work has been started on the development of hybrids resistant to herbicides from the *tribenuron-methyl* group of sulfonylureas. Two resistance sources have been discovered.

The first one was derived from *SU*-resistant wild *Helianthus annuus* plants collected from the same area in Kansas where *IMI* resistance was found in 2002. The USDA-ARS (NDSU) research group incorporated this genetic resistance into cultivated sunflower and released public lines maintainer SURES-1 (Reg. no. GS-28, PI 633749) and restorer SURES-2 (Reg. no. GS-29, PI 633750) [18].

The second *SU* resistance was detected by DuPont within an artificial mutagenesis project conducted in the early 1990s. This material was reselected, purified, and tested by Pioneer/DuPont during 1998–2000. Several mutation events were evaluated, and selectivity to the sunflower mutation event SU7 was confirmed for a narrow range of SU herbicides. Pioneer/DuPont and the Institute of Field and Vegetable Crops (IFVC), Novi Sad, Serbia, were first to place *SU*-resistant hybrids on the market. First observations from commercial production indicated that although it was the case of a single dominant gene, it was necessary for both parents to possess the *SU* gene. When resistance is incorporated in only one parental line (*Rf*), a problem of how to produce 100% tolerant hybrid seeds arose and farmers often had susceptible plants in commercial crops [19].

### 2.3 Enzymes and genes

Imidazolinone-tolerant plants with altered acetohydroxyacid synthase (*AHAS*) *genes* and *enzymes* have been discovered in many species. *IMI* and *SU* herbicides are the specific inhibitors of acetohydroxyacid synthase (*AHAS*, EC 2.2.1.6). Species differ in herbicide susceptibility and can develop resistance to different classes of *AHAS* inhibitors. With few exceptions, resistances to *AHAS*-inhibiting herbicides, in otherwise susceptible species, are caused by point mutations in genes encoding *AHAS* that reduce the sensitivity of the enzyme to herbicide inhibition.

Acetolactate synthase (*ALS*), also called acetohydroxyacid synthase (*AHAS*), is the first enzyme in the biosynthesis of three vital amino acids in plants: valine, leucine, and isoleucine. Four different classes of herbicides inhibit *ALS*, thus causing the herbicidal effect. The most common are imidazolinones (*IMI*) and sulfonylureas (*SU*). They have been widely used since their introduction in the early 1980s, and now, they constitute one of the major weed control mode-of-action classes for many crops. Resistant (tolerant) plants rapidly metabolize the herbicide in herbicidally inactive form. Sensitivity is likewise due to the lack of metabolic detoxification. Advantages of *ALS*-inhibiting herbicides are as follows: very low application rate, broad spectrum of weed control (broadleaf and grassy weed species), broad range of crop selectivity, etc.

Their disadvantages may be the following:

a. Their extensive use and genetic mutability of the trait have led to the development of resistance in many species.

b. More than one gene may be involved in resistance, and they may not be totally dominant. In these cases, both parental lines have to be resistant in order to reach a commercially accepted resistance of the hybrid (more costly breeding process).

c. Because of cross resistance, careful herbicide management is required to ensure their long-time usefulness. Cross resistance refers to plant resistance to multiple herbicides that have the same mechanism of action [19].

*IMI resistance* in sunflower is controlled by *two genes* with a semidominant type of gene action—a major gene having a semidominant type of gene action (*Imr1*), and a second gene (*Imr2*) with a modifier effect when the major gene is present [16]. Malidža et al. reported that resistance to imidazolinone herbicides was controlled by a *single gene* with partial dominance [15]. These different findings regarding the mode of inheritance of *IMI* resistance could perhaps be explained by mutations being present at multiple loci in the original population of wild *Helianthus annuus*.

On the other hand, Jocić et al. reported that the sunflower *resistance to tribenuron-methyl* is controlled by a *single dominant gene* [20].

The three *loci* LG 2 (*AHAS3*), LG 6 (*AHAS2*), and LG 9 (*AHAS1*), flanked by mapped *SSR* markers, were shown on the public sunflower map of simply inherited traits [21] (**Table 1**). Eleven more sunflower *AHAS ESTs* were found. When the DNA sequences of various amplicons were aligned, three paralogous *AHAS* genes were discovered and named *AHAS1, AHAS2,* and *AHAS3.* Single-nucleotide polymorphism (*SNP*) markers were developed for *AHAS1* and *AHAS3*, and single-strand conformation polymorphism (*SSCP*) markers were created for a six-base-pair *INDEL* in *AHAS2* and a G/A *SNP* in *AHAS3.* In addition, an *SSR* marker was developed for *AHAS1* based on the poly-Thr repeat in the putative transit peptide of *AHAS1* [22].

In the study of Kolkman et al., DNA polymorphisms were not found between herbicide-susceptible and herbicide-resistant inbred lines in the *AHAS2* and *AHAS3* coding sequences, but two mutations in the sunflower *AHAS1* gene were identified, i.e., an Ala205Val mutation and a Pro197Leu mutation, conferring resistance to *AHAS*-inhibiting

| Trait | Gene/locus | Linkage group (LG) | Population/s, line/s [23] |
|---|---|---|---|
| Herbicide resistance | *AHAS3* | LG 2 | RHA280 × RHA801, NMS373 × ANN1811, (HA425 × HA89) × HA89, IMISUN-2 × ZENB9 |
| | *AHAS2* | LG 6 | |
| | *AHAS1* | LG 9 | |

**Table 1.**
*The mapped loci in relation to the SSR markers.*

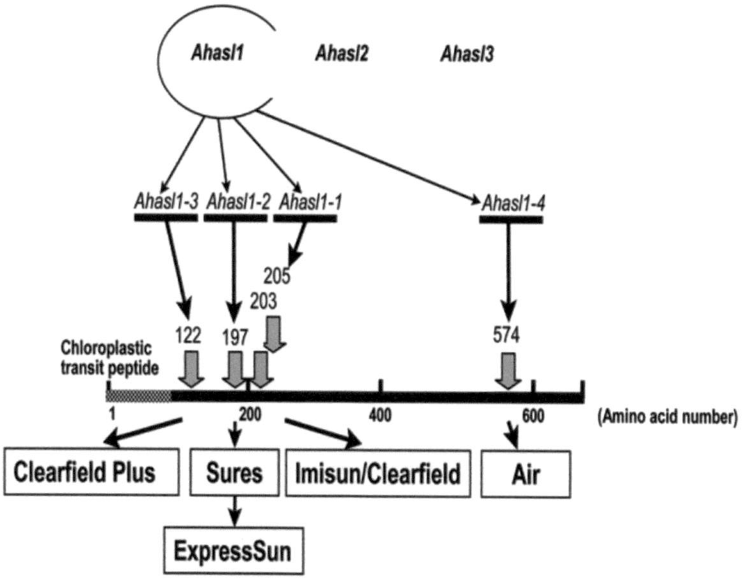


**Figure 2.**
*Genetics of AHAS-inhibitor herbicide tolerance in sunflower [8].*

herbicides (mutation codons in the acetohydroxyacid synthase gene *AHAS1* that confers resistance to sulfonylurea (*SU*) and imidazolonone (*IMI*) herbicides). Pro197 and Ala205 are conserved amino acids in *AHAS* enzymes in numerous species [9]. The mutation of Pro197 is one of the most common mutations found in plants resistant to *AHAS*-inhibiting herbicides but mutations of Ala205 in inhibitor-resistant plants had, thus far, only been reported in cocklebur, *Arabidopsis*, and sunflower [23].

There exist three genes encoding for catalytic subunits of the AHAS enzyme: AHASL1, AHASL2, and AHASL3 (**Figure 2**). Known mutations for herbicide tolerance so far described were located in AHASL1. Formally described alleles of this gene, the site of the aminoacidic substitution controlling tolerance in each case (following Arabidopsis thaliana nomenclature), and the herbicide tolerance trait developed from each allele/mutant are provided.

## 3. Sunflower breeding for tolerance to herbicides: Our results

In the past decade, our company also achieved significant results in sunflower breeding for resistance to some herbicides from the imidazolinones and sulfonylureas classes.

### 3.1 Genetic sources and geneplasm

Two *wild species* (**Figure 3**) were used—*Helianthus annuus* (two different accessions from ND and MN, USA) and *Helianthus argophyllus*, *sources* of herbicide resistance [24]. *H. petiolaris,* which was also used, showed partial tolerance, too.

According to Olson et al. [25], several populations of wild sunflower (*H. annuus* and *H. petiolaris*) from the USA and Canada have been screened for resistance to imazethapyr and imazamox herbicides. Eight percent of 50 wild sunflower populations had some resistance to imazamox and 57% had some resistance to tribenuron in the central USA. In additional, according to Miller and Seiler, in Canada, 52% out of 23 wild *H. annuus* populations had some resistance to tribenuron [26].

Several lines have been used as a source of herbicide resistance genes from *cultivated sunflower*: oilseed maintainer HA425 (Reg. no. GP-254, PI 617098) and restorer RHA 426 (Reg. no. GP-255, PI 617099) [27] and SURES-1. The resistance of every plant from the lines was identified in advance by treatment with herbicides.

A variety of lines obtained by hybridization and mutagenesis for herbicide resistance were also studied [28–41]. Different plant responses to treatment with a number of herbicides have been reported, but the main point is that with increasing doses, plant breakage at the base of the stem is observed. Imidazolinone-tolerant materials were susceptible to sulfonylurea herbicide. Some of the lines were also included in the subsequent breeding program since they have suitable parental base line *geneplasm* (*B* and *R*) for hybrid's registration.

### 3.2 New herbicide-resistant lines were obtained by using the breeding method

Herbicide resistance from wild species and resistant lines was transferred to cultivated sunflower by *hybridization* (*interspecific* and *intergeneric*). Self-pollination and yearly treating of the selected material were carried out. A high percentage of resistant plants was obtained from different crosses.

When treated with Pulsar 40 + Stomp 330 EK in 2009, 17 plants were killed and 4 were slightly affected out of 21 plants in total from hybrid material of the cross L. 1607 × *Matricaria* sp. For three of them, a normal seed set was obtained [24]. Since then, a new elite inbreed line (**Figure 4**) has been develop, and it can be deployed in the making of herbicide resistant hybrids.

A. *H. argophyllus* B. *H. petiolaris* C. *H. annuus* (w.f.).

**Figure 3.**
*A.* H. argophyllus. *B.* H. petiolaris. *C.* H. annuus *(w.f.).*

**Figure 4.**
*Elite line from cross* H. annuus × Matricaria sp.

**Figure 5.**
*(A) Susceptible plants and (B) resistant.*

The introduction of genes for resistance to *IMI* from wild species or from resistant genotype into elite *B* or *R* lines is done by backcrossing accompanied with continual resistance screening and elimination of sensitive and yellow flash plants.

During the first 2 weeks, the phenotypical distinction of plants is observed: resistant, intermediate with less yellow flash, dead plants (susceptible) (**Figure 5**) and intermediate with severe "yellow flash" (**Figure 6**).

Heterozygous plants are less tolerant than homozygous ones. Therefore, different herbicide concentrations are needed for screening and selecting phenotypically the genotypes, without injuring the tolerant or killing the heterozygous plants.

This is clearly shown in field conditions (**Figure 7**). However, there arises a new issue: stress. Thus, plants develop at different rates during the vegetation (growing) season.

Now, we have 30 more inbreed lines and 150 forms in different generation (from hybridization and mutagenesis), resistant to herbicides (Pulsar or Express). Some of

**Figure 6.**
*Effect of imidazolonone herbicide treatments on the produced crop: 15 days after treatment the effect of "yellow flash" phenomenon is observed on some plants (the less resistant to completely susceptible plants).*

**Figure 7.**
*Breeding for tolerance to herbicides by hybridization. (A) Screening for resistance totribenuron-methyl (SU-res) and imidazolonone (IMI-res) herbicide in field trial conditions—susceptible (dead) and resistant (normally developing) plants. (B) One resistant from many plants and (C) plants at different rates of vegetation.*

them have specifically morphological traits: *mutation* as *white pollen color*, *fasciation*, *wrinkled leaf*, *zig-zag stem*, and other (**Figure 8**). Many of the lines were with good combining ability, increased 1000-seed weight, high seed oil content, early maturity, and resistance to some diseases. Some of the lines could directly be used as parent forms of sunflowers for human food.

Except through *hybridization*, *genetic variability* in cultivated sunflower can be increased by *mutations*. Induced mutations are caused by humans, by treating plants with various physical or chemical agents. Mutagens create a wide range of heritable changes in sunflower. Mutations are most frequently observed in morphological traits, oil quality, resistance to herbicides, resistance to low or high temperatures, and other traits. In fact, some of the new traits can be used as morphological markers.

There were some *deformations* of sunflowers (**Figure 9**) after treatment with herbicides. These defects often are a result of the impact of the stress factor and the

**Figure 8.**
*Mutagenesis. 1. Zig-zag stem. 2. White pollen color. 3. Fasciation of stem, inflorescence, and leaves. 4. Wrinkled leaves. 5. Yellow spotting. 6. Altered shape of inflorescence with densely spaced disk flowers. 7. Different colors of the ray flower and center of the head. 8. Funnel-shaped ray flower.*

inability of the plant to adapt to it. Some of the extreme cases are of fertile disk flowers (*f*) in the sterile inflorescence (*st*) or vice versa, but more often plants simply do not develop normally or cannot produce a next generation.

### 3.3 A new line with resistance to herbicides Pulsar and Express was developed

The aim of the first experiment was to establish genetic variability (*geneplasm*) and sources (wild species and cultivated sunflower forms and lines) of herbicide resistance. The field trial was separated in three parts: the first part for treatment with herbicide Pulsar 40 (120 mL/dka)—*P*, the second part used as a control (not treated)—*K*, and the third part for treatment with Express (5 g/dka)—*SU* (**Figure 10**). The materials were sprayed during the optimum time as to test them for resistance to IMI herbicides is the stage of three to five pairs of permanent leaves. The dose of Pulsar was found to

1. *Branching (A), hardening (B), elongation (C) and breaking at the base (D) of the stem;*

**Figure 9.**
*Deformations of sunflower after treatment with herbicides. 1. Branching (A), hardening (B), elongation (C) and breaking at the base (D) of the stem. 2. Deformation of the inflorescence. 3. Fertile disk flowers (f) in the sterile inflorescence (st) or vice versa.*

effectively control several weeds and have a highly effective control of parasite broomrape (*O. cumana*), but it was not so capable of controlling *Convolvulus arvensis* properly.

The 2020 and 2021 placement of the crops is laid out in the same way, but the treatment doses are increased—165 mL/dka for Pulsar and 6.2 g/dka for Express

**Figure 10.**
P *group,* K *group, and* SU *group.*

**Figure 11.**
P *group,* K *group, and* SU *group in experiment, 2020.*

**Figure 12.**
*Drought impact.*

(**Figure 11**). This increase is in order to clear the heterozygous plants faster, i.e., to leave only those fully resistant (homozygous) to herbicides.

The drought during (**Figure 12**) this period was an additional stress factor besides the increased concentration of the chemicals. But, despite everything, we still have new elite *B* lines.

**Figure 13.**
*Mutant line with new morphological traits, 2015. A and B: Mutant plant. C: Leaf. D: Inflorescence and leaf.*

Herbicide resistance (IMI and SU) was successfully transferred from the different sources into elite *R* and *B* lines. Two years ago, we, *Mihsan* breeding group, reported a new form of sunflower with a new type of combined resistance to herbicides [42] (**Figure 13**). The study continued, and now, we have a line in homozygous state that is suitable for molecular analysis, which is the only thing that can show exactly what has happened.

The initial cross was between our mutant line and one American line, IMI type. The aim was to transfer IMI resistance into the morphologically specific line. The first treatment was done on the $F_2$ generation. Seeds from IMI resistant $F_2$ plants were divided into three groups. Every one of two groups was treated separately with herbicide Pulsar (*P* group) or Express (*SU* group). The received result was a very high percentage (60–95%) of alive plants in both the groups. After self-pollination, in $F_4$ hybrid generation, four numbers (all from seeds of only one plant from the previous year) from the *SU* group and five numbers from the *P* group were with 100% alive plants. The original $F_2$ plants are the initial parents of two of these numbers. The mentioned two numbers from the *P* group were used as donors of pollen for hybridization of seven not-resistant to herbicides B lines after emasculation. The aim was to understand how this new trait transmits (the mode of inheritance). $F_1$ and $F_2$ plants were treated with Pulsar. A different percentage (20–80%) of alive plants was received in every one of crosses. $F_2$ plants were received from alive isolated $F_1$ plants. The results were varying—numbers with all dead plants, numbers with all alive plants, and numbers with different percentage (7–92%) alive plants.

In 2020 and 2021, the treatment doses were increased for both detergents. As a result, in 2021, four numbers with plants in the sixth hybrid generation were homozygous for trait resistance to both herbicides. These were already *new elite B lines*. Out of the treated 11 lines, in one, all the numbers were dead in both the groups of treatment. In this same sixth generation, and with increasing treatment doses, a different and very interesting result was obtained, namely, in three lines, all numbers showed complete resistance to herbicide Express, but when treated with Pulsar, some of the plants were affected. It is also interesting to note that 95% of the affected plants die after the Pulsar treatment, and up to 2% of the plants produce only a few seeds. On the other hand, the plants affected by Express branch and deform their inflorescence, but about 80–90% of them survive. In a small proportion of these plants, however, no seed set is obtained after isolation and self-pollination. Three numbers of F4 plants from crosses between not-resistant B lines and plants with this new trait—resistance to both herbicides—show complete resistance to both herbicides.

## 4. Conclusion

In conclusion, it can be said that sunflower breeders use various methods in order to get new *genetic variability* (by interspecific and intergeneric hybridization and mutagenesis) in new elite lines that after having their combining abilities examined are later used for making new hybrids. The choice of methods is closely connected to the breeding goals set in advance, as well as the available staff, equipment, inheritance of the most significant agronomic traits, available genetic resources, and other factors. Molecular marker-assisted selection (MAS) can also be used to check herbicide tolerance.

The chemical industry is constantly creating new pesticide agents that are highly selective, slightly poisonous to humans, decompose quickly, and do not contaminate crop production. In this way, many of the disadvantages of the chemical method are avoided, and the possibility of widespread application is improved. Despite the increased use of biofertilizers and various biological control methods (entomopathogenic bacteria and fungi, predatory insects, parasitic insects, etc.), *the creation of resistant varieties* (both to the pests themselves and to the used chemicals), and many more, which make it possible to reduce the use of pesticides, the question arises: Which is the best and safest option for us, humans? Everyone has to answer for themselves, finding the balance between all the options, while trying to preserve the environment.

## Acknowledgements

I would like to thank *Mihsan* Ltd. for giving me the opportunity to present these results and photographs.

## Author details

Miroslava Hristova-Cherbadzhi
University of Forestry, Sofia, Bulgaria

*Address all correspondence to: miroslavaxristova75@gmail.com

## References

[1] Blamey FPC, Zollinger RK, Schneiter AA. Sunflower production and culture. In: Schneiter AA, editor. Sunflower Technology and Production, Agronomy №. 35. Madison, Wisconsin, USA: American Society of Agronomy; 1997. pp. 595-670. ISBN 08-911-81350

[2] Farkas D, Horotán K, Orlóci L, Neményi A, Kisvarga S. New methods for testing/determining the environmental exposure to glyphosate in sunflower (*Helianthus annuus* L.) plants. Sustainability. 2022;**14**:588. DOI: 10.3390/su14020588

[3] Berville A, Delbut J, Bedergat R. Mutagenic treatments performed on seeds of a sunflower hybrid variety with the purpose of obtaining bifenox or glyphosate resistant mutants. Helia. 1992;**15**(16):53-58

[4] Škorić D. Desired model of hybrid sunflower and the newly developed NS-hybrids. Helia. 1980;**3**:19-24

[5] Škorić D. Sunflower breeding. Uljarstvo. 1988;**25**(1):1-90

[6] Škorić D. Achievement and future directions of sunflower breeding. Field Crops Research. 1992;**30**:231-270

[7] Alonso LC, Rodriguez-Ojeda MI, Fernandez-Escobar J, Lopez-Ruiz-Calero G. Chemical control of broomrape (*Orobanche cernua* Loefl.) in sunflower (*Helianthus annuus* L.) resistant to imazethapyr herbicide. Helia. 1998;**21**(29):45-54

[8] Sala CA, Bulos M, Altieri E, Ramos ML. Genetics and breeding of herbicide tolerance in sunflower. Helia. 2012;**35**(57):57-70

[9] Tranel PJ, Wright TR. Resistance of weeds to AHAS inhibiting herbicides: What have we learned? Weed Science. 2002;**50**:700-712

[10] Yuan JS, Tranel PJ, Stewart N Jr. Non-target site herbicide resistance: A family business. Trends in Plant Science. 2006;**12**(1):6-13

[11] Breccia G, Gil M, Vega T, Zorzoli R, Picardi L, Nestares G. Effect of cytochrome P450s inhibitors on imidazolinone resistance in sunflower, Proceedings of the 18th International Sunflower Conference; Mar del Plata-Balcarce, Argentina2012. pp. 507-512

[12] Sala CA, Bulos M, Altieri E, Ramos ML. Imisun tolerance is the result of the interaction between target and non-target tolerance mechanisms. In: Proceedings of the 18th International Sunflower Conference; Mar del Plata-Balcarce, Argentina. 2012. pp. 551-556

[13] Al-Khatib K, Beungurtner JR, Peterson DE, Currie RS. Imazethapyr resistance in common sunflower (*Helianthus annuus* L.). Weed Science. 1998;**46**:403-407

[14] Al-Khatib K, Miller JF. Registration of four genetic stocks of sunflower resistant to imidazolinone herbicides. Crop Science. 2000;**40**(3):869-870

[15] Malidža G, Škorić D, Jocić S. Imidazolinone resistant sunflower (*Helianthus annuus* L.): Inheritance of resistance and response towards selected sulfonyl urea herbicides. In: Proceedings of the 15th–17th International Sunflower Conference, Vol. 2; 12–15 June 2000; Toulouse, France. 2000. pp. 42-47

[16] Bruniard JM, Miller JF. Inheritance of iImidazolinone-herbicide resistance in sunflower. Helia. 2001;**24**(35):11-16

[17] Sala CA, Bulos M, Echarte AM. Genetic analysis of an induced mutation conferring imidazolinone resistance in sunflower. Crop Science. 2008;**48**: 1817-1822

[18] Miller JF, Al-Khatib K. Registration of two oil seed sunflower genetic stocks, SURES1 and SURES-2, resistant to tribenuron herbicide. Crop Science. 2004;**44**(3):1037-1038

[19] Škorić D. Sunflower breeding. In: Škorić D, editor. Sunflower Genetics and Breeding. International Monography. Branch in Novi Sad: Serbian Academy of Sciences and Arts; 2012. pp. 165-354

[20] Jocić S, Miklič V, Malidža G, Hladni N, Gvozdenović S. New sunflower hybrids tolerant of tribenuron-methyl. In: Proceedings of the 17th International Sunflower Conference, Vol. 2; 8–12 June 2008; Cordoba, Spain. 2008. pp. 505-508

[21] Tang S, Slabaugh MB, Shintani DK, Knapp SJ. Simple sequence repeat map of the sunflower genome. Theoretical and Applied Genetics. 2002;**105**:1124-1136

[22] Hahn V, Wieckhorst S. Mapping and tagging of simply inherited traits. In: Hu J, Seiler G, Kole Ch, editors. Genetics, Genomics and Breeding of Sunflower. Boca Raton, Florida, USA: CRC Press; 2010. pp. 111–134. ISBN 978–1–57808-676-4

[23] Kolkman JM, Slabaugh MB, Bruniard JM, Berry S, Bushman BS, Olungu C, et al. Acetohydroxyacid synthase mutations conferring resistance to imidazolinone or sulfonylurea herbicides in sunflower. Theoretical and Applied Genetics. 2004;**109**:1147-1159

[24] Christov M, Hristova-Cherbadzi M, Entcheva V, Shindrova P, Kiryakov I, Valkova D, et al. Creating new sunflower forms and lines, resistant to diseases, the broom rape parasite and certain types of herbicides applying the interspecific and intergeneric hybridization. In: Proceedings of the International Sunflower Symposium "Breeding of Sunflower on Resistance to Diseases"; 23–24 June 2010; VNIIMK, Krasnodar. 2010. pp. 72-84

[25] Olson B, Al-Khatib K, Aiken RM. Distribution of resistance to imazamox and tribenuron-methyl in native sunflower. In: Proceedings of the 26th Sunflower Research Workshop; 14–15 January 2004; Fargo, ND, USA. 2004

[26] Miller JF, Seiler GJ. Tribenuron resistance in accessions of wild sunflower collected in Canada. In: Proceedings of the 27th Sunflower Research Workshop; 12–13 Jan 2005; Fargo, ND, USA. 2005

[27] Miller JF, Al-Khatib K. Registration of imidazolinone herbicide-resistant sunflower maintainer (HA425) and fertility restorer (RHA426 and RHA427) germplasms. Crop Science. 2002;**42**(3): 988-989

[28] Christov M. A new sunflower mutant form. Helia. 1996;**19**(24):39-46

[29] Christov M. Hybridization of cultivated sunflower and wild *Helianthus* species. In: Caligari PDS, Hind DJN, editors. Compositae: Biology & Utilization, Vol. 2. Kew: Royal Botanic Gardens; 1996. pp. 603-670

[30] Christov M. Investigation on the inheritance of several traits in mutant sunflower form 674. Scientific Reports, Dobrich. 2001;**3**:23-27

[31] Christov M. Changes in sunflower morphology as a result of applying of the experimental mutagenesis. In: Proceedings of the International Conference "Conventional and

Molecular Breeding of Field and Vegetable Crops"; 24–27 November 2008; Novi Sad, Serbia. 2008. pp. 184-186

[32] Christov M. Contribution of interspecific and intergeneric hybridization to sunflower breeding. Helia. 2013;**36**(58):1-18

[33] Christov M, Nikolova V. Increasing of the sunflower genetic diversity by mutagenesis. In: Proceedings of the 14th International Sunflower Conference, Vol. 1; 12–20 June 1996; Beijing/Shenyang, China. 1996. pp. 19-30

[34] Christov M, Sabev G, Valkova D, Hristova-Cherbadzi M. New sunflower forms and hybrids, resistant to herbicides. In: Proceedings of the International Conference "Conventional and Molecular Breeding of Field and Vegetable Crops"; 24–27 November 2008; Novi Sad, Serbia. 2008, 2008. pp. 329-331

[35] Hristova-Cherbadzi M, Christov M. Hybridization between cultivated sunflower *Helianthus annuus* L. and wild perennial species *Helianthus pumilus* Nuttall. In: Proceedings of the 17th International Sunflower Conference, Vol. 2; 8–12 June 2008; Cordoba, Spain. 2008. pp. 673-678

[36] Hristova-Cherbadzi M, Christov M. Characterization of hybrids from crosses between cultivated *Helianthus annuus* L. and subspecies *rydbergii* (Britton) Long of perennial diploid *Helianthus nuttallii*. In: Proceedings of the 17th International Sunflower Conference, Vol. 2; 8–12 June 2008; Cordoba, Spain. 2008, 2008. pp. 691-696

[37] Hristova-Cherbadzi M, Christov M. Sunflower lines and forms, obtained from the intergeneric hybridization. In: Proceedings of the International Conference "Conventional and Molecular Breeding of Field and Vegetable Crops"; 24–27 November 2008; Novi Sad, Serbia. 2008. pp. 158-160

[38] Hristova-Cherbadzhi M. Characterization of hybrids, forms and lines, obtained from interspecific hybridization of cultivated sunflower *Helianthus annuus* L. with wild species of genus *helianthus*. Biotechnology and Biotechnological Equipment. 2009; **23**(2):112-116

[39] Hristova-Cherbadzhi M. Study of new forms of sunflower received by distant hybridization. In: Breeding and Genetics of Cultivated Sunflower—Methods, New Lines, New Crosses, New CMS Source. Saarbrücken, Germany: Lambert Academic Publishing; 2012. ISBN 978–3–659-13617-7

[40] Hristova-Cherbadzhi M. Intergeneric hybidization of sunflower (*Helianthus annuus* L.) with spiny plumeless thistle (*Carduus acanthoides* L.). Helia. 2020; **43**(72):83-98. ISSN: 2197–0483. DOI: 10.1515/helia-2020-0008

[41] Christov M, Hristova-Cherbadzhi M. Hybridization between sunflowers (*Helianthus annuus* L.) and Less Stem Rosette (*Carlina acanthifolia* All.). Characterization of received intergeneric forms. In: Proceedings of the 19th International Sunflower Conference; 29 May–3 June 2016; Edirne, Turkey. 2016. pp. 593-597

[42] Christov M, Hristova-Cherbadzhi M. New form cultivated sunflower (*Helianthus annuus* L.) with resistance to the herbicides Pulsar and Express. Helia. 2020;**43**(73):185-189. eISSN 2197–0483, ISSN 1018–1806. DOI: 10.1515/helia-2020-0007

Chapter 3

# Management Strategies and Alternatives for Fungicidal Resistance in Potato

*Rahul R. Bakade, S. Sundaresha and Mehi Lal*

### Abstract

Fungicides have been used for over 200 years to protect plants from damage by fungi, but today fungicidal resistance is very common among potato pathogens and it is difficult to control. The best and intensively studied example is metalaxyl resistance in *Pytophthora infestans*. Causes are many to get pathogen resistance against the fungicides like intensive use or misuse of it, repeated application of same fungicides, etc. Hence, it is today's need to find out the different strategies like different cultural practices, use of bio-agents, use of green chemicals, elimination of disease source, etc. to manage this fungicidal resistance. There are also alternative ways like increasing host resistance, use of new molecules, etc. that can be adopted to reduce the risk of fungicidal resistance.

**Keywords:** fungicide, resistance, potato diseases, management, *phytophthora*

## 1. Introduction

Development of human civilization has been closely associated with the cultivation of crops, and plant diseases have been a concern to mankind probably since plants were cultivated more than 10,000 years ago [1]. The record of severe epidemics, which threatened crops, is reported in early Greek and Roman literature in 500 BC, and in the Roman world, Robigus, the God for cereals, was worshipped to prevent crop failure. Chemical plant protection schemes were developed at the beginning of the twentieth century with copper and sulphur as antifungal agents to control downy and powdery mildews. There was the multi-billion-dollar industry that has modern fungicides belonging to various chemical classes, differing in their mode of action against pathogens and characteristics of uptake and distribution within the plant.

Later, numerous cases of fungicide resistance have occurred worldwide because of release of fungicides with several target sites (so-called multi-side inhibitors) to the market, like organomercurials, regardless of their human toxicity and environmental pollution.

Along with that, systemic fungicides are also released which were taken up by the plant and are subsequently distributed within the entire plant, protecting also newly formed tissues. Thus, these fungicides were curative and allowed to control pathogens after infection had occurred. In the progress highly specific modern fungicides, which block only one target in the pathogen (monospecific fungicides or single-site inhibitors), were developed. Examples of single-site inhibitors are strobilurins, phenylamides and benzimidazoles, which were released to the market in the late 1970s and in the mid-1990s. Surprisingly, after 2 years, the apple scab fungus *Venturia inaequalis* and the *polyphagous* grey mould fungus *Botrytis cinerea* developed resistance against benzimidazoles. Also, in other single-site inhibitors such as the phenylamides and the strobilurins, resistant strains got developed within 2 years after the compounds, were introduced to the market and widely used.

## 2. Evolution of resistance

The life cycle of fungus is so small that evolution happens frequently. Its interaction with fungicide forces it to modify itself for its survival. Basically, fungicides disrupt metabolism and threaten the fungal survival, and as a result, pathogenic fungi can initiate mechanisms to resist lethal effects. Fungal genomes are very unstable and may contain thousands of polymorphisms [2]. Fungicides target a specific biochemical step, and a single point mutation causing one amino acid change can rapidly and effectively block fungicide binding within the target site (single-site inhibitors) and generally cause high levels of resistance. Fungicides of multisite inhibitors that target many biochemical steps require a combination of many mutations and so resistance evolves slowly.

The basic process of development of fungicidal resistance is depicted in **Figure 1**. The fungal spore populations are having the genetic potential to resist the disease and here the resistance can be developed initially (represented by the filled circles in the figures). When a newly formed fungicide is applied, maximum of the fungal spores are killed but very few get survived and that became the resistant spores. These spores are extremely low in numbers and that will be the start of the process. Along with the resistance spores, some sensitive spores also survived, because they 'escaped' the fungicide treatment and were not got exposed to the applied fungicides. The survived spores get developed and started the disease activity in favourable environmental conditions and produce a new crop of spores. This new crop of spores has a higher percentage of resistant spores because of its survival in the previous crop of spores.

The fungicide with the same mode of action is the core feature of fungicide resistance and hence the same specific resistance mechanism, show cross-resistance,

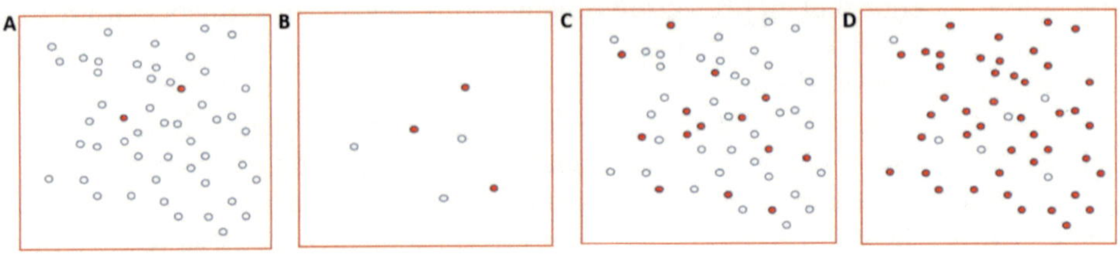


**Figure 1.**
*Process of fungicidal resistance development. (A) Population of spores before fungicide use. Most spores are sensitive (open circles), but sometimes a very low number are genetically resistant to the fungicide (filled circles). (B) After a fungicide application, the number of surviving spores is greatly reduced and only the resistant spores survived the treatment. Also, some sensitive spores (open circles) escaped the treatment. (C) If environmental conditions favour a new cycle of disease activity, the next generation of spores will have a higher percentage of resistant spores. Continued use of the fungicide selects for these resistant spores. (D) Multiplication of resistant spores in the next generation and spread.*

but not resistance to other modes of action. This resistance may generate resistance between products with different modes of action, known as multidrug resistance (MDR), especially in laboratory assays.

Sometimes the interaction of fungicides of the same mode of action group (FRAC) may be different to a particular change in the target site, which differs in different resistance levels of fungi resulting in the evolution of resistance within pathogen populations [3]. For example, the interaction of Prothioconazole, with the haem component of the target-site sterol 14α-demethylase (CYP51), was differently from other azoles [4], showing lower resistance but still effective control of some cereal diseases. In *Mycosphaerella graminicola*, the cause of wheat leaf blotch, azole resistance was shown to different target-site mutations alone, or in combination, generate different cross-resistance patterns [5] and indeed improved the performance of prochloraz [6].

## 3. Genetics of fungicidal resistance

Development of fungicidal resistance in fungal plant pathogens is a challenge in modern crop protection. Because of the short life cycle, fungi are indeed very able to adapt to changing environmental conditions, like the introduction of a new fungicide in the agricultural practice. These changed environmental conditions forced the genetic material to change and several genetic mechanisms happen in fungus and influence the chance and time of its appearance and spreading in fungal populations.

Acquired resistance develops in fungi in their wild-type form are sensitive and may develop resistance after their exposure to new fungicides. This resistance is due to genetic modifications transmissible to the progeny so that a chemical, which was once effective against the organism, is no longer effective. Fortunately, till the late 1960s, fungicides used in crop protection were sulphur, copper derivatives, dithiocarbamates and these were multisite inhibitors, affecting multiple target sites and hence interfering with many metabolic processes of the pathogen. Afterwards, the single-site fungicides were introduced and as a consequence of their frequent and repeated use, fungicidal resistance has become a major concern in modern crop protection seriously threatening effectiveness of several fungicides [7]. Hence, Fungicidal resistance is a result of adaptation of a fungus to a fungicide due to a stable and inheritable genetic change, leading to the appearance and spread of mutants with reduced fungicide sensitivity [8].

### 3.1 Genetic bases of fungicide resistance

Borck and Braymer [9] have analysed the genetics of fungicide resistance and found four important factors responsible for the resistance, these are (1) involvement of number of loci, (2) the number of allelic variants at each locus, (3) the existence and relevance of dominant or recessive relationship between resistant and wild-type alleles and (4) the additive or synergistic interactions between resistance genes.

These resistance genes may be located either in the nucleus on chromosome or in the cytoplasm on extrachromosomal genetic determinants and that can be differentiated by their inheritance patterns. Chromosomal genes typically show disomic inheritance in sexual crosses, in which one allele from each parent is received by the zygote and in cytoplasmic extrachromosome a uniparental (usually maternal) transmission [10]. In cytoplasmic genes, resistance stability gets affected by vegetative segregation and intracellular selection [11, 12].

In pathogenic fungus, mutation occurs, as a result of fungicidal resistance, in single major genes or from additive or synergistic interactions between several mutant genes [13].

Mostly this resistance is the result of mutations in major genes and these genes conferring resistance to fungicides having different modes of action may also occur in the same isolate, causing multiple resistance. Laterally, these genes have an appreciable influence on the phenotype which results in qualitative change in response to a fungicide, with the appearance in the field of new fungicide-resistant sub-populations well distinguishable from the wild-type sensitive ones. In the case of oligogenic resistance, many different major genes are involved, any one of which can mutate to cause an increase in resistance to the same fungicide. Like, in *Pyricularia oryzae* the resistance against kasugamycin as well as resistance to the two fungicides ethirimol and triadimenol in *Blumeria graminis* f. sp. *hordei* may be controlled by three different loci where a resistance allele at any one locus confers resistance [14].

### 3.2 Ploidy level

Cellular ploidy is the number of complete sets of chromosomes in a cell. Many eukaryotic species have two (diploid) or more than two (polyploid) sets of chromosomes [15]. The difference in ploidy level affects the number of alleles at each locus, which constitutes an Fmajor genomic trait that results in the evolution of fungicidal resistance. This ploidy level directly affects the frequency of mutations which may arise in single individuals as a result of the different numbers of mutational targets [16]. Most of the plant pathogenic fungi are in haploid state for a major part of their life cycle. On the contrary, *Oomycetes* typically show a diploid life cycle, and the haploid phase is restricted to the gametes [17]. Moreover, polyploids have been frequently identified among *Oomycetes*, such as *Plasmopara viticola* and *Phytophthora* spp. [18, 19].

CAA (carboxylic acid amide) fungicides, which inhibit cellulose biosynthesis in *Oomycete* phytopathogens, are considered at low to medium resistance risk depending on the fungal species. Classic genetic analysis showed that resistance to all CAA fungicides co-segregates and has the same genetic basis [20, 21]. The intrinsic risk of resistance is estimated to be significantly higher than CAA due to their genetic differences; however, no cross-resistance exists between CAA and other fungicides currently available against *Oomycetes*, such as phenylamides and QoI fungicides. In phenylamides, the resistance is a monogenic trait conferred by a semi-dominant chromosomal gene [22, 23], while QoI resistance is due to mutations in the mitochondrial *cytb* gene [24].

The probability of resistance is significantly increased by the occurrence of gene recombination, although, to express it phenotypically and for making resistance fixed several sexual reproduction cycles are required.

### 3.3 Heterokaryosis and nuclear number

Heterokaryosis is the association of genetically distinct nuclei coexisting in a common hyphal compartment and is a process involved in the generation of fungal variation occurs frequently in some fungal taxa and is a potential source of genetic variation. Heterokaryosis permits changes in the proportions of different nuclei in *Ascomycetes*, for selection and is essential for parasexual recombination [25]. In *Basidiomycetes*, the stable dikaryotic state is established in which two distinct parental haploid nuclei coexist (heterothallic) without fusion in each cell, which is genetically equivalent to a diploid. They both, heterokaryons and dikaryons, provide the chances for genes to complement each other (genetic complementation). In heterokaryons, fungicidal resistant and fungicide-sensitive genes may be able to develop in the presence or absence of fungicides [25].

It is very rare that the resistant mutants gain competitiveness under the selection force of fungicide sprays and are selected to frequencies at which disease control becomes unsatisfactory [26]. Mutation for resistance occurs at different rates depending on the number of genes conferring resistance. A rapid shift towards resistance may occur in monogenic resistance, leading to a discrete resistant sub-population. In polygenic resistance, mutation for resistance occurs slowly, leading to a reduced sensitivity of the entire population. Two types of selection pressures are able to keep resistant and wild-type sub-populations in a dynamic equilibrium, (1) the disruptive selection (directional selection), which develops because of repeated sprays of fungicides having the same mode of action and favours resistant sub-population(s), and (2) stabilizing selection, is developed because of a negative pleiotropic effect of resistance mutations leading to reduced fitness and favours the wild-type sensitive populations.

## 4. Management strategies and alternatives

Potato is considered a poor man's food and carries many diseases. The healthy potato will have a direct impact on people's food security and increase the income in potato growing countries. Worldwide, efficient use of land, water and nutrients can be improved by achieving healthy potato tubers but practically it is not possible. To reduce the disease loss, lots of fungicides get spread on the crop, which creates the cause of resistance development. This is one of the reasons for fungicidal resistance in potato pathogens and it can be reduced with the adaptation of good management practices and strategies, which are discussed below.

To delay the development of fungicidal resistance is the primary goal of resistance management rather than managing resistant fungal strains and a management strategy should be implemented before resistance becomes a problem. In this way, resistance can be prevented from becoming economically important. Also, minimizing the use of at-risk fungicides helps to avoid the development of fungicidal resistance without sacrificing disease control. This can be accomplished by using the at-risk fungicide with other fungicides and with non-chemical control measures, such as an integrated disease management program and the use of disease-resistant varieties. Also, the use of resistant cultivars, growing the crop in pathogen-free areas, lengthening of crop rotation, disease forecasting tools, proper use of fungicides (repeated use of same fungicide, dose, time, place of fungicide), creating unfavourable environment helps to avoid the development of fungicidal resistance without sacrificing disease control. Elimination of disease source will be the alternative with which development of fungicidal resistance can be avoided. Anyhow, it is critical to use an effective disease management program to delay the build-up of resistant strains. The larger the pathogen population exposed to an at-risk fungicide, the greater the chance a resistant strain will develop. However, in the broad aspect the management strategies and alternatives for fungicidal resistance in potatoes can be exploiting host resistance; exploiting host resistance; exploitation of race-nonspecific resistance; enhancement of natural disease resistance in potatoes; biotechnology approach; biological approach; use of botanical and many more.

### 4.1 Exploiting host resistance

To avoid the development of fungicidal resistance the cultivar resistance could be exploited to reduce fungicide input while achieving an acceptable control of potato disease, especially in late blight, in both foliage and tubers. Nærstad et al. [27], in 2007, confirmed that the host resistance against the pathogen is responsible to avoid the development of fungicidal resistance (i) by spraying at the right time with the recommended dose in cultivars with low field resistance to blight; (ii) the fluazinam dose can be reduced to 80% of the recommended dose by exploiting medium foliar resistance at high disease pressure and to approximately 40% at low disease pressure, by applying fungicides at right time, when field resistance to tuber blight is high; (iii) exploiting a high level of foliar resistance carries a high risk when the level of field resistance to tuber blight is low because a light foliar infection can provide enough spores to cause a high frequency of infected tubers; (iv) The application intervals may also be extended at high levels of field resistance to blight.

### 4.2 Enhancement of natural disease resistance in potatoes by chemicals

It is possible to enhance the existing host resistance against potato pathogen by exogenous application of some chemicals like acetylsalicylic acid (ASA), acibenzolar-S-methyl (BTH), 2,6-dichloroisonicotinic acid (INA), DL-3-aminobutyric acid (BABA), etc. The expression of the pathogenesis-related (PR) gene was observed by the spray of INA to tomato [28, 29], ASA and BTH to tobacco [30], and benzothiazole to potato plants and results in disease resistance, but the level of resistance and the set of PR-proteins induced are highly plant-specific [28, 31].

BABA induces the accumulation of high levels of three PR-protein families, PR-1, PR-2 and PR-5 in potatoes, and protects against late blight caused by *Pytophthora infestans* [32]. BABA has also been reported in the partial protection of potato plants against *P. infestans* in field experiments [32]. The fosetyl aluminium (aluminium tri(ethyl hydrogen phosphonate)) is a systemic fungicide, which has acropetal and basipetal mobility and is active against *Oomycetes*. Its mode of action showed that it can act directly on the fungus and indirectly by activating disease resistance mechanisms, such as phytoalexin production in tomato, tobacco, capsicum and grapevine plants [33].

### 4.3 Bio-technological approach

This approach is one of the most promising approaches for avoiding the development of fungicidal resistance and getting disease-free potato tubers by making the host more compatible to fight with pathogens.

This approach develops the host resistance by understanding the knowledge of molecular biology and genetics of inter-action between plant and *Oomycetes* which helps in discovering many resistance genes, numerous effector proteins and their mode of action [34]. Mainly two approaches are there in biotechnology aspect i.e., Cis-genic and Trans-genic.

#### *4.3.1 Cis-genic approach*

In this, resistant genes naturally occurring in the plant itself or from other related species are used and it is mainly based on the availability of resistant genes in potato crop. This approach is ethically and socially more acceptable to the public [35].

The start and end product during this programme is potato varieties which consist of potato genes (resistant) only. In this programme, no new varieties were developed and only point is that in the old variety resistant genes of wild potato species were incorporated. This cis-genic modification approach with potato's own gene is societally acceptable and also results in simplification in the legislation on the use of cis-genic modification approach [34]. To develop the durable resistance in potato crop, the DuRPh (Durable Resistance against *Pytophthora infestans*) programme was made in which cloning, transformation and selection of desired resistance were involved. Surprisingly, no markers are used in this approach, so the variety obtained will be made free, and to confirm the presence of resistance gene for *P. infestans*, PCR (polymerase chain reaction) technique is used.

#### *4.3.2 Transgenic approach*

This includes detection, isolation, cloning and transformation of gene from wild species or any other species into existing varieties through a bacterial vector (*Agrobacterium tumefaciens*). The mutate plantlets regenerated through callus culture and are screened to assess for resistance. Importantly, the mutant should have the same phenotype as the wild variety into which resistance genes are introduced. In this, there are two sub approaches, genetic engineering and RNAi technology. Plant genetic engineering is the act of inserting one or more agriculturally important genes into the genome of a plant by in vitro techniques. The genes inserted by genetic engineering are called transgenes that may (partly) originate from other organisms (such as bacteria or fungi) or non-crossable plant species. The first transgenic potato was developed about 20 years ago, and many of the transgenic potato plant products with enhanced characteristics are to be commercialized in the present decade [36]. Another important class of transgenes is based on RNAi for silencing existing traits coding for starch composition, processing traits or other quality traits.

Transgenics for late blight resistance: The disease caused by the *Oomycetous* fungus, has a history of causing catastrophic famine in Ireland where people depended heavily on this crop. In recent years, India and China emerged as the global leaders in potato production together contributing about 27% of world production. Occurrence of both A1 and A2 mating types of *P. infestans* resulting in sexual reproduction and survival through resilient oospores have been reported that may give rise to immense variability in the pathogen population, thereby endangering durability of a cultivar. Moreover, this population is gradually becoming tolerant to higher doses of prophylactic fungicides. As a consequence of this, a hidden but serious population shift in *P. infestans* has succumbed to this disease, in Kufri Jyoti, the most popular Indian cultivar, after a sustained performance for about 30 years. The other popular cultivar Kufri Bahar does not have any resistance to *P. infestans*. Together, these two cultivars occupy >60% of the potato area in India creating an imminent danger under our nose.

Race-specific, major genes from the wild potato species *Solanum demissum* have been extensively used in resistance breeding programmes throughout the world including India. However, the efficacy of such major genes had been too short-lived to justify their deployment. Because of this stress, late blight breeding has now moved to deployment of multi-gene, horizontal resistance. Although, identifying the genes responsible for horizontal resistance and their pyramiding is a difficult task. Recently, a new gene has emerged, i.e., the RB gene, which behaves like non-host resistance and is effective against all known races of *P. infestans.* This gene has been mapped and cloned by two independent groups in the USA and The Netherlands. The potato cultivar Katahdin, Transgenic clones of RB gene, showed late blight resistance at Toluca valley, the centre of origin of *P. inefstans*. The Agricultural Biotechnology Support Project-II operating from The Cornell University, USA has initiated a programme to popularize the use of RB gene in South and South-East Asia.

Bacterial wilt resistance: Bacterial wilt is another chronic disease problem that does not have any reliable resistance source. Therefore, an antimicrobial peptide gene, bovine enteric beta defensin (EBD) is being used for conferring bacterial wilt resistance in potatoes. Transgenic lines of Kufri Badshah showed a very high level of resistance to bacterial wilt in glass house screening. In India, the gene has now been transferred to two commercial potato cultivars Kufri Giriraj and Kufri Jyoti. Kufri Giriraj was selected because of its popularity in Shimla and Nilgiri hills where bacterial wilt is prevalent. Kufri Jyoti is a popular variety in eastern plains where bacterial wilt is endemic. Twenty-seven putative transgenic lines of Kufri Giriraj and 12 lines of Kufri Jyoti have been developed that are being characterized at present.

Viruses are also important pathogens which are ubiquitous and cause 80% losses in potato yield. Potato has been infected by more than 40 viruses and 2 viroids [37]. So far, only 9 viruses and 2 viroids are of economic significance for the growing potato industry. These potato viruses include potato viruses A, M, S, V, X, and Y (PVA, PVM, PVS, PVV, PVX, and PVY), potato leafroll virus (PLRV), tobacco rattle virus (TRV) and potato mop-top virus (PMTV). PLRV and PVY are currently considered the most dangerous viruses [38]. There are numerous variable factors i.e., plant genetic diversity, biology, lifecycle of the host plant/pathogen, vector species, biotype and environmental conditions that affect the incidence and severity of viral diseases. Potato is clonally propagated by planting tuber, which enhances the risk of accumulation of viruses in the next crop and tuber generations. Viral infection on potato (either individual or mixed infection) results in varied tuber infections i.e., spraying (TRV); necrotic ringspots (PVY NTN), net necrosis (PLRV) and deformed tubers (potato spindle tuber viroid) that render the tubers unsaleable [39]. PLRV is among the most prevalent viral diseases of potato in India, which almost causes 50–80% loss in potato yield and produces only a few, small to medium tubers [40]. Mineral oil and pesticide spray (chemical spray) are partial protection techniques that are not effective and efficient means to control viral diseases. The generation of resistant cultivars is considered the most economic and environmentally acceptable way of controlling viral diseases in potatoes [38]. Transgenic development by using pathogen-derived resistance is, therefore, being pursued for their management. The molecular technique involves two sets: (i) cellular technique which involves transformation and regeneration and (ii) includes identification, isolation and specific genes coding for interesting traits.

Coat protein mediated virus resistance: In 1986, Abel et al. reported the first example of resistance derived from coat protein. Transgenic potatoes expressing the PVY CP gene were found to be highly resistant to PVY and PLRV [41]. The PVYo strain was collected from field infected samples. The CP gene has been amplified by RT-PCR, cloned and sequenced. Sense, antisense and hairpin constructs have been designed and cloned in pDrive. The constructs were sub-cloned in the binary vector pBinAR. The vectors were then mobilized into *A. tumefaciens* strain EHA 105. Co-cultivation was done with sense construct and 11 putative transgenic lines of Kufri Bahar regenerated so far. Further screening of these 2 lines for PLRV resistance is to be undertaken in the glass house at CPRI, Shimla.

Generation of virus resistance through RNA silencing: Resistance to PVY in potatoes was done by induction of RNA silencing an ectopically expressed dsRNA, conserved between different PVY strains. PVY strains express Hc-Pro suppressor protein that interferes with the plant host defence. The expression of virus-derived dsRNA from transgenes can fully suppress viral infection through RNA silencing, thus overcoming viral suppressors [42–44]. Coat protein of PVY was cloned in pT3T7 vector followed by subcloning in binary vector pART7/27 and transferred into *A. tumefaciens* strain LBA4404. Transgenic lines generated were found resistant to viral infection as confirmed by ELISA measurements, northern hybridizations and RT-PCR. The transgenic lines generated have not been yet tested under field conditions, which would be necessary for further use of these lines [45].

Engineering virus resistance using a modified potato gene: Virus resistance genes have recessively inherited that function in a passive manner, whereby host factors evolve to avoid an interaction, which is essential for an invading virus to complete its lifecycle [46]. Using this theory, a study was carried out by [47] where natural mutations in translation initiation factor eIF4E confer resistance to potyviruses in potato plants. elf4E from potato cultivar Russet Burbank strain 'Ida' was cloned in TOPO cloning vector and further subcloned in a plant cloning vector pBI121 further, *Agrobacterium*-mediated transformation was performed on potato stem internode segments. All control plants (wild-type 'Russet Burbank' and transgenic lines over-expressing the1+ or GUS genes) developed typical PVY symptoms and tested positive for DAS-ELISA. To determine eIF4E expression, Northern blot and cDNA sequencing analysis were conducted on transgenic and non-transgenic plants.

Movement protein-mediated resistance: Resistance to virus movement in plants reduces the initial infection site [48]. Potato transformed with sequence from PLRV open reading frame (ORF) 4, encode a protein (pr17) i.e., phloem specific movement protein [49]. The mutant pr17 binds to the plasmodesmata and inhibiting cell to cell movement of unrelated viruses. Transgenic plants showed reduced accumulation of PLRV on secondary infection (operating at RNA level) and were also found resistant to PVY and PVX virus infection (protein-mediated resistance).

Fungicide Resistance Action Committee (FRAC) were developed to facilitate managing resistance by categorising fungicidal group and coding them, which designate chemical group of that fungicide and mentioned on the front of label. The fungicides with the same mode of action have categorised into one group. For managing resistance, it is critically important to know the group code for the fungicides being used to avoid alternating among chemically similar fungicides. Currently, there are 48 numbered FRAC Group Codes plus NC (not classified), 7 numbered with a 'P' (for host plant defence), 7 numbered with a 'U' (for unknown mode of action), and 12 numbered with an 'M' (for multi-site contact activity).

The codes for all fungicide active ingredients are based on a common name that can be downloaded from the FRAC website and can be used to spray the effective fungicides against a particular disease.

### 4.4 Use of bio-agents

In biological control, living micro-organisms provide disease protection through the production of antibiotics, competition for food and space, induced plant resistance, etc. This helps to avoid the use of fungicides which directly reduce the chances of development of fungicidal resistance. Various fungi and bacteria were tested against *P. infestans* in potato crop [50] and results in suppressed blight infection in leaflets [51]. Daayf et al. [52] also studied on biological control of potato late blight by detached leave method, whole planting testing system and *in vitro*.

*Trichoderma* (formulation) @ 10 g/l and *Pseudomonas* (formulation) @ 10 g/l found antagonistic behaviour and best results against late blight disease [53]. Bio-agent *Xenorhabdus* spp. gave most consistent results of biological control against late blight disease. Application of *Steinernema feltiae* was also studied against late blight both in vivo and in vitro [54].

In search of antagonistic against *P. infestans,* lot of work has been done and *Burkholderia* spp., *Streptomyces* spp., *Pseudomonas* spp., and *Trichoderma* spp. were obtained from leaves, stems, tubers and rhizoplane of potato plants were tested. The efficacy of these bio-agents to A1 and A2 mating type of *P. infestans* was assessed in greenhouse; field and on potato leaves in moist chamber and all three found to reduce the *P. infestans* infection applied individually or in combination [55].

The bio-control agents in combination with products such as neem oil could be effective to manage late blight severity [56] and it could be another option to reduce crop losses caused by the pathogen. Among the seven potato phylloplane fungi, only three fungi viz., *Fusarium* spp., *Trichoderma* spp., *Aspergillus* spp. showed antagonistic potential against *P. infestans* [53].

Systemic acquired resistance: Induction of SA (salicylic acid) is elicited by both *bacilli* and *pseudomonad* PGPR strains but ethylene and jasmonic acid dependent [57].

Bio-fungicide: *Chaetomium* mycofungicide found to reduce incidence of late blight and reduce its population in the soil with significant reduction the potato late blight [58]. A significantly reduction in *P. infestans* sporangial germination was observed with the spray of *T. viride* and *P. viridicatum* formulation and has potential to control potato late blight under control condition [59].

Rhizobacteria: Kim and Jeun [60] reported the drenching with plant growth promoting rhizobacteria isolates increased the total weight of tubers per potato plants, in addition to effectively controlling late blight. Yang et al. [61] also reported the *Bacillus pumilus* and *Pseudo-monas fluorescens* induced resistance to *P. infestans* and there was reduction in zoospore formation and germination.

### 4.5 Green chemicals

Biological origin pesticides, especially extract and natural substances originating from plants, microorganisms, algae and animals, are called green pesticides, like botanicals, essential oil, etc., also called ecological pesticides, which are considered environmentally friendly and are causing less harm to human and animal health and to habitats and the ecosystem are gaining a lot of interest for the integrated management of fungal diseases. Botanicals are one of them which is a substance obtained or derived from a plant such as a plant part or extract used for many purposes and can be used against the pathogens. Green plants are a huge reservoir of various effective chemotherapeutics and could serve as an environmentally friendly natural alternative to fungicides and directly the risk of fungicidal resistance development can be avoided.

Eleven extracts from different plant species were tested for antibacterial activity against potato soft rot bacteria, *E. carotovora* subsp. *carotovora* (Ecc) P-138, under in vitro and storage conditions and found effective. These are bael (*Aegle marmelos* L.), mander (*Erythrina variegata*), chatim (*Alstonia scholaris* L.), marigold (*Tagetes erecta*), garlic (*Allium sativum* L.), onion (*Allium cepa*), lime (*Citrus aurantifolia*), turmeric (*Curcuma longa* L.), jute (*Corchorus capsularis* L.), cheerota (*Swertia chirata* Ham.) and neem (*Azadirachta indica*) [62]. The work on botanicals with anti-oomycetes activity has increased over the years and the efficacy of botanicals against pathogens has also been demonstrated. Several preliminary *in vitro* studies have been conducted in China and India [63, 64]. Few plant extracts from different plant materials were tested for controlling effects against the infection of *P. infestans* on potato tuber slices, seedlings and detached leaves and *Galla chinensis* showed the best inhibiting effect among *Terminalia chebula, Sophora flavescens, G. chinensis, Rheum rhabarbarum, Potentilla erecta* and *Salvia officinalis* [65]. Cao and Van Bruggen, (2001) observed the treatments of garlic cloves extract at 1 or 2 percent completely inhibit the zoospore formation and colony growth of pathogen [63].

Essential oils are used to manage potato diseases which are obtained from plants through fermentation, enfleurage, extraction and steam distillation. Essential oils are used because of two prominent features, i.e., low toxicity for people and the environment due to their natural properties and low risk for resistance development by pathogenic micro-organisms [66]. The antifungal activity of essential oil obtained from three medicinal plants, i.e., *Zataria multiflora*, *Thymus vulgaris* and *Thymus kotschyanus* against phytopathogenic fungi were reported [67].

## 5. Conclusion

Today, food security is dependent on crop protection and fungicide-based plant protection is indispensable for efficient and large-scale crop production. As it takes a long time to develop new fungicides, it is necessary to take proper care to avoid the development of fungicidal resistance. In single-target fungicidal resistance, development may occur within a few years and fungicides with novel modes of action are found rarely. The earlier strategy was to apply single-target fungicides, which impose a strong selection pressure for the development of resistant mutants. The fungicides with more than one target are not easily overcome by mutations. Hence, there is a need to explore and enhance the existing host resistance against the potato pathogen. Among all, transgenic approach is the most promising to avoid the risk of development of fungicidal resistance; use of alternatives like green chemicals, bio-agents and explore and enhance the host resistance.

## Author details

Rahul R. Bakade[1]*, S. Sundaresha[2] and Mehi Lal[3]

1 ICAR-Indian Institute of Natural Resins and Gums, Ranchi, India

2 ICAR-Central Potato Research Institute, Shimla, HP, India

3 ICAR-Central Potato Research Institute, Meerut, Uttar Pradesh, India

*Address all correspondence to: rahulpathology@gmail.com

## References


[1] Deising HB, Reimann S, Peil A, Weber WE. Disease Management of Rusts and Powdery Mildews. In: Kempken F (eds). Agricultural Applications. Berlin, Heidelberg: The Mycota, Springer. 2002;**11**. DOI: 10.1007/978-3-662-03059-2_13

[2] Cuomo CA, et al. The Fusarium graminearum genome reveals a link between localized polymorphism and pathogen specialization. Science. 2007;**317**(5843):1400-1402. DOI: 10.1126/science.1143708

[3] Forsyth W (1802). A treatise on the culture and management of fruit trees. Nichols, London FRAC (Fungicide Resistance Action Committee). FRAC list of plant pathogenic organisms resistant to disease control agents. 2014. Available online: www.frac.info

[4] Parker JE, et al. Mechanism of binding of prothioconazole to *Mycosphaerella graminicola* CYP51 differs from that of other azole antifungals. Applied and Environmental Microbiology. 2011;**77**(4):1460-1465. DOI: 10.1128/AEM.01332-10

[5] Cools HJ, et al. Impact of recently emerged sterol 14α-demethylase (CYP51) variants of *Mycosphaerella graminicola* on azole fungicide sensitivity. Applied and Environmental Microbiology. 2011;**77**(11):3830-3837. DOI: 10.1128/AEM.00027-11

[6] Leroux P, Walker AS. Multiple mechanisms account for resistance to sterol 14α-demethylation inhibitors in field isolates of *Mycosphaerella graminicola*. Pest Management Science. 2011;**67**(1):44-59. DOI: 10.1002/ps.2028

[7] Brent KJ, Hollomon DW. Fungicide resistance: the assessment of risk. FRAC monograph no. 2. Global Crop Protection Federation, Brussels; 2007

[8] Delp CJ, Dekker J. Fungicide resistance: Definitions and use of terms. EPPO Bulletin. 1985;**15**(3):333-335. DOI: 10.1111/j.1365-2338.1985.tb00237.x

[9] Borck K, Braymer HD. The genetic analysis of resistance to benomyl in Neurospora crassa. Journal of General Microbiology. 1974;**85**(1):51-56. DOI: 10.1099/00221287-85-1-51

[10] Griffiths AJF. Mitochondrial inheritance in filamentous fungi. Journal of Genetics. 1996;**75**(3):403-414. DOI: 10.1007/BF02966318

[11] Birky CW. The role of knox genes in plant development. Annual Review of Genetics. 2001;**35**:c125/148

[12] Ziogas BN, Markoglou AN, Tzima A. A non-Mendelian inheritance of resistance to strobilurin fungicides in Ustilago maydis. Pest Management Science. 2002;**58**(9):z08-z916. DOI: 10.1002/ps.543

[13] Kalamarakis AE, de Waard MA, Ziogas BN, Georgopoulos SG. Resistance to fenarimol in Nectria haematococca var. cucurbitae. Pesticide Biochemistry and Physiology. 1991;**40**(3):12-220. DOI: 10.1016/0048-3575(91)90092-Z

[14] Taga M. Identification of three different loci controlling Kasugamycin resistance in *Pyricularia oryzae*. Phytopathology. 1979;**69**(5):463. DOI: 10.1094/phyto-69-463

[15] Otto SP, Whitton J. Polyploid incidence and evolution. Annual Review

of Genetics. 2000;**34**(1):401-437. DOI: 10.1146/annurev.genet.34.1.401

[16] Otto SP, Whitton J. Pollyploid incide and evolution. Genetic Resources and Crop Evolution. 2000;**34**:401-437

[17] Fincham JRS, Day Peter R, Radford A. Fungal Genetics. 4th edition. Botanical Monographs. New York: USA. **41979**; ISBN: 05-200-38185

[18] Rumbou A, Gessler C. Particular structure of *Plasmopara viticola* populations evolved under Greek island conditions. Phytopathology. 2006;**96**(5):501-509. DOI: 10.1094/PHYTO-96-0501

[19] Bertier L, Leus L, D'hondt L, De Cock AWAM, Höfte M. Host adaptation and speciation through hybridization and polyploidy in phytophthora. PLoS One. 2013;**8**(12):1-15. DOI: 10.1371/journal.pone.0085385

[20] Young DH, Kemmitt GM, Owen J, Dehne HW, Gisi U, Kuck KH, et al. A comparative study of XR-539 and other oomycete fungicides: similarity to dimethomorph and amino acid amides in its mechanism of action. 2005

[21] Gisi U, Waldner M, Kraus N, Dubuis PH, Sierotzki H. Inheritance of resistance to carboxylic acid amide (CAA) fungicides in *Plasmopara viticola*. Plant Pathology. 2007;**56**(2):199-208. DOI: 10.1111/j.1365-3059.2006.01512.x

[22] Gisi U, Cohen Y. Resistance to phenylamide fungicides: A case study with *Phytophthora infestans* involving mating type and race structure. Annual Review of Phytopathology. 1996;**34**:549-572. DOI: 10.1146/annurev.phyto.34.1.549

[23] Knapova G, Schlenzig A, Gisi U. Crosses between isolates of *Phytophthora infestans* from potato and tomato and characterization of F1 and F2 progeny for phenotypic and molecular markers. Plant Pathology. 2002;**51**(6):698-709. DOI: 10.1046/j.1365-3059.2002.00762.x

[24] Gisi U, Sierotzki H. Fungicide modes of action and resistance in downy mildews. European Journal of Plant Pathology. 2008;**122**(1):157-167. DOI: 10.1007/s10658-008-9290-5

[25] de miccolis, Milvia & Pollastro, Stefania & Faretra, Francesco. Genetics of Fungicide Resistance. 2015. DOI: 10.1007/978-4-431-55642-8_2

[26] Hobbelen PHF, Paveley ND, Van Den Bosch F. The emergence of resistance to fungicides. PLoS One. 2014;**9**(3):1-14. DOI: 10.1371/journal.pone.0091910

[27] Nærstad R, Hermansen A, Bjor T. Exploiting host resistance to reduce the use of fungicides to control potato late blight. Plant Pathology. 2007;**56**(1):156-166. DOI: 10.1111/j.1365-3059.2006.01491.x

[28] Kessmann H, Staub T, Ligon JIM, Oostendorp M, Ryals J. Activation of systemic acquired disease resistance in plants SAR: Biological phenomenon. European Journal of Plant Pathology. 1994;**100**:359-369

[29] Metraux JP, et al. Induced systemic resistance in cucumber in response to 2,6-dichloro-isonicotinic acid and pathogens. In: Hennecke H, DPS V, editors. Advances in Molecular Genetics of Plant-Microbe Interactions Vol. 1: Proceedings of the 5th International Symposium on the Molecular Genetics of Plant-Microbe Interactions, Interlaken, Switzerland, September 9-14, 1990. Dordrecht: Springer Netherlands; 1991. pp. 432-439

[30] Romero AM, Ritchie DF. Systemic acquired resistance delays race

shifts to major resistance genes in bell pepper. Phytopathology. 2004 Dec;**94**(12):1376-82. DOI: 10.1094/PHYTO.2004.94.12.1376. PMID: 18943709

[31] Summermatter K, Sticher L, Metraux JP. Systemic Responses in Arabidopsis thaliana Infected and Challenged with Pseudomonas syringae pv syringae. Plant Physiol. 1995 Aug;**108**(4):1379-1385. DOI: 10.1104/pp.108.4.1379. PMID: 12228548; PMCID: PMC157515

[32] Cohen YR. B-aminobutyric acid-induced resistance against plant pathogens. Plant Disease. 2002;**86**(5):448-457. DOI: 10.1094/PDIS.2002.86.5.448

[33] Jiang Y. Alterations in the host-parasite interface between *Phytophthora infestans* and tomato leaf cells after treatment with fosetyl-Al. Plant Science. 1990;**71**(2):267-276. DOI: 10.1016/0168-9452(90)90017-I

[34] Haverkort AJ, Struik PC, Visser RGF, Jacobsen E. Applied biotechnology to combat late blight in potato caused by *Phytophthora infestans*. Potato Research. 2009;**52**(3):249-264. DOI: 10.1007/s11540-009-9136-3

[35] Jochemsen H. An ethical assessment of cisgenesis in breeding late blight resistant potato. Potato Research. 2008;**51**(1):59-73. DOI: 10.1007/s11540-008-9090-5

[36] Khurana SM, Chakrabarti S. Genetically engineered improvement of root and tuber crops. Biotechnology in Horticultural and Plantation Crops. New Delhi: Malhotra publication house. 2000. P. 422-451

[37] Jeffries C, Barker H, Khurana SMP. Potato Production, Improvement and Post-Harvest Management. New York, USA: The Haworth's Food Products Press; 2005

[38] Solomon-Blackburn RM, Barker H. Breeding virus resistant potatoes (Solanum tuberosum): A review of traditional and molecular approaches. Heredity (Edinb). 2001;**86**(1):17-35. DOI: 10.1046/j.1365-2540.2001.00799.x

[39] Salazar LF, Querci M, Bartolini I, Lazarte V. Aphid transmission of potato spindle tuber viroid assisted by potato leaf roll virus. Fitopatologia. 1995;**30**(1):56-58

[40] Jeevalatha A, Kaundal P, VenkatasalamEP, ChakrabartiSK, SinghBP. Uniplex and duplex PCR detection of geminivirus associated with potato apical leaf curl disease in India. Journal of Virological Methods. 2013;**193**(1):62-67. DOI: 10.1016/j.jviromet.2013.05.012

[41] Abel PP, Nelson RS, De B, Hoffmann N, Rogers SG, Fraley RT, et al. Delay of disease development in transgenic plants that express the tobacco mosaic virus coat protein gene. Science. 1986 May 9;**232**(4751):738-743. DOI: 10.1126/science.3457472. PMID: 3457472

[42] Kalantidis K, Psaradakis S, Tabler M, Tsagris M. The occurrence of CMV-specific short RNAs in transgenic tobacco expressing virus-derived double-stranded RNA is indicative of resistance to the virus. Molecular Plant-Microbe Interactions. 2002;**15**(8):826-833. DOI: 10.1094/MPMI.2002.15.8.826

[43] Smith NA, Singh SP, Wang M-B, Stoutjesdijk PA, Green AG, Waterhouse PM. Total silencing by intron-spliced hairpin RNAs. Nature. 2000;**407**(6802):319-320. DOI: 10.1038/35030305

[44] Wang MB, Abbott DC, Waterhouse PM. A single copy of a virus-derived transgene encoding

hairpin RNA gives immunity to barley yellow dwarf virus. Molecular Plant Pathology. 2000;**1**(6):347-356. DOI: 10.1046/j.1364-3703.2000.00038.x

[45] Missiou A, Kalantidis K, Boutla A, Tzortzakaki S, Tabler M, Tsagris M. MITRA missiou2004.pdf, 185-197, 2004

[46] Fraser RSS. The genetics of resistance to plant viruses. Annual Review of Phytopathology. 2003;**28**:179-200. DOI: 10.1146/annurev.py.28.090190.001143

[47] Cavatorta J, Perez K, Gray S, Eck J, Yeam I, Jahn M. Engineering virus resistance using a modified potato gene. Plant Biotechnology Journal. 2011;**9**:1014-1021. DOI: 10.1111/j.1467-7652.2011.00622.x

[48] Hong J-S, Ju H-J. The plant cellular systems for plant virus movement. Plants Pathology Journal. 2017;**33**(3):213-228. DOI: 10.5423/PPJ.RW.09.2016.0198

[49] Tacke E, Salamini F, Rohde W. Genetic engineering of potato for broad-spectrum protection against virus infection. Nature Biotechnology. 1996;**14**(11):1597-1601. DOI: 10.1038/nbt1196-1597

[50] Ghorbani R, Wilcockson S, Leifert C. Alternative treatments for late blight control in organic potato: Antagonistic micro-organisms and compost extracts for activity against *Phytophthora infestans*. Potato Research. 2005;**48**(3-4):181-189. DOI: 10.1007/BF02742375

[51] Lamsal K, Kim SW, Kim YS, Lee YS. Biocontrol of late blight and plant growth promotion in tomato using rhizobacterial isolates. Journal of Microbiology and Biotechnology. 2013;**23**(7):897-904. DOI: 10.4014/jmb.1209.09069

[52] Daayf F, Adam L, Fernando WGD. Comparative screening of bacteria for biological control of potato late blight (strain US-8), using invitro, detached-leaves, and whole-plant testing systems. Canadian Journal of Plant Pathology. 2003;**25**(3):276-284. DOI: 10.1080/07060660309507080

[53] Pundhir VS. Efficacy of fungicides and bio-agents against late blight severity, infection rate and tuber yield of potato. Journal of Plant Disease Sciences. 2008;**3**(1):4-8

[54] Kaya HK, et al. Status of entomopathogenic nematodes and their symbiotic bacteria from selected countries or regions of the world. Biological Control. 2006;**38**(1):134-155. DOI: 10.1016/j.biocontrol.2005.11.004

[55] Lozoya-Saldaña H, Coyote-Palma M, Ferrera-Cerrato R. Microbial antagonism against *Phytophthora infestans* (Mont) de Bary. Agrociencia. 2006;**40**:491-499

[56] Diniz LP, Maffia LA, Dhingra OD, Casali VWD, Santos RHS, Mizubuti ESG. Quantification of the efficacy of alternative products for tomato late blight control. Phytopathol. bras. 2006;**31**(2):171-179. DOI: 10.1590/S0100-4158200600020000 8

[57] Yan Z, Reddy MS, Ryu CM, McInroy JA, Wilson M, Kloepper JW. Induced systemic protection against tomato late blight elicited by plant growth-promoting rhizobacteria. Phytopathology. 2002;**92**(12):1329-1333. DOI: 10.1094/PHYTO.2002.92.12.1329

[58] Soytong K, Ratanacherdchai K. Application of mycofungicide to control late blight of potato. Journal of Agricultural Technology. 2005;**1**:19-32

[59] Gupta H, Singh BP, Mohan J. Biocontrol of late blight of potato. Potato Journal. 2004;**31**:39-42

[60] Kim H-J, Jeun Y-C. Resistance induction and enhanced tuber production by pre-inoculation with bacterial strains in potato plants against *Phytophthora infestans*. Mycobiology. 2006;**34**(2):67. DOI: 10.4489/myco.2006.34.2.067

[61] Yang XF, Zhang ZM, Yang HW, Jian H. Inhibition of metabolites from Xenorhabdus nematophilus against Phytophthora infestans. Journal of Hebei Agricultural University. 2001;**24**(2):65-68

[62] Rahman MM, Khan AA, Ali ME, Mian IH, Akanda AM, Abd Hamid SB. Botanicals to control soft rot bacteria of potato. Scientific World Journal. 2012;**2012**. DOI: 10.1100/2012/796472. Epub 2012 May 22. PMID: 22701096; PMCID: PMC3366214. Article ID 796472, 6 pages

[63] Cao K, van Bruggen A. Inhibitory efficacy of several plant extracts and plant products on *Phytophthora infestans*. Journal of Agricultural University of Hebei. 2001;**24**:90-96

[64] Deepa MA, Narmatha Bai V, Basker S. Antifungal properties of Pseudarthria viscida. Fitoterapia. 2004;**75**(6):581-584. DOI: 10.1016/j.fitote.2004.04.008

[65] Wang S, Hu T, Zhang F, Forrer HR, Cao K. Screening for plant extracts to control potato late blight. Frontiers of Agriculture in China. 2007;**1**(1):43-46. DOI: 10.1007/s11703-007-0007-x

[66] Daferera DJ, Ziogas BN, Polissiou MG. GC-MS analysis of essential oils from some Greek aromatic plants and their fungitoxicity on Penicillium digitatum. Journal of Agricultural and Food Chemistry. 2000;**48**(6):2576-2581. DOI: 10.1021/jf990835x

[67] Amini M, Safaie N, Salmani MJ, Shams-Bakhsh M. Antifungal activity of three medicinal plant essential oils against some phytopathogenic fungi. Trakia Journal of Sciences. 2012;**10**(1):1-8

Chapter 4

# Biomimetic and Hemisynthetic Pesticides

*Ahissan Innocent Adou, Garrick Bibian, Odile Bordelais, Léa Farouil, Muriel Sylvestre, Sarra Gaspard, Marie-Noëlle Sylvestre and Gerardo Cebrián-Torrejón*

## Abstract

Pests are responsible for most losses associated with agricultural crops. In addition, due to the indiscriminate use of synthetic pesticides, several problems have arisen over the years, such as pest resistance and contamination of important planetary sources such as water, air and soil. This awareness regarding pest problems and environment has led to the search for powerful and eco-friendly pesticides that degrade after some time, avoiding pest persistence resistance, which is also pest-specific, non-phytotoxic, nontoxic to mammals, and relatively less expensive in order to obtain a sustainable crop production Biodegradable biomimetic pesticides can be a potential green alternative to the pest industry.

**Keywords:** biopesticides, biomimetic, phytochemistry

## 1. Introduction

Chlordecone (CLD, **Figure 1**), a chlorinated insecticide, with a homocubane structure was used in Guadeloupe and Martinique (French West Indies (FWI)) to control the banana weevil, *Cosmopolites sordidus* from 1971 to 1993. Larvae of this insect are the most destructive stage, and they use their strong mandibles to escavate and create tunnels or galleries in the rhizome of banana trees [1]. To fight against this insect in the FWI, CLD was marketed in France from 1981 to 1993 as a formulation called Curlone®. The authorization for CLD was withdrawn by the French Ministry of Agriculture in 1990 but used in the FWI until September 1993. The estimated chlordecone amount applied over this time is 300 tons [2]. CLD is a very stable compound due to its high persistence; consequently, the entire environment (soil, surface, ground water and costal marine waters) and food chain remain contaminated. Therefore, animals, raised in banana production areas, are affected by this molecule [3, 4].

In banana cultivated areas of Guadeloupe, CLD concentrations between 0.1 and 37.4 $mg.kg^{-1}$ can be found in topsoil and up to 10 $\mu g.L^{-1}$ in aquatic systems [5]. Following the contamination of foodstuffs, the population of Guadeloupe and Martinique is exposed to chlordecone contamination through the consumption of

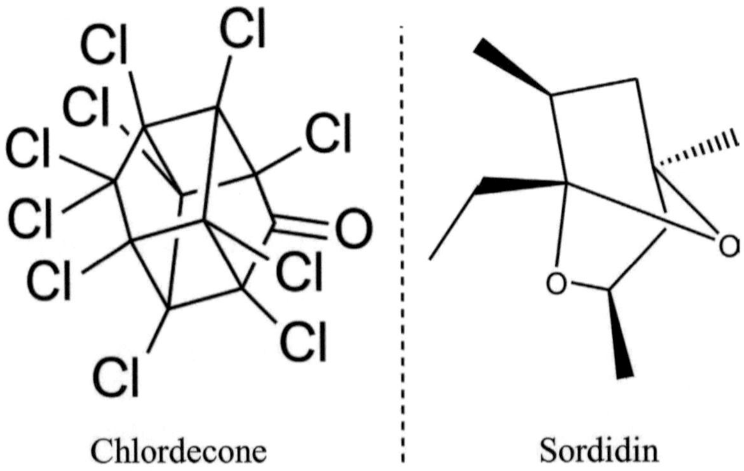


**Figure 1.**
*Chlordecone (left) and sordidin (right) structures.*

contaminated food and drinking water [6]. As a consequence, 92.5% of Martinicans and 94.9% of Guadeloupeans have detectable concentrations of CLD in their blood [7]. Several epidemiological studies were conducted to determine the health impact of this exposure. A correlation between pre- and postnatal chlordecone exposure and short-term memory and fine motor skills in young infants in the TIMOUN study [8–10]. CLD presents also endocrine disrupting properties [11] and is associated to type 2 diabetes [12]. Recent studies show that CLD exposure may be associated with altered epigenetic marks [13] and autoimmune diseases [14].

When CLD was prohibited in 1993, a weevil trap containing a pheromone, sordidin (**Figure 1**) was used [15]. Sordidin is a male-produced aggregation pheromone of banana weevil, related to ketal pheromones from Scolytids [15]. The production of this pheromone was first evidenced by Budenberg *et al.* in 1993. It was then identified and isolated in 1995 by Ducrot *et al.* as a major pheromone. Sordidin was first synthesized using the regioselective Baeyer-Villiger reaction of 2,6-disubstituted cyclohexanon as a key step, giving a mixture containing 4 stereoisomers. Trap system using this hormone is employed in the FWI and the Canary Islands. This trap supposes an interesting alternative mimicking the natural hormone allowing to substitute the use of CLD. A study of two plantations of over 200 hectares each shows a reduction in corm damage of 62.86% after the implantation of this biomimetic strategy [16].

The environmental problems produced by the CLD have led to the search for powerful and eco-friendly biomimetic pesticides as sordidin. These biomimetic pesticides should be: biodegradable (avoiding pest persistence); pest-specific; non-toxic to mammals and plants; and relatively less expensive in order to obtain a sustainable crop production.

Biomimetic compounds, as biopesticides, are obtained by synthetic routes which tend to transpose enzymatic reactions within the framework of synthetic organic chemistry. The concept of biomimetic synthesis of natural products was introduced by Robinson, following his straightforward synthesis of tropinone reported in 1917 [17, 18]. Several years later, the different ideas and the philosophy covering the biomimetic or biogenetic type synthesis was proposed by Van Tamelen in his work in 1961 [19]. Biomimetic synthesis can also describe a sequence of reactions carried out to support a biogenetic hypothesis which is generally accepted with succeeded reactions [20]. Poupon, Nay and coworkers have compiled the biomimetic syntheses of several families of organic compounds including alkaloids [21], terpenoids, polyphenols and polyketides (as sordidin) [21]. Over past decades, numerous publications contain the biomimetic term associated with organic synthesis but also sensoring particularly in the pollution control field. For example, we may notice an increasingly use of molecular imprinted polymers as recognition elements in mimicking molecular/ionic recognition by natural receptors [22, 23]. Khadem *et al.* have designed an electrochemical selective sensor to determine the dicloran by modifying the working electrode with molecular imprinted polymer [24]. Liu *et al.* have developed a biomimetic absorbent containing the lipid triolein embedded in the cellulose acetate spheres to remove persistent organic pollutants from water [25]. More recently, Sicard *et al.* have proposed a strategy for the decontamination of organic pollutants combining pesticides and drugs based on the use of nucleolipids, polymer-free bioinspired materials. The advantage of using the latter lies in their degradation providing nontoxic natural biomolecules [21], such as nucleosides, phosphates, and lipids [26].

The present chapter shows a compilation of biomimetic and hemisynthetic pesticides, classified by several different mechanisms affecting one or more biological systems, including:

1. Pesticides targeting nervous system.

2. Pesticides targeting endocrine system.

3. Pesticides targeting digestive system.

4. Pesticides targeting different cellular structures.

## 2. Pesticides targeting nervous system

Insects have a simple nervous system with a brain linked to a ventral nerve cord that consists of paired segmental ganglia running along the ventral midline of the thorax and abdomen (**Figure 2**). An insect's brain is a complex of six fused ganglia located dorsally within the head capsule. These ganglia can be separated in 3 pairs:

- Protocerebrum, associated with vision.

- Deutocerebrum, processing sensory information collected by the antennae.

- Tritocerebrum, which innervate the labrum and integrate sensory inputs from proto-and deutocerebrums while also linking the brain with the rest of the ventral nerve cord.

Below the brain another complex of fused ganglia, the subesophageal ganglion innervates mandibles, maxillae, labium, the hypopharynx, salivary glands, and neck muscles. In the thorax, three pairs of thoracic ganglia control locomotion by innervating the legs and wings. Thoracic muscles and sensory receptors are also associated with these ganglia. Similarly, abdominal ganglia control movements of abdominal muscles [28]. In some insects, the thoracic ganglia fuse to form a single ganglion.

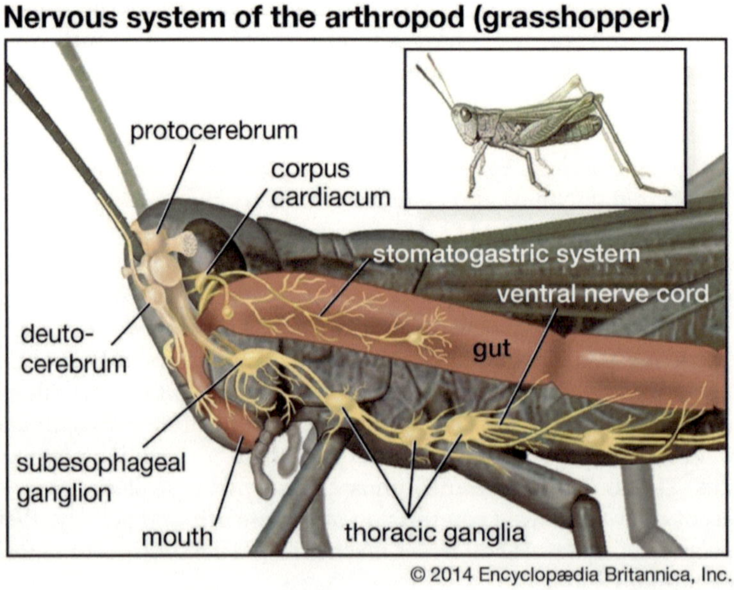


**Figure 2.**
*Nervous system of the grasshopper [27].*

Similarly, sometimes most of the abdominal ganglia are fused to form a single compound ganglion as in the blood sucking bug.

From molecular point of view, several receptors can be explored targeting the nervous system at different levels. These receptors are: i. Glutamate-gated chloride channels, ii. Voltage-gated sodium channels, iii. Transient receptor potential vanilloid channels, iv. Gamma-amino butyric acid receptors, v. Octopaminergic system and vi. Nicotinic acetylcholine receptors.

### 2.1 Glutamate-gated chloride channels (GluCls)

Glutamate-gated chloride channels (GluCls) are found only in protostome invertebrate phyla. Their functions include: the control and modulation of locomotion, the regulation of feeding, and the mediation of sensory inputs [29]. This channel is composed of 5 adjacent subunits. Each subunit is a polypeptide chain large extracellular N-terminal domain (for ligand binding) and four transmembrane domains (1–4) (**Figure 3**) [30].

#### *2.1.1 Compounds acting via glutamate-gated chloride channels*

- Avermectins and milbemycins:

Avermectins and milbemycins (**Figure 4**) are two families of hemisynthetic macrolides that have been widely used as pesticides in agriculture. The hemisynthesis of these molecules produces an increase of the chemical diversity starting from the natural ones produces by bacteria from the order Actinomycetes (as for example *Streptomyces avermitilis*). As an example of these molecules, we can mention aglycone milbemycin (**Figure 4**), abamectin, emamectin, doramectin,

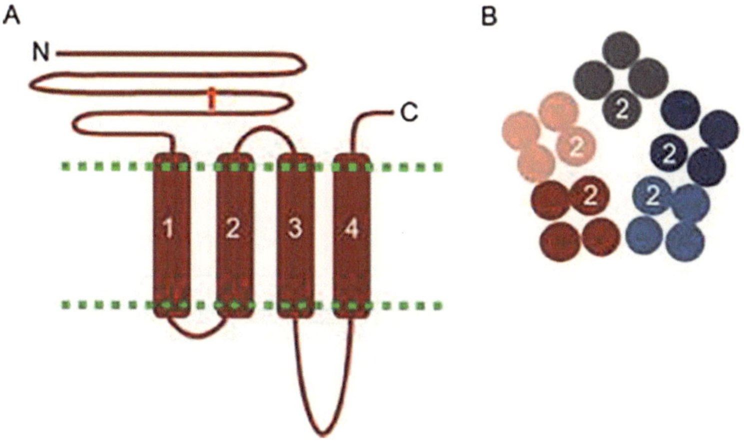


**Figure 3.**
*Glutamate-gated chloride channels [31]. (A) Transmembrane scheme; (B) Scheme of top-view of channel.*

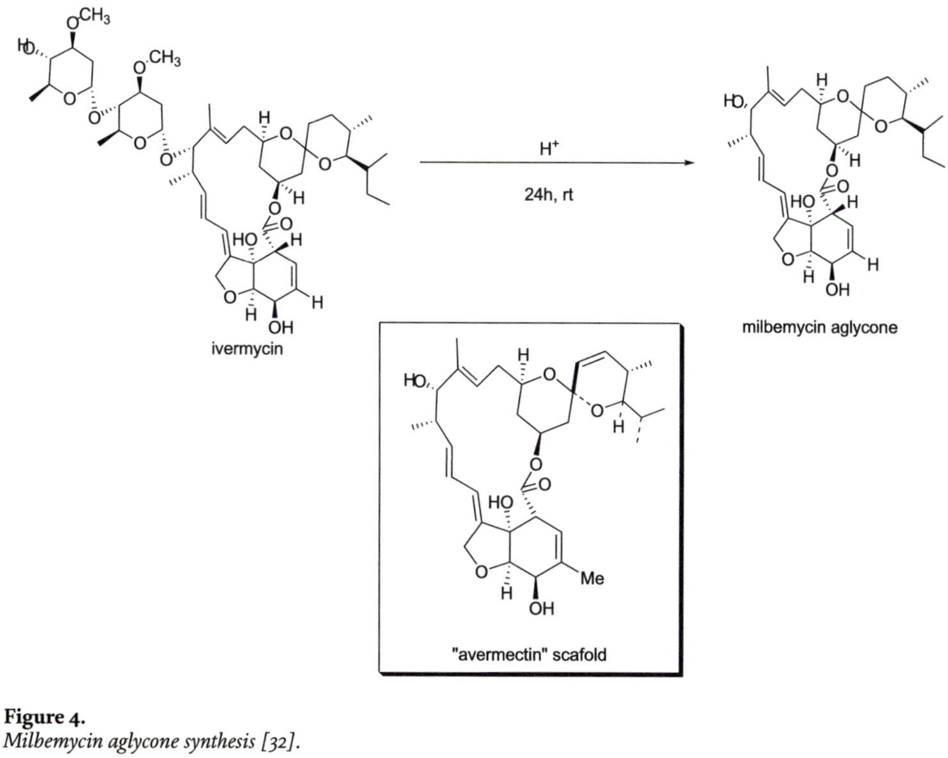


**Figure 4.**
*Milbemycin aglycone synthesis [32].*

milbemycin oxime, latidectin etc. These molecules attack the nervous system of insects [33]. This macrocyclic lactones exert their parasiticide (anthelmintic) and insecticidal effects mainly by potentiating the agonistic action of glutamate on GluCls or by directly activating GluCls, as in the case of *Drosophila melanogaster* [31], [34]. They act by linking to glutamate-dependent chloride channels common to invertebrate nerve and muscle cells. This binding causes the opening of the channels, increasing the flow of chloride ions and hyperpolarizing the cell membranes, paralyzing and killing the invertebrate. **Figure 4** shows the hemisynthesis of aglycone milbemycin by solubilizing ivermectin in a concentrated sulfuric acid solution [32]. However, the milbemycins have characteristics that are harmless to the environment.

## 2.2 Voltage-gated sodium channels (VGSC)

The voltage-gated sodium channel (VGSC) mediates the increase in sodium conductance during the rapid depolarization phase of the membrane action potential (high concentration of sodium ions ($Na^+$) and a low concentration of potassium ions ($K^+$)). Therefore, this channel represents a key structural element that controls cellular excitability in biological systems [35]. Mammalian sodium channels are composed of a pore-forming α-subunit and one or more β-subunits. Sodium channel α-subunits have four homologous repeat domains (I–IV), each possessing six α-helical transmembrane segments (**Figure 5**). There are no orthologs of mammalian β-subunits in insects. Instead, the non-orthologous proteins TipE and three to four TipE-homologs (TEH1–4) seem to serve as auxiliary subunits of sodium channels *in vivo*. Structurally, both TipE and TEH1 have intracellular N- and C-termini and two transmembrane segments connected by a large extracellular loop [37]. In physiological function, the flow of sodium ions into and out of the insect synapse occurs through the sodium channel present on the cell membrane of the neuron. This flow is controlled by the normal movements of the insect's muscles. When the sodium channel is open, the muscle is activated, when it is closed, the muscle can relax.

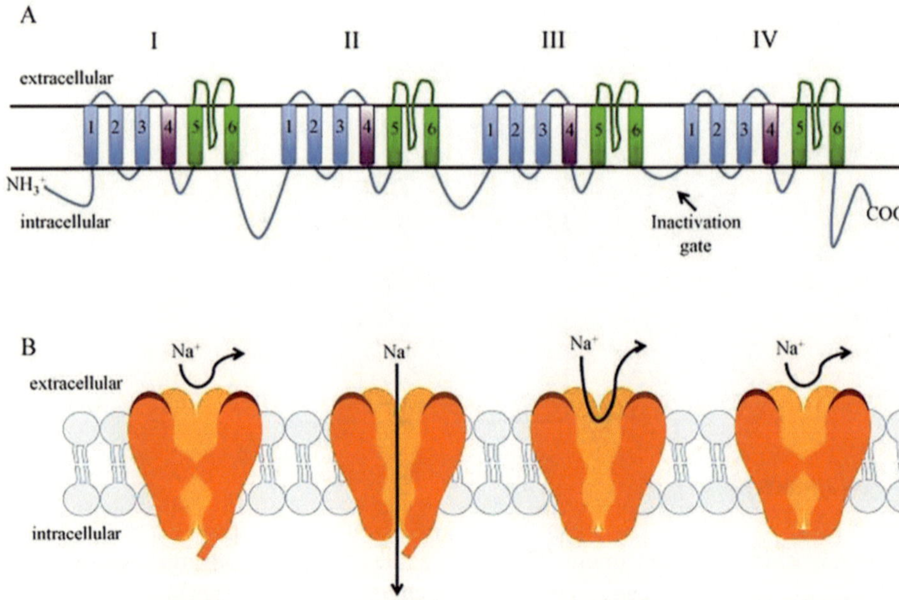


**Figure 5.**
*The α subunit of the voltage-gated sodium channel. A. Structure of the subunit (four domains (I-IV) and their six transmembrane segments (1–6)). B. Four states of the VGSC (channel closed: at resting membrane potentials; activated or opened channel: During the rising phase of an action potential; inactivated channel: Falling phase; deactivated channel: During the undershoot phase prior to returning to the closed phase) [36].*

*2.2.1 Compounds acting via Voltage-Gated Sodium Channels*

- Pyrethroids

Pyrethroids are biomimetic molecules adapted from natural pyrethrins isolated from the flowers of *Chrysanthemum cinerariifolium* [38]. Cypermethrin and deltamethrin are two examples of pyrethroid compounds. Their synthetical pathway consists of a cyclopropanation reaction of an α,β-unsaturated ester derived from D-glyceraldehyde, giving a hemicaronaldehyde, which subsequently leads to deltamethrin (**Figure 6**) [39].

Pyrethroids have been used in pest control as the main insecticides. The mode of action of pyrethroids consists in the binding and modulation of the activity of the VGSC, leading to a prolonged opening of sodium channels, and a continuous firing of action potential [40]. This neurotoxic action produces in the insect hyperactivity and convulsions, followed by lethargy, paralysis and death [41, 42].

## 2.3 Transient receptor potential vanilloid (TRPV) channels

The transient receptor potential vanilloid (TPRV) channel is a subfamily of 6 cationic channels. They are tetrameric and each subunit is composed of 6 transmembrane domains with 3 to 5 N-terminal ankyrin repeats and a TRP box in

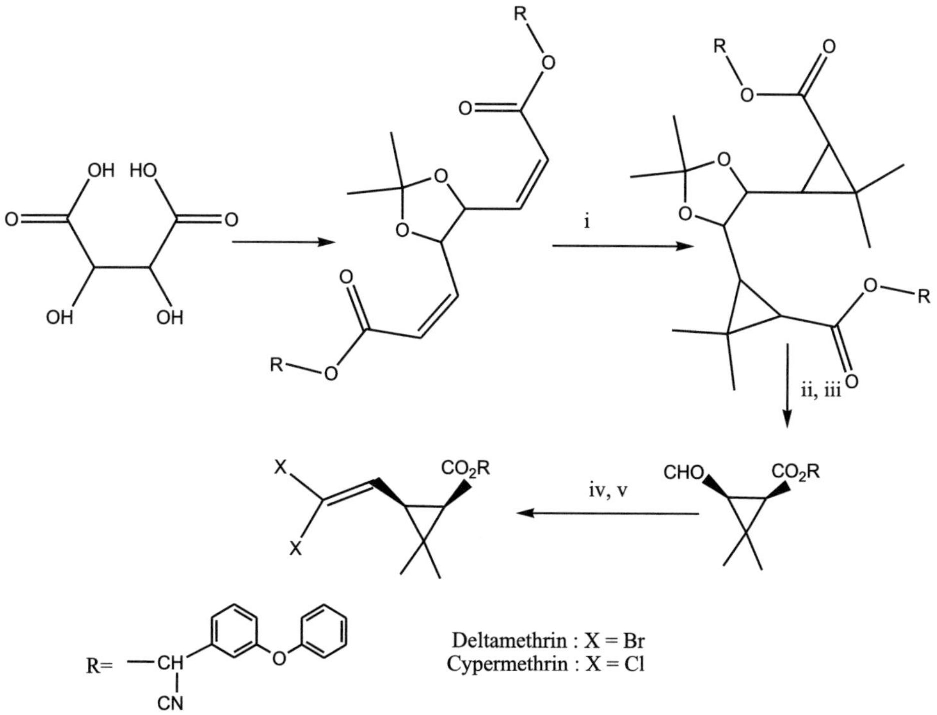


(i) 2,5 equiv. $Ph_2S{=}C(Me)_2$; DME .-78°C; (ii) 4 equiv. 2Naq, $HClO_4$, THF,20°C,6h (iii) 1,5 equiv, $NaIO_4$, MeOH, Phosphate buffer PH=7,2, 20°C, 1h. (iv) KOH, MeOH then aq,HCl (v) $CBr_4$, $PPH_3$

**Figure 6.**
*Deltametrin synthesis.*

their C-terminal (**Figure 7**). The activation of TRPV1 primarily permits an influx of extracellular $Ca^{2+}$, which is involved in a number of essential physiological functions, such as neurotransmitter release, membrane excitability, and muscle cell contraction [44].

### *2.3.1 Compounds acting via TRPV Channels*

- Afidopyropens

Afidopyropens (for example the keto-pyripyropene A) are new hemisynthetic insecticides derived from pyripyropene A (**Figure 8**). This family of molecules presents a strong insecticidal activity against aphids. These molecules modulate TRPV channels in the chordotonal organs of insects [46]. It is a class of ester molecules that are marketed under the common name of afidopyropen [47] including keto-pyripyropene A. The biomimetic hemisynthesis of keto-pyripyropene A is done as

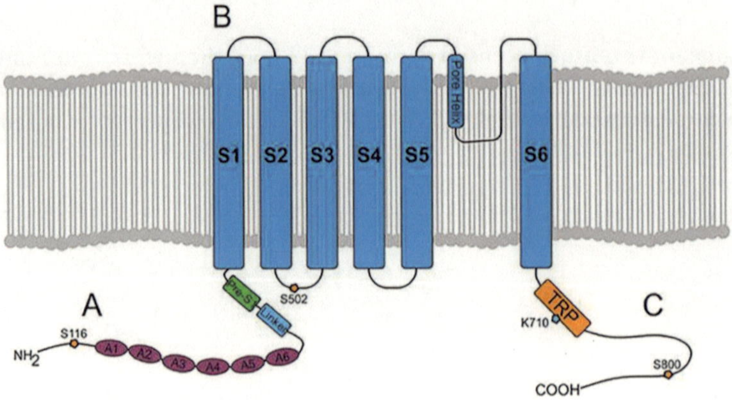


**Figure 7.**
*Structure of a TRPV1 subunit. A. N-terminus containing 6 ankyrin subunits (A1–A6) and a linking region consisting of a linker and a pre-S1 helix segment. B. Transmembrane region with 6 helical segments (S1–S6). C. C-terminus containing a TRP domain and binding sites for protein kinase A, C, phosphatidylinositol-4,5-bisphosphate (PIP₂), and calmodulin [43].*

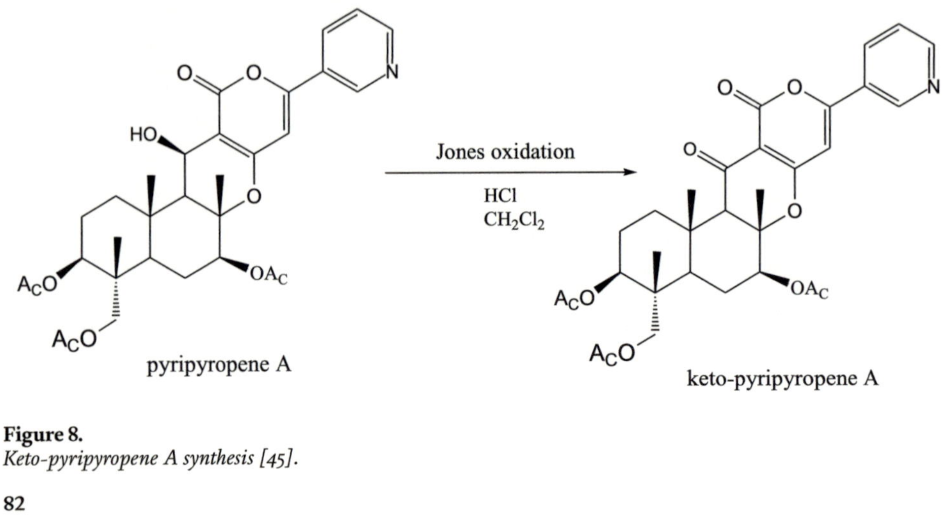


**Figure 8.**
*Keto-pyripyropene A synthesis [45].*

follows, a modification of pyripyropene A through a Jones oxidation allows to obtain keto-pyripyropene A (**Figure 8**) [45].

### 2.4 Gamma-amino butyric acid (GABA) receptors

The GABA receptors are located in the nervous system of many insects. It is an oligomer of 5 subunits (**Figure 9**), each being polypeptide with a large domain in their N-terminal and 4 transmembrane domains [49]. The binding of GABA on its receptors leads to the inhibition of the nerve impulse. GABA acts by binding to its specific transmembrane receptors (GABA-gated chloride channels) present in the plasma membrane of neurons, opens the chloride ($Cl^-$) channels to allow the flow of $Cl^-$ into the neurons. This results in a negative charge on the transmembrane potential causing hyperpolarization and a reduction in membrane entry resistance. Pesticides by binding to insect GABA receptors, decrease or increase $Cl^-$ influx into neurons, and kill insects by causing excessive excitation or inhibition of the nervous system (hyperactivity, hyperexcitability, convulsions, production of prolonged high frequency discharges, etc.) [50, 51].

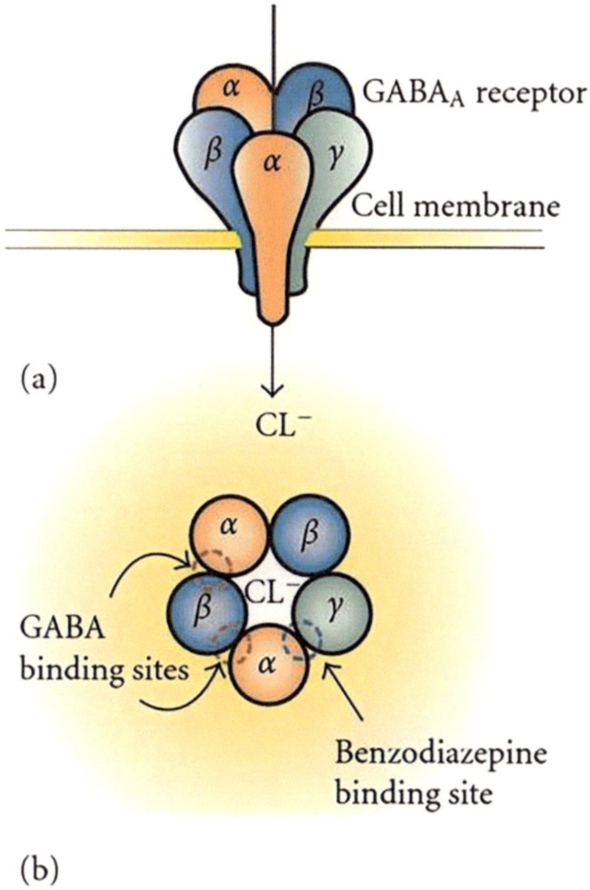


**Figure 9.**
*Representation of the GABAA receptor structure. (a). The inhibitory GABAA receptor consists of five subunits that together form a ligand-gated chloride ($Cl^-$) channel. (b). The most common subtype is a pentamer with 2α,2β, and 1 γ-subunit [48].*

### *2.4.1 Compounds acting via GABA receptors*

- Avermectins

The avermectins are hemisynthetic pesticides acting on glutamate-gated chloride channels (GluCls) and gamma-amino butyric acid (GABA) receptors, and then causes neuromuscular paralysis that eventually leads to death [52]. Several derivates can be synthetized starting from avermectins. For example, the hydrogenation of avermectin

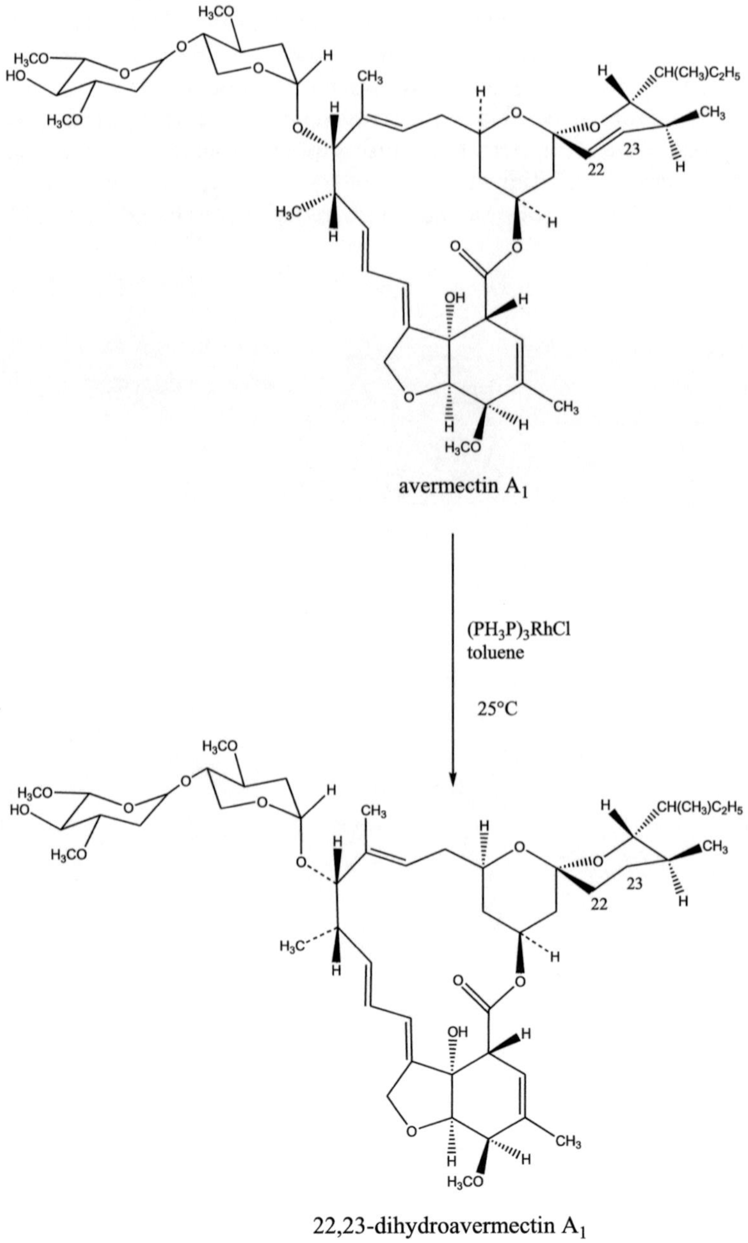


**Figure 10.**
*Dihydroavermectin $A_1$ synthesis.*

$A_1$ with Wilkinson's catalyst ($(PH_3P)_3RhCl$) [53] yields to the dihydroavermectin $A_1$ (**Figure 10**).

- Emamectin

Emamectin (**Figure 11**) is another hemisynthetic pesticide targeting the same receptors (GABA and GluCls) that the avermectins. The emamectin causes neuromuscular paralysis that eventually leads to death [23]. Emamectin activates chloride channels by stimulating high-affinity GABA receptors and GluCls channels, which increases membrane permeability to $Cl^-$ and disrupts nerve signals in arthropods. This results in hyperpolarization and removal of signal transduction in the insect nervous system, which reduces neurotransmission [54]. The insect larva stops feeding after exposure and becomes irreversibly paralyzed which leads to death within 3 or 4 days [55].

### 2.5 Octopaminergic (OA) system

Octopamine is a neurohormone (released in the hemolymph for lipid mobilizing during flight and long-lasting motor behaviors), a neuromodulator and a neurotransmitter present in relatively high concentrations in every invertebrate tissue [56]. Octopamine binds to a specific G protein-coupled membrane receptor. The binding of octopamine to these receptors leads to the activation of the enzyme adenylyl cyclase. It transforms ATP to cAMP and causes an increase in the cAMP level, which is a signaling molecule, activating the protein kinase A (PKA). The G protein also activates phospholipase C (**Figure 12**). It leads to the release of calcium from deposits in the endoplasmic reticulum and to the elevation of its intracellular level as well as to the activation of the calcium-dependent protein kinase C (PKC). Protein kinases phosphorylate several enzymes and receptors leading to the modulation of their activity. This phosphorylation produces important changes in cell functions [57].

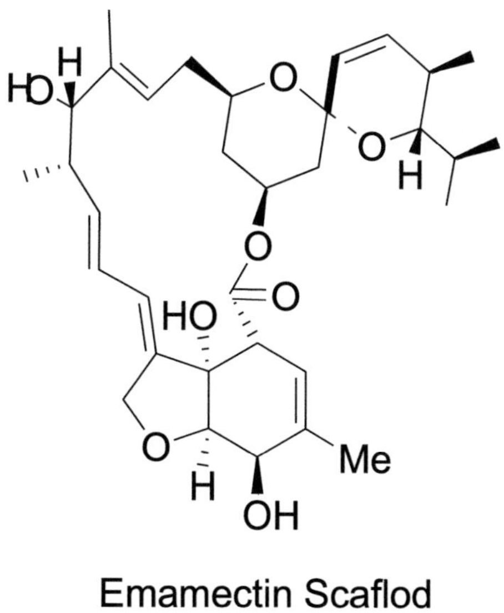


**Figure 11.**
*Structure of emamectin.*

### 2.5.1 Compounds acting via the Octopaminergic (OA) System

- Phenylpropanoids

Hemisynthetic phenylpropanoids derivatives can interfere with the octopaminergic system. Their binding to the octopamine receptor causes its blockage which leads to decreased cAMP levels within cells, thus resulting in antifeedant and larvicidal effect [58]. For example, the dillapiole, a phenylpropanoid isolated from the essential oil of leaves of *Piper aduncum* L. and several hemisynthetic derivatives present an activity against *Aedes aegypti* L. being several derivates more actives than the dillapiole [59]. Following a similar approach, Sinha's team designed a hemisynthetic method to obtain cinnamic esters from the oxidation of cinnamaldehydes (**Figure 13**).

Eugenol is a phenylpropanoid and a major constituent of clove oil (*Syzygium aromaticum* L.) with many applications in the pharmaceutical, food, agricultural and cosmetic industries [60]. Eugenol can mimic octopamine by increasing intracellular calcium levels in cloned brain cells of *Periplaneta americana* L. and *Drosophila melanogaster* Meigen [61].

## 2.6 Nicotinic acetylcholine receptors (nAChRs)

Nicotinic acetylcholine receptors (nAChRs) are ion channels that mediate fast neurotransmission in the central and peripheral nervous systems. nAChRs are

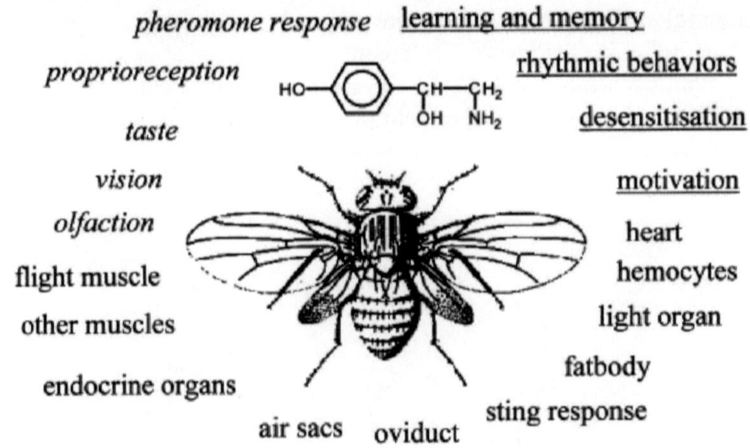


**Figure 12.**
*Effects of OA on different tissues of invertebrates. Sense organs (italic); central systems (underlined) modulated by OA [56].*

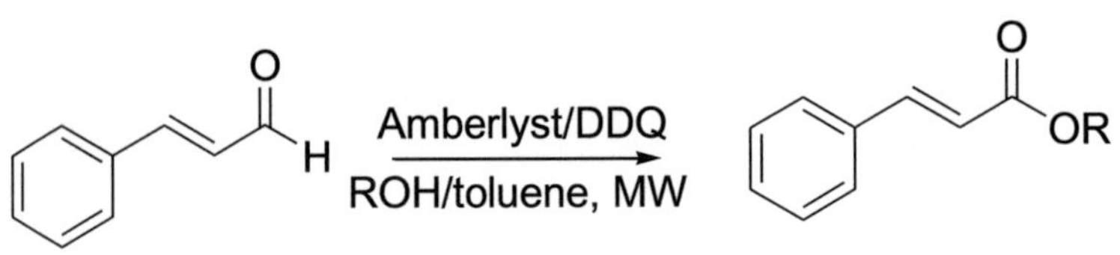


**Figure 13.**
*Hemisynthesis of phenylpropanoids derivatives.*

formed by the assembly of 5 transmembrane subunits (**Figure 9**) among 17 different nAChR subunits. nAChRs regulate the flow of mainly sodium, potassium and calcium ions across the cell membrane (**Figure 14**) [62].

### *2.6.1 Compounds acting via nicotinic acetylcholine receptors*

- Flupyradifurone

Flupyradifurones are a class of synthetic butenolide insecticides, mimic of natural neonicotinoids, active against various pests and suckers with an excellent safety profile. It acts reversibly as an agonist on the nicotinic acetylcholine receptors of insects. It binds to the nAChR blocking it. Flupyradifurone is a

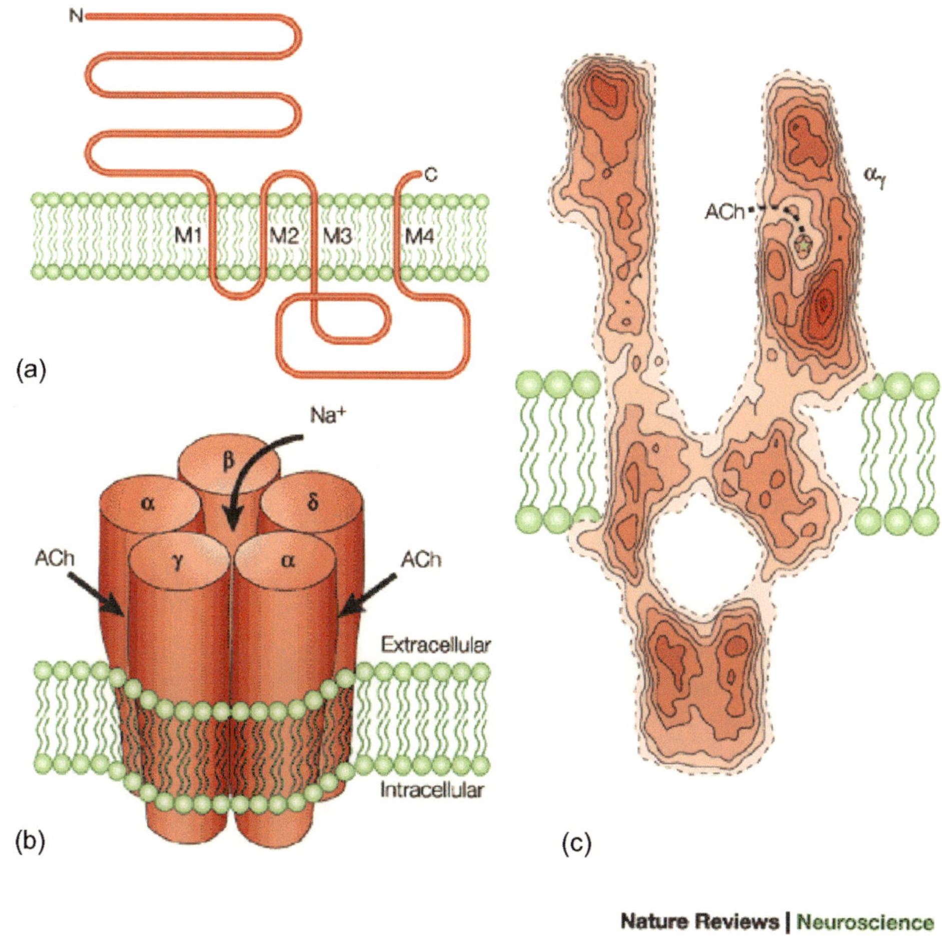


**Figure 14.**
*Structure of the nicotinic acetylcholine receptors. (a). The threading pattern of receptor subunits through the membrane. (b). A schematic representation of the quaternary structure, showing the arrangement of the subunits in the muscle-type receptor, the location of the two acetylcholine (ACh)-binding sites (between an α- and a γ-subunit, and an α- and a δ-subunit), and the axial cation-conducting channel. (c). A cross-section through the 4.6-Å structure of the receptor [63].*

novel butenolide insecticide that is also systemic and a nicotinic acetylcholine receptor (nAChR) agonist. Tosi and Nieh [64] provide the first demonstration of adverse synergistic effects on honeybee survival and behavior (*Apis mellifera* L.) (poor coordination, hyperactivity, apathy). Two different pathways for the synthesis of flupyradifurone are presented in **Figure 15**. Starting from tetronic acid, one approach consists in two consecutive reactions. Where the tetronic acid reacts firstly with a difluoroethane-1-amine and secondly with 2-Chloro-5-(chloromethyl)pyridine (Method A, **Figure 15**). And the other approach, where the tetronic acid reacts with difluoroethane-1-amine derivative in the presence of 4-toluenesulfonic acid in a "one pot" approach (Method B, **Figure 15**) to yield flupyradifurone [65].

- Triflumezopyrim

Triflumezopyrim is biomimetic mesoionic insecticide, containing domains characteristics of natural betaines that have shown excellent control of sucking insects. Mesoionic insecticides bind to the orthosteric site of the nACHR and act primarily by inhibition of the binding site. A method for the synthesis of these pyrimidones is described as follows: 2-aminopyridine reacted with pyrimidine-5-carbadehyde to form imine, the imine was exposed to reductive amination conditions to generate amine which reacts with malonic chloride to form triflumezopyrim (**Figure 16**) [66].

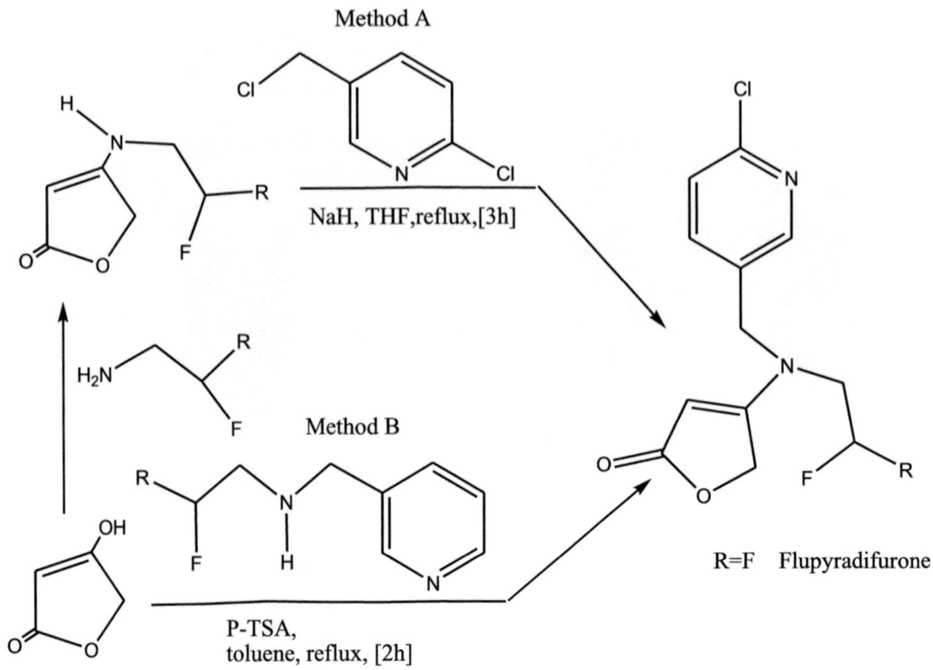


**Figure 15.**
*Flupyradifurone synthesis.*

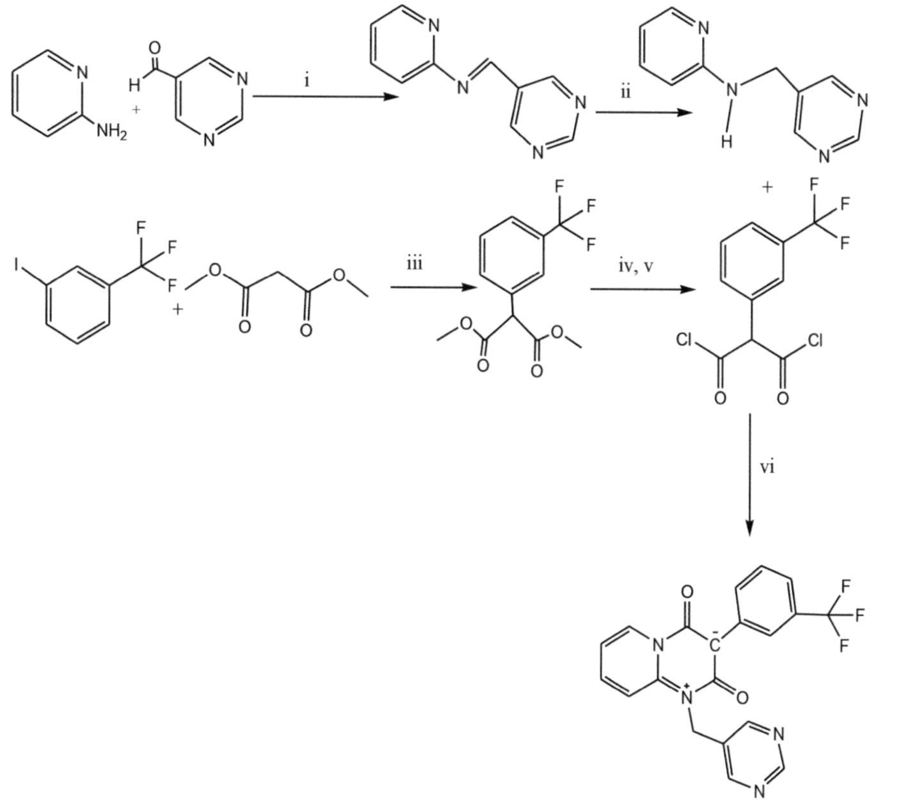


Triflumezopyrim

Reagents and conditions: i) $CHCl_3$, 25°C ,99%; ii) $NaBH_4$, MeOH, THF,rt, 64% ; iii) $CS_2CO_3$, CuI (cat), 1,4-dioxane,90°C, 58%; iv) aq, KOH, 2-25°C, drying, 97%, v) $(COCl)_2$, DMF, (cat), toluene, 3°C to rt; vi) TEA, toluene, 3-25°C, 50% for 2 steps starting from the malonate salt

**Figure 16.**
*Triflumezopyrim synthesis.*

## 3. Pesticides targeting endocrine system

Complementary to the nervous system, the endocrine system ensures the functioning of the organism thanks to the production and the transport of various hormones trough the body. There are types of endocrine glands: neurosecretory cells within the central nervous system whose secretions act on effector organs or on other endocrine glands and specialized endocrine glands, *corpora cardiaca*, *corpora allata*, and the prothoracic glands [67].

### 3.1 Apoptosis

Apoptosis is a programmed cellular death occurring under regulated conditions. At the end of the process, the cell divides in many apoptotic bodies that will be phagocyted. Caspases (cysteine aspartate-specific proteinases) are a family of cysteine proteases that serve as both the initiators and the executioners of apoptosis. They are crucial mediators of apoptosis, and their activation is carefully controlled by a death program. An unbalance in this program can lead to deleterious apoptosis [68]. Caspases are frequently considered synonymous with apoptotic cell death [69], but the review of Accorsi, 2015 [70] prove that these proteases may exert their activities in non-apoptotic functions (developmentally regulated autophagy during insect metamorphosis, neuroblast self-renewal and the immune response).

*3.1.1 Compounds acting via apoptosis*

- Phenylpropanoids

Several amino-alcohols biomimetic derivates of the phenylpropanoid eugenol are insecticides that act against *Spodoptera frugiperda* Smith and increase the activity of caspases leading to apoptosis [71]. The amino-alcohols are derivatives obtained by a hemisynthetic reaction of eugenol. In this reaction eugenol was converted to the corresponding epoxide with *m*-chloroperoxybenzoic acid (*m*-CPBA) in dichloromethane and then reacted with a series of nucleophilic amines to give the corresponding β-amino alcohols (**Figure 17**) [71].

### 3.2 Ecdysteroids

Steroid hormones play indispensable roles in modulating a broad range of biological processes in nearly all multicellular organisms. Once produced, steroid hormones are circulated in hemolymph and are easily transported to target cells to act as ligands for the nuclear receptor family of transcription factors. In insects, the major steroid hormones are ecdysteroids, also known as molting hormones (**Figure 18**). They play essential roles in coordinating developmental transitions, such as larval molting and metamorphosis (**Figure 13**) [73].

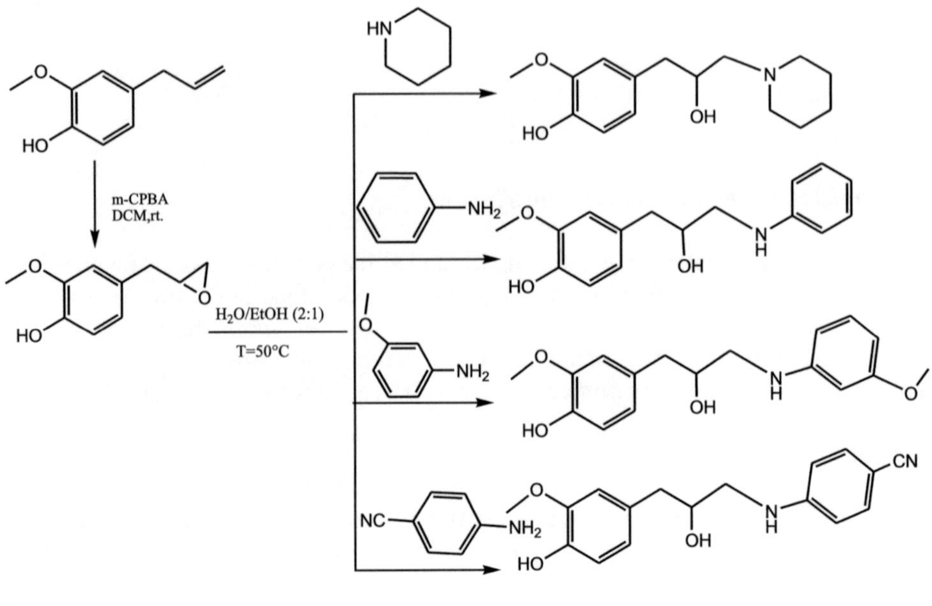


**Figure 17.**
*Hemisynthesis of eugenol alcohols.*

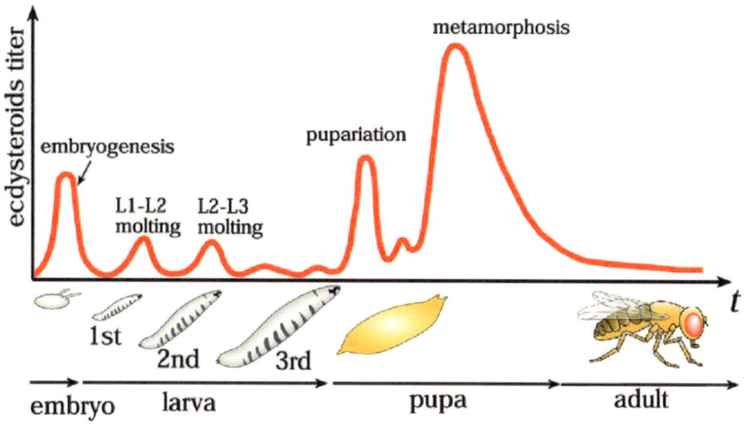


**Figure 18.**
*The developmental stages and ecdysteroid titers in D. melanogaster [72].*

The different metamorphoses undergone by insects to pass from one stage to another (larval stage to metamorphosis) are called molts (ecdysis). Molting takes place under the control of steroid hormone (ecdysone) responsible for molting [74] and the juvenile hormone (JH), responsible for inhibiting the steroid hormone to maintain the insect in its larval state and thus avoid premature ecdysis [75]. Activation and release of ecdysone into the hemolymph are controlled by the prothoracic hormone (PPTH), produced by the *corpora cardiaca*, and the insect insulin.

### *3.2.1 Compounds acting via ecdysteroids*

- Azadirachtin-A derivatives

Tetrahydroazadirachtin, alongside with other azadirachtin-A analogues like 22,23-dihydroazadirachtin; 3-tigloylazadirachtol; 11-methoxydihydroazadirachtin and 22,23-bromoethoxydihydroazadirachtin are hemisynthetic pesticides disrupting the endocrine system. By blocking the release of neurosecretory peptides which regulate synthesis of ecdysteroids and juvenile hormone they provoke molt disruption leading to death [76].

Azadirachtin causes a slowdown in the synthesis and release of prothoracicotropic hormone (PPTH), which affects the functioning of the nucleus of secretory neurons and endocrine glands and the insect can no longer molt. Azadirachtin also modifies the production and stop of the growth functions [77].

Azadirachtin is a synthetic insecticide that belongs to the triterpenoid class of limonoids. One method of synthesis of azadirachtin starts with the selective acetylation in C3 of the triol to give the acetate derivate which by a series of reactions gives the triglate intermediate. Cleavage of the benzyl ether from the triglate intermediate occurred to provide lactol. Then the conversion of methyl acetal into phenyl sulfide during a treatment with thiophenol and catalytic PPTS (pyridinium toluene-P-sulfonic acid) in toluene followed by an oxidation with dimethyldioxirane (DMDO) followed by a pyrolysis to obtain azadirachtin (**Figure 19**) [78].

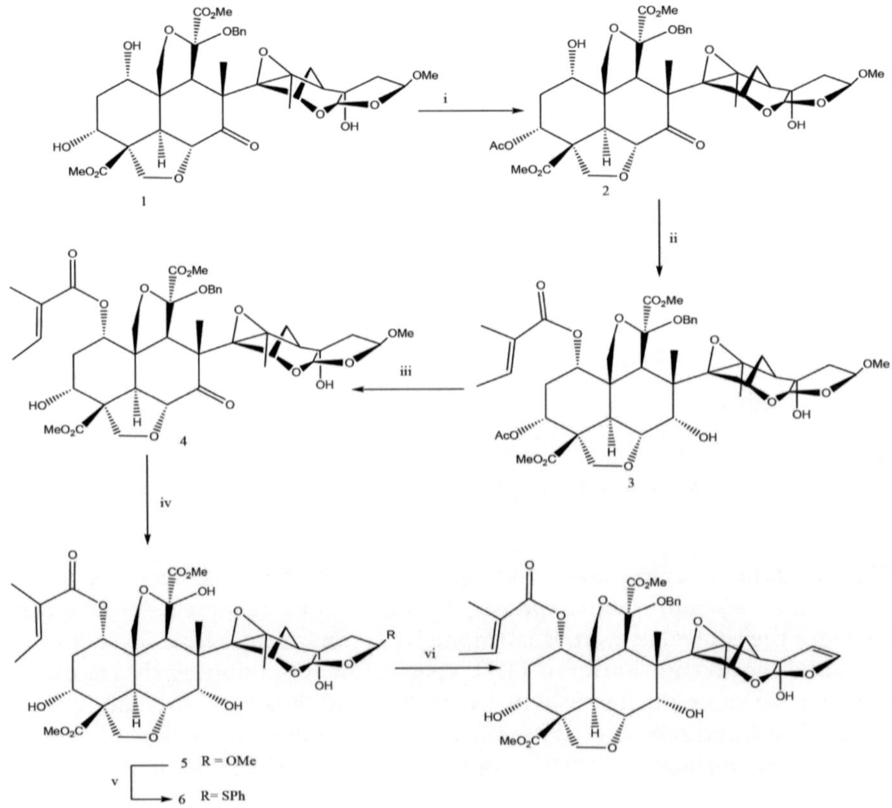

Reagents and conditions : (i) $Ac_2O$. $Et_3N$, DMAP, $CH_2Cl_2$, r.t.; (ii) $CH_3CH(CH_3)(CO)O(CO)C_6H_2Cl_3$, $CS_2CO_3$, toluène, reflux, (iii)$NaBH_4$, $CeCl_3$,$7H_2O$, MeOH, 0°C. (iv) Pd/C, $H_2$, MeOH, r.t.; (v) PhSH, PPTS, $CH_2ClCH_2Cl$, 80°C, (vi) DMSO, $CH_2Cl_2$, -78°C, toluène reflux.

**Figure 19.**
*Azadirachtin synthesis.*

## 4. Pesticides targeting cellular structure

### 4.1 Chitin metabolism

Chitin is one of the most important natural biopolymers. It is mainly produced by fungi, arthropods, and nematodes. In insects, it supports the cuticles of tissues like the epidermis or the trachea. As for bones in the vertebrates, chitin is constantly synthesized and degraded. This balance is strictly controlled by the production of chitin synthases and chitinolytic enzymes to ensure a correct growth [79]. Chitin is widely distributed in the fungal kingdom since nearly all fungi have significant amounts of chitin in their cell wall (**Figure 14**). Cell wall architecture is well documented and it was described several decades ago that inhibition of chitin synthesis produces cell death [80]. Regarding the importance of chitin in growth and development of insects and in fungi cell wall, its synthesis is an interesting target for a pesticide. Chitin plays a key role in the insect's water system. It controls water homeostasis. The loss of this impermeable layer leads to transpiration which is fatal for the insect (**Figure 20**) [81].

### *4.1.1 Compounds acting via chitin metabolism*

- N-amino-maleimide

N-amino-maleimide derivatives containing a hydrazone group are imides mimicking the synthesis of linderone and methyllinderone which were isolated from *Lindera erythrocarpa* M. Makino. They are fungicides that inhibit chitin synthase B-1,3-glucan synthase, leading to an alteration of the cell walls of fungi. A hemisynthetic method is described as follows: various aryl-substituted unsaturated ketones were synthesized and reacted with N-amino-maleimide under reflux of dry ethanol with a catalytic amount of *p*-toluenesulfonic acid to produce a variety of N-amino-maleimide derivatives containing a hydrazone group (**Figure 21**) [83].

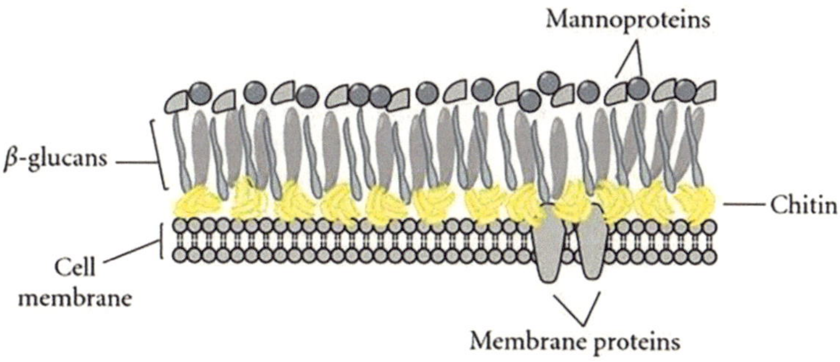


**Figure 20.**
*Fungal cell wall components [82].*

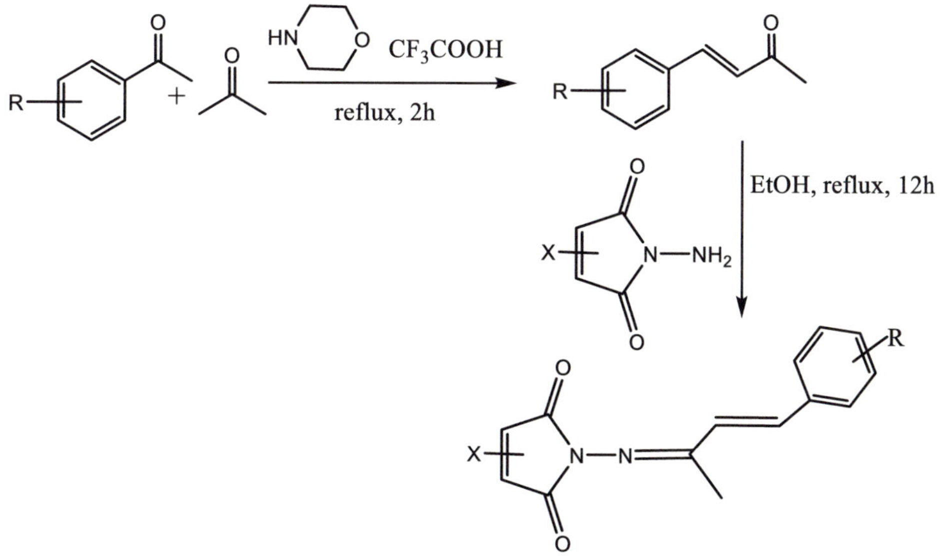


**Figure 21.**
*Synthesis of N-amino-maleimide derivatives containing a hydrazone group.*

## 4.2 Cell membrane

In any living being, the cell membrane ensures the smooth-running of the exchanges between the cytoplasm and the extracellular matrix. It is composed of a lipid bilayer containing proteins, glycoprotein, glycolipids and sterols. The latter are important component of the cell membrane, they regulate its fluidity and the enzymes in it (like the chitin synthases) [84].

### *4.2.1 Compounds acting via the cell membrane*

- Spiroxamine

Spiroxamine is a synthetic fungicide mimic of the class of natural or synthetic morpholines such as fenpropidin, tridemorph, fenpropimorph etc. It inhibits both delta-14 reductase and delta-7–delta-8 isomerase, which leads to the formation of carbocation sterols, and strongly affects hyphae and mycelium development [85]. One method of synthesis of this molecule is as follows, tert-butyl cyclohexanone is first reacted with 3-chloro-1,2-propanediol. Formation of the ketal under acidic conditions leads to 8-tert butyl-1,4-dioxanspiro[4,5]decan-2-ylmethyl chloride, which is reacted with ethyl propylamine to form spiroxamine following nucleophilic substitution (**Figure 22**) [86].

- Prochloraz

Nitrogen compounds such as prochloraz (imidazole), fenarimol (pyrimidine), epoxyconazole, fluzilazole, and tebuconazole (triazole) are synthetic fungicides that act on essential fungal functions. They are inhibitors of the α-methylation of sterols [84]. Prochloraz for example is a fungicide of the imidazole family which can be obtained in several steps. Initially 2,4,6-trichlorophenol is alkylated with 1,2-dibromomethane in a Williamson ether synthesis. The following reaction with propylamine provides a secondary amine which is reacted with phosgene. This acid chloride of a carbamic acid is finally reacted with imidazole to give prochloraz (**Figure 23**) [87].

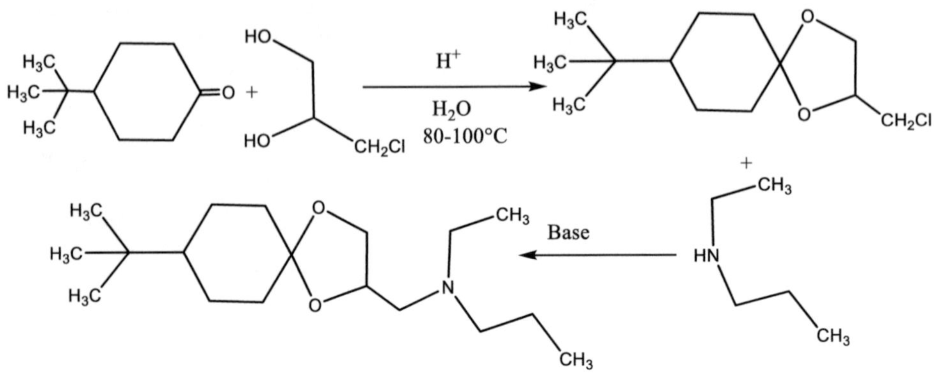


**Figure 22.**
*Spiroxamine synthesis.*

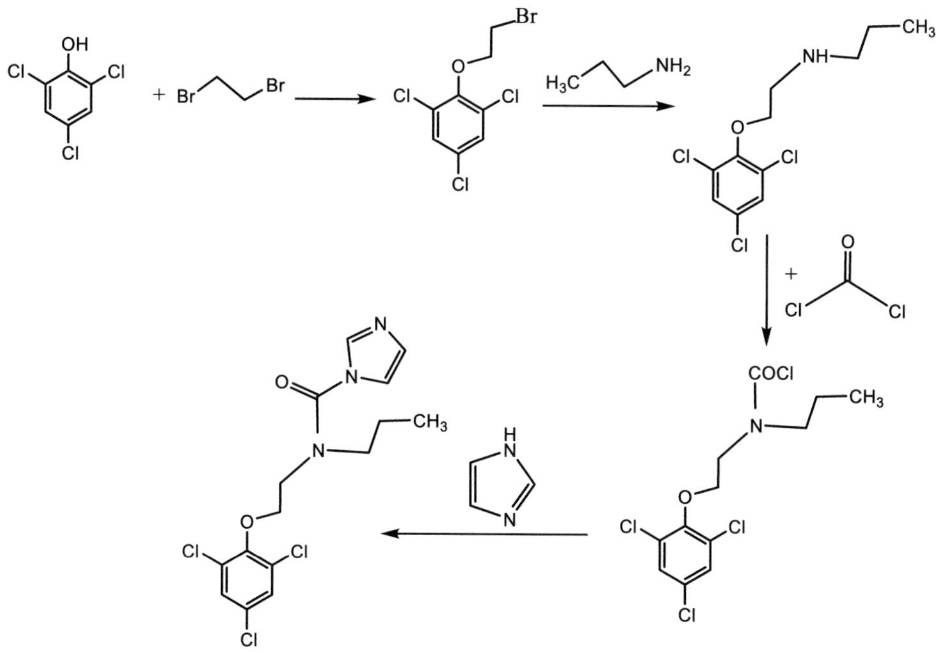


**Figure 23.**
*Prochloraz synthesis.*

### 4.3 Cytoskeleton

The cytoskeleton and membrane systems of eukaryotic cells play key roles in the intracellular transport of vesicles, organelles, and macromolecules. The actin cytoskeleton is mainly composed of globular actin (G-actin), which is monomeric actin able to self-assemble into filamentous actin (F-actin) [88]. The actin cytoskeleton is subjected to alterations and organizations to promote cellular dynamic and particle transport within and between cells. Eukaryotic cells polymerize actin filaments to provide mechanical integrity and motility force for a wide range of cellular mechanisms [89]. Microtubules are the main components of the cytoskeleton and the spindle apparatus (the cytoskeletal structure separating the sister chromatids during cell division). They are formed through α−/β-tubulin heterodimers assembling into cylindric filaments (**Figure 24**). The plus-ends of these filaments grow pointing towards the plasma membrane into protrusions, while their minus ends are anchored at microtubule-organizing centers (MTOCs) such as the centrosome. This polarity allows selective directional long-range cargo transport at the cell periphery [88]. Any substance able to impair with the formation or functioning of those microtubules blocks cell division in general and hyphae in fungi [84].

#### *4.3.1 Compounds acting via the cytoskeleton*

- Carbendazim

Carbendazim is a biomimetic benzimidazole that inhibits microtubule assembly and therefore blocks cell division in fungi. This effect appears to be related to their β-tubulin

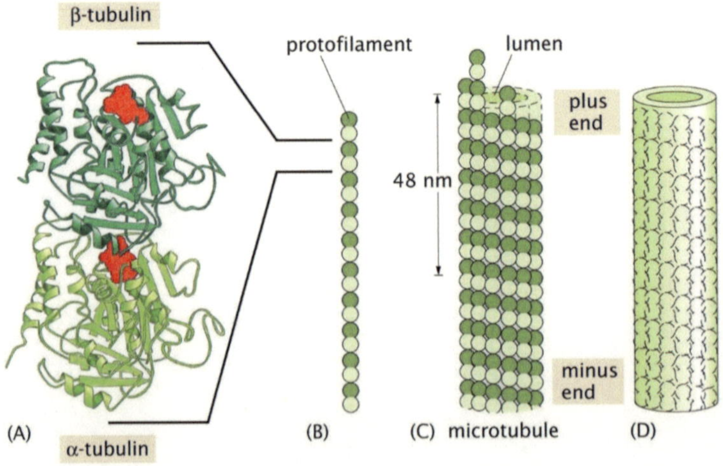


**Figure 24.**
*Microtubules formation: A. Tubulin dimerization; B. Tubulin dimers polymerization; C. Protofilament association; D. Formed microtubule [90].*

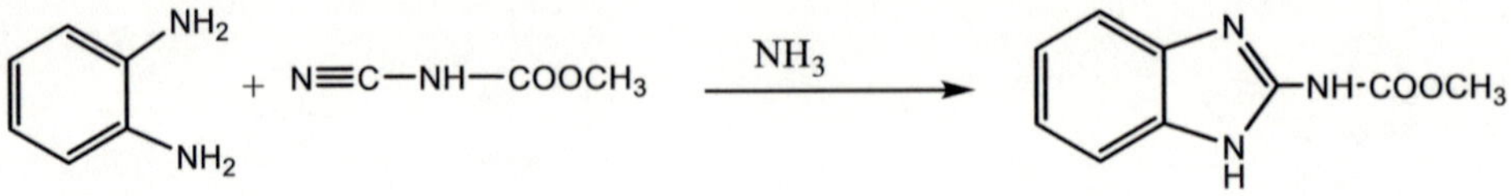


**Figure 25.**
*Carbendazim synthesis.*

binding, the main component of microtubulins. Carbendazim binds to β-tubulin and prevents tubulin formation [84]. A simple way to synthesize this molecule was realized by the condensation of orthophenylene diamine with an ester of aminonitrile in the presence of ammonia according to the figure below (**Figure 25**) [91].

Carbendazim is a widely used broad-spectrum fungicide that inhibits mitotic microtubule formation and cell division. The use of proteomics approaches, suggest that carbendazim is an environmental risk factor that likely weakens honeybees (*Apis mellifera*) colonies, partially due to reduced expression of major royal jelly proteins, which may be potential causes of colony collapse disorder [92].

- Ethylicin

Biomimetic organosulfur compounds have received considerable attention in recent years. Among various organosulfur compounds have shown a broad spectrum of biological activity such as fungicidal activity [93, 94]. They can block the normal metabolism of microorganisms by sulfenylation of the thiol groups of enzymes [95, 96]. Ethylicin is therefore a biomimetic organosulfur fungicide with a broad spectrum for plants. It can inhibit the growth of *Pseudomonas syringae* pv. *actinidiae* and prevent cancer in the plant stem [97, 98]. It is a bionic organosulfur pesticide (S-ethyl ethanethiosulfonate) that mimics the natural allicin obtained from garlic (*Allium sativum* L.). It was first prepared and studied in the laboratory during the synthetic research of allicin and its homologs in 1958 and developed as a broad spectrum biomimetic fungicide in China [99]. Because of the widespread application

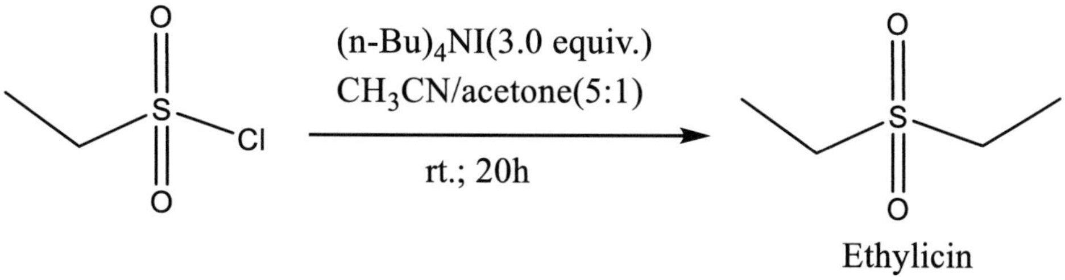


**Figure 26.**
*Ethylicin synthesis.*

of thiosulfonates, considerable effort has been made to develop synthetic methods for these compounds. Therefore, one of the synthesis methods used is the reduction of sulfonyl chlorides (**Figure 26**) [100].

## 5. Pesticides targeting digestive system

All insects have a complete digestive system in the form of a tube-like enclosure. Named the alimentary canal and running lengthwise through the body from mouth to anus, it consists of three regions: the foregut or stomodaeum, the midgut or mesenteron, and the hindgut or proctodaeum (**Figure 21**). An insect's mouth, located centrally at the base of the mouthparts, is a sphincter that marks the "front" of the foregut. Then goes the pharynx, from which food passes into the esophagus, a simple tube that connects the pharynx to the crop, a food-storage organ where food remains until it can be processed through the remaining sections of the alimentary canal. In some insects, the crop opens posteriorly into the proventriculus, which grinds and pulverizes food particles before they reach the stomodeal valve, a sphincter regulating the flow of food from the stomodeum to the mesenteron. The midgut begins just past the stomodeal valve. Near its anterior end, finger-like projections (usually from 2 to 10) diverge from the walls of the midgut. Gastric caecae provide extra surface area for secretion of enzymes or absorption of water (and other substances) from the alimentary canal. The rest of the midgut is called the ventriculus — it is the primary site for enzymatic digestion of food and absorption of nutrients. Digestive cells lining the walls of the ventriculus have microscopic projections (microvilli) that increase surface area for nutrient absorption. The posterior end of the midgut is marked by another sphincter muscle, the pyloric valve. It serves as a point of origin for dozens to hundreds of Malpighian tubules. These long, spaghetti-like structures extend throughout most of the abdominal cavity where they serve as excretory organs, removing nitrogenous wastes from the hemolymph (analog of blood in arthropods). The rest of the hindgut plays a major role in homeostasis by regulating the absorption of water and salts from waste products in the alimentary canal (**Figure 27**) [101].

### 5.1 Compounds acting *via* the digestive system

- Triterpenic derivatives

Balanced nutritional intake is essential to ensure that insects undergo adequate larval development and metamorphosis. Terpenes are a class of hydrocarbons,

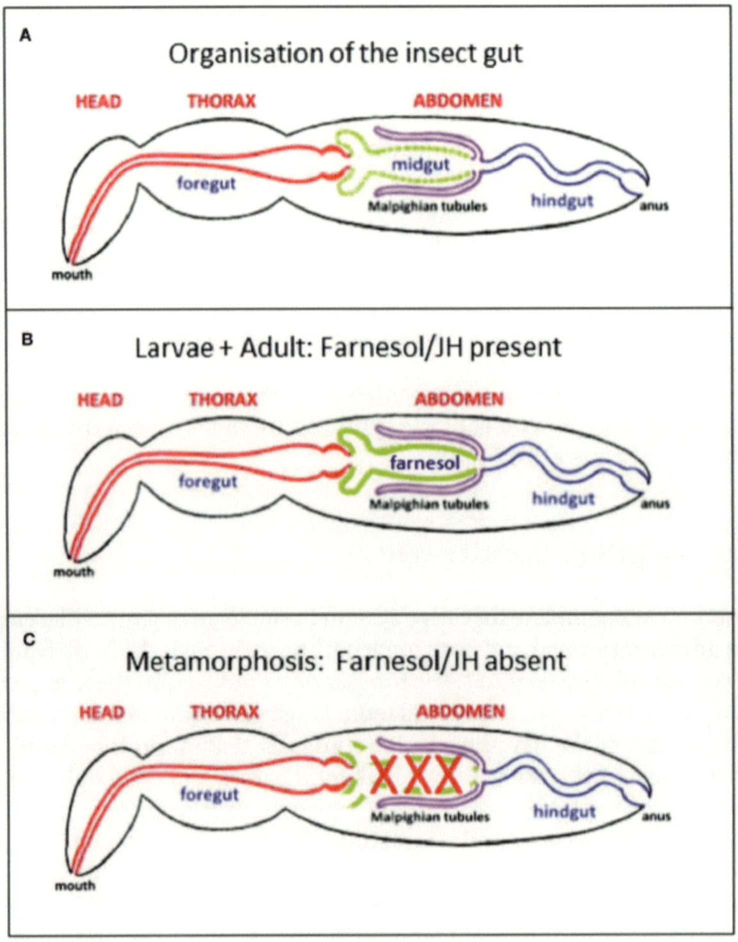


**Figure 27.**
*Generalized insect digestive system illustrating the three main regions at different stages of development [102] (JH = juvenile hormone).*

produced by many plants. The aim is to optimize insecticidal triterpene derivatives by biomimetic oxidation with hydrogen peroxide and iodosobenzenes catalyzed by porphyrin complexes. Therefore, the hemisynthesis of the derivatives were made from 31-norlanosterol an insecticide isolated from the latex of *Euphorbia officinarum* L. and were subjected to oxidation with hydrogen peroxide ($H_2O_2$) and iodosobenzene (PhIO) catalyzed by porphyrin complexes following a biomimetic strategy. Main transformations were epoxidation of double bonds and hydroxylations of non-activated C-H groups as shown in **Figure 12** [103]. These compounds caused a decreased digestive enzyme secretion and histolysis of intestinal tissues and led to indigestion, nutritional deficiency and decreased body weight of larvae. This prevented the larvae from reaching a critical weight and a normal population [103]. Similarly, work carried out on the development of *Chlosyne lacinia* caterpillars fed on *Heliantheae* leaves showed that the main discriminating metabolites of these leaves, diterpenes, caused a delay in the complete development of the caterpillars to the adult phase and that the latter showed a higher rate of diapause (**Figure 28**) [104].

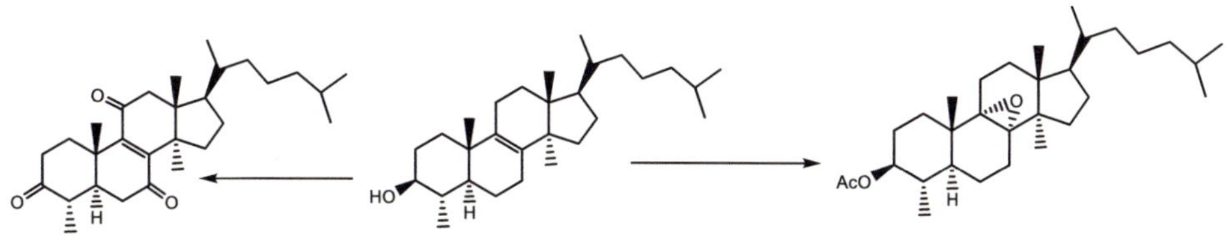


**Figure 28.**
*Triterpenic derivatives synthesis.*

## 6. Electron transport chain

Mitochondria regulate critical cellular processes, from energy production to apoptosis; within these organelles, sugars and long chain fatty acids are broken down, ADP is recycled back into ATP, steroids and lipids are synthesized, ancient DNA is replicated, transcribed and proteins are translated, along with numerous other reactions that are essential for human life [105]. Characteristic properties of all insect mitochondria are their low stability, their exceptionally high respiratory and phosphorylative activity with their physiological substrates, their relatively poor rate of oxidation of Krebs-cycle intermediates and the low P:0 ratios accompanying these slow oxidations. The phosphorylating respiratory chain of insect mitochondria strongly resembles that of mammalian mitochondria [106]. The electron transport chain is a mitochondrial pathway in which electrons move across a redox span of 1.1 V from NAD/NADH to $O_2/H_2O$. Three complexes are involved in this chain, namely, complex I, complex III, and complex IV. Some compounds like succinate, which have more positive redox potential than NAD/NADH, can transfer electrons via a different complex—complex II (**Figure 29**) [107].

### 6.1 Compounds acting *via* electron transport chain

- Cyazofamid

Cyazofamid is a cyanoimidazole fungicide particularly effective on *Oomycota*. This molecule, inspired in natural imidazoles, inhibits ATP production in cells by

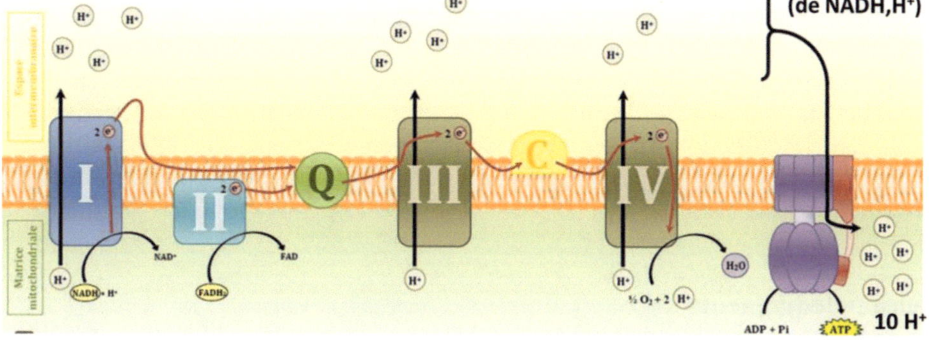


**Figure 29.**
*Electron transport chain [108].*

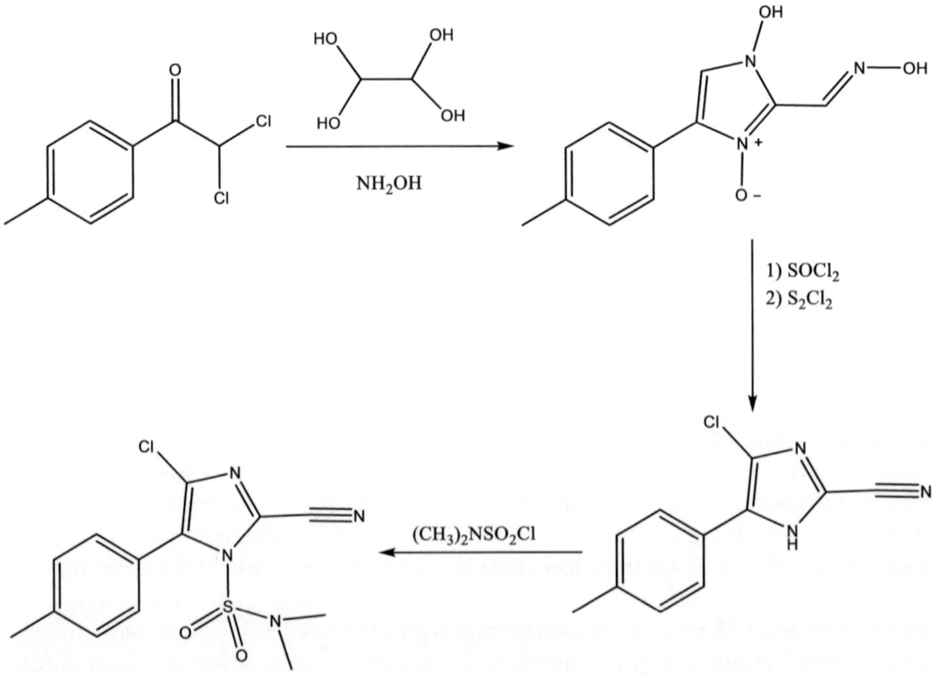


**Figure 30.**
*Cyazofamid synthesis.*

inhibiting the complex III of the respiratory chain of the mitochondria [84]. It is a synthetic fungicide whose synthesis is described as follows; an acetophenone derivative was treated with aqueous glyoxal and hydroxylamine to form an oxime substituted imidazole ring system. This intermediate was treated with thionyl chloride and disulfide dichloride to convert the oxime to a cyano group chlorinating the imidazole in the position near the phenyl ring. Finally, treatment with dimethylsulfamoyl chloride gave cyazofamide (**Figure 30**).

## 7. Conclusion

In summary, pesticides are a major environmental issue. Alternative strategies need to be explored, based on phytocompounds and natural extracts. Following this idea, the synthetic compounds mimicking the structure of natural products and modifying this molecules by several approaches as for example hemisynthesis or total synthesis will enhance the molecular diversity. These innovative biomimetic modified pesticides will open new perspectives in the fight against pests, improving crop efficiency and decreasing food crisis while maintaining sustainability.


## Acknowledgements

This research was funded by "Plan Chlordecone III en Guadeloupe et Martinique" in the framework of DICHOAL-MAREC (DIREC) project.

## Author details

Ahissan Innocent Adou[1†], Garrick Bibian[1,2†], Odile Bordelais[1], Léa Farouil[1], Muriel Sylvestre[1], Sarra Gaspard[1], Marie-Noëlle Sylvestre[2*] and Gerardo Cebrián-Torrejón[1,3*]

1 Department of Chemistry, COVACHIM-M2E Laboratory EA 3592, University of the French West Indies, Pointe-à-Pitre, Guadeloupe, France

2 Institut de Systématique, Evolution, Biodiversité (ISYEB), Muséum National d'Histoire Naturelle, CNRS, Sorbonne Université, EPHE, UFR des Sciences Exactes et Naturelles, Université des Antilles, Pointe-à-Pitre, Guadeloupe, France

3 C$^3$MAG, UFR des Sciences Exactes et Naturelles, Université des Antilles, Pointe-à-Pitre, Guadeloupe, France

*Address all correspondence to: marie-noelle.sylvestre@univ-antilles.fr and gerardo.cebrian-torrejon@univ-antilles.fr

† These authors contributed equally.

## References


[1] Okolle NJ. Alternatives to synthetic pesticides for the management of the banana borer weevil (*Cosmopolites sordidus*) (Coleoptera: Curculioniidae). CAB Reviews: Perspectives in Agriculture, Veterinary Science, Nutrition and Natural Resources. 2020;**15**(26)

[2] Procaccia C, Le Deaut JY. Impacts de l'utilisation de la chlordécone et des pesticides aux Antilles: bilan et perspectives d'évolution. Available from: https://www.senat.fr/rap/r08-487/r08-487.html [Accessed: April 14, 2022]

[3] Cabidoche Y-M, Achard R, Cattan P, Clermont-Dauphin C, Massat F, Sansoulet J. Long-term pollution by chlordecone of tropical volcanic soils in the French West Indies: A simple leaching model accounts for current residue. Environmental Pollution. 1987;**157**(5):1697-1705

[4] Devault DA, Laplanche C, Pascaline H, Bristeau S, Mouvet C, Macarie H. Natural transformation of chlordecone into 5b-hydrochlordecone in French West Indies soils: Statistical evidence for investigating long-term persistence of organic pollutants. Environmental Science and Pollution Research International. 2016;**23**(1):81-97. DOI: 10.1007/s11356-015-4865-0

[5] Mottes C et al. Hydrological and pesticide transfer modeling in a tropical volcanic watershed with the WATPPASS model. Journal of Hydrology. 2015;**529**:909-927. DOI: 10.1016/j.jhydrol.2015.09.007

[6] Dubuisson C et al. Impact of subsistence production on the management options to reduce the food exposure of the Martinican population to Chlordecone. Regulatory Toxicology and Pharmacology. 2007;**49**(1):5. DOI: 10.1016/j.yrtph.2007.04.008

[7] Dereumeaux C et al. Chlordecone and organochlorine compound levels in the French West Indies population in 2013-2014. Environmental Science and Pollution Research International. 2020;**27**(33):41033-41045. DOI: 10.1007/s11356-019-07181-9

[8] Boucher O et al. Exposure to an organochlorine pesticide (chlordecone) and development of 18-month-old infants. Neurotoxicology. 2013;**35**:162-168. DOI: 10.1016/j.neuro.2013.01.007

[9] Dallaire R et al. Cognitive, visual, and motor development of 7-month-old Guadeloupean infants exposed to chlordecone. Environmental Research. 2012;**118**:79-85. DOI: 10.1016/j.envres.2012.07.006

[10] Multigner L et al. Chlordecone exposure and risk of prostate cancer. Journal of Clinical Oncology: Official Journal of the American Society of Clinical Oncology. 2010;**28**(21):3457-3462. DOI: 10.1200/JCO.2009.27.2153

[11] Multigner L, Kadhel P, Rouget F, Blanchet P, Cordier S. Chlordecone exposure and adverse effects in French West Indies populations. Environmental Science and Pollution Research International. 2016;**23**(1):3-8. DOI: 10.1007/s11356-015-4621-5

[12] Han X et al. Exposure to organochlorine pesticides and the risk of type 2 diabetes in the population of East China. Ecotoxicology and Environmental Safety. 2020;**190**:110125. DOI: 10.1016/j.ecoenv.2019.110125

[13] Legoff L et al. In utero exposure to chlordecone affects histone

modifications and activates LINE-1 in cord blood. Life Science Alliance. 2021;**4**(6)

[14] Khan I, Saeed K, Khan I. Nanoparticles: Properties, applications and toxicities. Arabian Journal of Chemistry. 2019;**12**(7):908-931. DOI: 10.1016/j.arabjc.2017.05.011

[15] Beauhaire J, Ducrot P-H, Malosse C, Rochat D, Ndiege IO, Otieno DO. Identification and synthesis of sordidin, a male pheromone emitted by Cosmopolites sordidus. Tetrahedron Letters. 1995;**36**(7):1043-1046. DOI: 10.1016/0040-4039(94)02437-G

[16] Alpizar D, Fallas M, Oehlschlager AC, Gonzalez LM. Management of Cosmopolites sordidus and Metamasius hemipterus in banana by pheromone-based mass trapping. Journal of Chemical Ecology. 2012;**38**(3):245-252. DOI: 10.1007/s10886-012-0091-0

[17] Robinson R. LXIII.—A synthesis of tropinone. Journal of the Chemical Society, Transactions. 1917;**111**:762-768. DOI: 10.1039/CT9171100762

[18] Robinson R. LXXV.—A theory of the mechanism of the phytochemical synthesis of certain alkaloids. Journal of the Chemical Society, Transactions. 1917;**111**:876-899. DOI: 10.1039/CT9171100876

[19] van Tamelen EE. Progress in the chemistry of organic natural products. Journal of the American Chemical Society. 1960;**82**(14):3804-3804. DOI: 10.1021/ja01499a099

[20] de la Torre MC, Sierra MA. Comments on recent achievements in biomimetic organic synthesis. Angewandte Chemie, International Edition. 2004;**43**(2):160-181. DOI: 10.1002/anie.200200545

[21] Poupon E, Nay B. Biomimetic Organic Synthesis. 1er édition ed. Wiley-VCH; 2011

[22] Rao TP, Prasad K, Kala R, Gladis JM. Biomimetic sensors for toxic pesticides and inorganics based on optoelectronic/electrochemical transducers—An overview. Critical Reviews in Analytical Chemistry. 2007;**37**(3):191-210. DOI: 10.1080/10408340701244664

[23] Schirhagl R, Latif U, Dickert FL. Atrazine detection based on antibody replicas. Journal of Materials Chemistry. 2011;**21**(38):14594-14598. DOI: 10.1039/C1JM11576F

[24] Khadem M, Faridbod F, Norouzi P, Foroushani A, Ganjali M, Shahtaheri S. Biomimetic electrochemical sensor based on molecularly imprinted polymer for dicloran pesticide determination in biological and environmental samples. Journal of the Iranian Chemical Society. 2016;**13**. DOI: 10.1007/s13738-016-0925-8

[25] Liu H, Qu J, Dai R, Ru J, Wang Z. A biomimetic absorbent for removal of trace level persistent organic pollutants from water. Environmental Pollution. 2007;**147**(2):337-342. DOI: 10.1016/j.envpol.2006.06.031

[26] Sicard M, Crauste-Manciet S, Dole F, Verget J, Thiéry A, Barthélémy P. Decontamination of organic pollutants from aqueous media using polymer-free bioinspired materials. ACS Sustainable Chemistry & Engineering. 2020;**8**(30):11052-11057. DOI: 10.1021/acssuschemeng.0c04329

[27] Arthropod - Nervous system and organs of sensation | Britannica. Available from: https://www.britannica.com/animal/arthropod/Nervous-system-and-organs-of-sensation [Accessed: March 22, 2022]

[28] "Nervous System," *ENT 425 - General Entomology*. Available from:

https://genent.cals.ncsu.edu/bug-bytes/nervous-system/ [Accessed: March 22, 2022]

[29] Wolstenholme AJ. Glutamate-gated chloride channels. The Journal of Biological Chemistry. 2012;**287**(48):40232-40238. DOI: 10.1074/jbc.R112.406280

[30] French-Constant RH, Williamson MS, Davies TGE, Bass C. Ion channels as insecticide targets. Journal of Neurogenetics. 2016;**30**(3-4):163-177. DOI: 10.1080/01677063.2016.1229781

[31] Ozoe Y. Chapter Four - γ-Aminobutyrate- and glutamate-gated chloride channels as targets of insecticides. In: Cohen E, editor. Advances in Insect Physiology. Vol. 44. Academic Press; 2013. pp. 211-286. DOI: 10.1016/B978-0-12-394389-7.00004-1

[32] Zhao J-H, Xu X-J, Ji M-H, Cheng J-L, Zhu G-N. Design, synthesis, and biological activities of milbemycin analogues. Journal of Agricultural and Food Chemistry. 2011;**59**(9):4836-4850. DOI: 10.1021/jf2001926

[33] Designed Biosynthesis of 25-Methyl And 25-Ethyl Ivermectin with Enhanced Insecticidal Activity by Domain Swap of Avermectin Polyketide Synthase | Microbial Cell Factories | Full Text. Available from: https://microbialcellfactories.biomedcentral.com/articles/10.1186/s12934-015-0337-y [Accessed: March 22, 2022]

[34] Wolstenholme A, Rogers A. Glutamate-gated chloride channels and the mode of action of the abamectin/milbemycin anthelmintics. Parasitology. 2005;**131**(Suppl):S85-S95. DOI: 10.1017/S0031182005008218

[35] Insect-Selective Spider Toxins Targeting Voltage-Gated Sodium Channels - PubMed. Available from: https://pubmed.ncbi.nlm.nih.gov/17223149/ [Accessed: March 22, 2022]

[36] Wakeling EN, Neal AP, Atchison WD. Pyrethroids and Their Effects on Ion Channels. IntechOpen; 2012. DOI: 10.5772/50330

[37] Silver KS et al. Voltage-gated sodium channels as insecticide targets. Advanced in Insect Physiology. 2014;**46**:389-433. DOI: 10.1016/B978-0-12-417010-0.00005-7

[38] Zhou J, Kang H-M, Lee YH, Jeong C-B, Park JC, Lee J-S. Adverse effects of a synthetic pyrethroid insecticide cypermethrin on life parameters and antioxidant responses in the marine copepods Paracyclopina nana and Tigriopus japonicus. Chemosphere. 2019;**217**:383-392. DOI: 10.1016/j.chemosphere.2018.10.217

[39] Krief A, Dumont W, Pasau P, Lecomte P. Stereoselective synthesis of methyl (1R) trans- and (1R) cis-hemicaronaldehydes from natural tartaric acid: Application to the synthesis of s-bioallethrin and deltamethrin insecticides§. Tetrahedron. 1989;**45**(10):3039-3052. DOI: 10.1016/S0040-4020(01)80130-7

[40] Soderlund DM. Molecular mechanisms of pyrethroid insecticide neurotoxicity: Recent advances. Archives of Toxicology. 2012;**86**(2):165-181. DOI: 10.1007/s00204-011-0726-x

[41] Awoyemi OM, Kumar N, Schmitt C, Subbiah S, Crago J. Behavioral, molecular and physiological responses of embryo-larval zebrafish exposed to types I and II pyrethroids. Chemosphere. 2019;**219**:526-537. DOI: 10.1016/j.chemosphere.2018.12.026

[42] Palmquist K, Salatas J, Fairbrother A. Pyrethroid Insecticides: Use, Environmental Fate, and Ecotoxicology. London, UK: IntechOpen; 2012. DOI: 10.5772/29495

[43] Christie S, Wittert GA, Li H, Page AJ. Involvement of TRPV1 channels in energy homeostasis. Frontiers in Endocrinology. 2018;**9**

[44] Du Q, Liao Q, Chen C, Yang X, Xie R, Xu J. The role of transient receptor potential Vanilloid 1 in common diseases of the digestive tract and the cardiovascular and respiratory system. Frontiers in Physiology. 2019;**10**:1064. DOI: 10.3389/fphys.2019.01064

[45] Obata R et al. Chemical modification and structure-activity relationships of pyripyropenes. 1. Modification at the four hydroxyl groups. Journal of Antibiotics (Tokyo). 1996;**49**(11):1133-1148. DOI: 10.7164/antibiotics.49.1133

[46] "Afidopyropen, a novel insecticide originating from microbial secondary extracts | Scientific Reports." Available from: https://www.nature.com/articles/s41598-022-06729-z [Accessed: March 22, 2022]

[47] Goto K et al. Synthesis and insecticidal efficacy of pyripyropene derivatives. Part II-Invention of afidopyropen. Journal of Antibiotics (Tokyo). 2019;**72**(9):661-681. DOI: 10.1038/s41429-019-0193-9

[48] Vinkers C, Olivier B. Mechanisms underlying tolerance after long-term benzodiazepine use: A future for subtype-selective GABA(A) receptor modulators? Advances in Pharmacological Sciences. 2012;**2012**:416864. DOI: 10.1155/2012/416864

[49] Hosie A, Sattelle D, Aronstein K, French-Constant R. Molecular biology of insect neuronal GABA receptors. Trends in Neurosciences. 1997;**20**(12):578-583. DOI: 10.1016/S0166-2236(97)01127-2

[50] Casida JE. Insecticide action at the GABA-gated chloride channel: Recognition, progress, and prospects. Archives of Insect Biochemistry and Physiology. 1993;**22**(1-2):13-23. DOI: 10.1002/arch.940220104

[51] Bloomquist JR. Chloride channels as tools for developing selective insecticides. Archives of Insect Biochemistry and Physiology. 2003;**54**(4):145-156. DOI: 10.1002/arch.10112

[52] Xu Z, Bai J, Li L, Liang L, Ma X, Ma L. Sublethal concentration of emamectin benzoate inhibits the growth of gypsy moth by inducing digestive dysfunction and nutrient metabolism disorder. Pest Management Science. 2021;**77**(9):4073-4083. DOI: 10.1002/ps.6432

[53] Chabala JC et al. Ivermectin, a new broad-spectrum antiparasitic agent. Journal of Medicinal Chemistry. 1980;**23**(10):1134-1136. DOI: 10.1021/jm00184a014

[54] Ishaaya I, Kontsedalov S, Horowitz AR. Emamectin, a novel insecticide for controlling field crop pests. Pest Management Science. 2002;**58**(11):1091-1095. DOI: 10.1002/ps.535

[55] Grafton-Cardwell E, Godfrey L, Chaney W, Bentley W. Various novel insecticides are less toxic to humans, more specific to key pests. California Agriculture. 2005;**59**(1). DOI: 10.3733/ca.v059n01p29

[56] Octopamine in Invertebrates - ScienceDirect. Available from: https://www.sciencedirect.com/science/

article/pii/S0301008299000167 [Accessed: March 22, 2022]

[57] Jankowska M, Rogalska J, Wyszkowska J, Stankiewicz M. Molecular targets for components of essential oils in the insect nervous system—A review. Molecules. 2018;**23**. DOI: 10.3390/molecules23010034

[58] Bhardwaj A, Kumar Tewary D, Kumar R, Kumar V, Kumar Sinha A, Shanker A. Larvicidal and structure-activity studies of natural phenylpropanoids and their semisynthetic derivatives against the tobacco armyworm Spodoptera litura (Fab.) (Lepidoptera: Noctuidae). Chemistry & Biodiversity. Jan. 2010;**7**(1):168-177. DOI: 10.1002/cbdv.200800345

[59] Pohlit A. Adulticidal activity of dillapiol and semi-synthetic derivatives of dillapiol against Aedes aegypti (L.) (Culicidae). Journal of Mosquito Research. 2012. DOI: 10.5376/jmr.2012.01.0001

[60] Cortés-Rojas DF, de Souza CRF, Oliveira WP. Clove (Syzygium aromaticum): A precious spice. Asian Pacific Journal of Tropical Biomedicine. 2014;**4**(2):90-96. DOI: 10.1016/S2221-1691(14)60215-X

[61] Enan EE. Molecular and pharmacological analysis of an octopamine receptor from American cockroach and fruit fly in response to plant essential oils. Archives of Insect Biochemistry and Physiology. 2005;**59**(3):161-171. DOI: 10.1002/arch.20076

[62] Ho TNT, Abraham N, Lewis RJ. Structure-function of neuronal nicotinic acetylcholine receptor inhibitors derived from natural toxins. Frontiers in Neuroscience. 2020;**14**

[63] Karlin A. Emerging structure of the nicotinic acetylcholine receptors. Nature Review in Neuroscience. 2002;**3**(2)

[64] Tosi S, Nieh JC. Lethal and sublethal synergistic effects of a new systemic pesticide, flupyradifurone (Sivanto®), on honeybees. Proceedings of the Royal Society B: Biological Sciences. 2019;**286**(1900):20190433. DOI: 10.1098/rspb.2019.0433

[65] Nauen R et al. Flupyradifurone: A brief profile of a new butenolide insecticide. Pest Management Science. 2015;**71**(6):850-862. DOI: 10.1002/ps.3932

[66] Zhang W et al. Mesoionic pyrido[1,2-a]pyrimidinones: Discovery of triflumezopyrim as a potent hopper insecticide1. Bioorganic & Medicinal Chemistry Letters. 2017;**27**(1):16-20. DOI: 10.1016/j.bmcl.2016.11.042

[67] "Endocrine System," *ENT 425 - General Entomology*. Available fro: https://genent.cals.ncsu.edu/bug-bytes/endocrine-system/ [Accessed: March 22, 2022]

[68] Sadowski-Debbing K, Coy JF, Mier W, Hug H, Los MJ. Caspases - Their role in apoptosis and other physiological processes as revealed by knock-out studies. Archivum Immunologiae et Therapiae Experimentalis (Warsz). 2002;**50**(1):19-34

[69] Cooper DM, Granville DJ, Lowenberger C. The insect caspases. Apoptosis. 2009;**14**(3):247-256. DOI: 10.1007/s10495-009-0322-1

[70] Accorsi A, Zibaee A, Malagoli D. The multifaceted activity of insect caspases. Journal of Insect Physiology. 2015;**76**:17-23. DOI: 10.1016/j.jinsphys.2015.03.007

[71] Pereira RB et al. Amino alcohols from eugenol as potential semisynthetic

insecticides: Chemical, biological, and computational insights. Molecules. 2021;**26**. DOI: 10.3390/molecules26216616

[72] Shimada-Niwa Y, Niwa R. Neural control of steroid hormone biosynthesis during development in the fruit fly Drosophila melanogaster. Genes & Genetic Systems. 2014;**89**:27-34. DOI: 10.1266/ggs.89.9

[73] Recent Progress in Understanding the Role of Ecdysteroids in Adult Insects: Germline Development and Circadian Clock in the Fruit Fly *Drosophila melanogaster* | SpringerLink." Available from: https://link.springer.com/article/10.1186/s40851-015-0031-2 [Accessed: March 22, 2022]

[74] Truman JW. The evolution of insect metamorphosis. Current Biology. 2019;**29**(23):R1252-R1268. DOI: 10.1016/j.cub.2019.10.009

[75] Riddiford LM. How does juvenile hormone control insect metamorphosis and reproduction? General and Comparative Endocrinology. 2012;**179**(3):477-484. DOI: 10.1016/j.ygcen.2012.06.001

[76] Mordue A et al. Actions of azadirachtin, a plant allelochemical, against insects. Pesticide Science. 1998;**54**:277-284

[77] Mordue AJ, Morgan ED, Nisbet AJ. Azadirachtin, a natural product in insect control. In: Comprehensive Molecular Insect Science. Elsevier; 2005. pp. 117-135. DOI: 10.1016/B0-44-451924-6/00077-6

[78] Veitch GE, Beckmann E, Burke BJ, Boyer A, Ayats C, Ley SV. A relay route for the synthesis of azadirachtin. Angewandte Chemie. 2007;**119**(40):7777-7779. DOI: 10.1002/ange.200703028

[79] Merzendorfer H, Zimoch L. Chitin metabolism in insects: Structure, function and regulation of chitin synthases and chitinases. The Journal of Experimental Biology. 2003;**206**(24):4393-4412. DOI: 10.1242/jeb.00709

[80] Roncero C. The genetic complexity of chitin synthesis in fungi. Current Genetics. 2002;**41**(6):367-378. DOI: 10.1007/s00294-002-0318-7

[81] Wigglesworth VB. Transpiration Through the Cuticle of Insects. The Journal of Experimental Biology. 1945;**21**(3-4):97-114. DOI: 10.1242/jeb.21.3-4.97

[82] Vega K, Kalkum M. Chitin, Chitinase responses, and invasive fungal infections. International Journal of Microbiology. 2012;**2012**:920459. DOI: 10.1155/2012/920459

[83] Molecules | Free Full-Text | Natural Product-Based Pesticide Discovery: Design, Synthesis and Bioactivity Studies of N-Amino-Maleimide Derivatives. Available from: https://www.mdpi.com/1420-3049/23/7/1521 [Accessed: March 22, 2022]

[84] Leroux P. Modes d'action des produits phytosanitaires sur les organismes pathogènes des plantes. Comptes Rendus Biologies. 2003;**326**(1):9-21. DOI: 10.1016/S1631-0691(03)00005-2

[85] Debieu D et al. Inhibition of Ergosterol Biosynthesis by Morpholine, Piperidine, and Spiroketalamine Fungicides in Microdochium nivale: Effect on Sterol Composition and Sterol $\Delta 8 \rightarrow \Delta 7$-Isomerase Activity. Pesticide Biochemistry and Physiology. 2000;**67**(2):85-94. DOI: 10.1006/pest.2000.2485

[86] Krämer W et al. Chemistry, stereochemistry and biological properties

of KWG 4168. Pesticide Science. 1999;**55**(5):610-614. DOI: 10.1002/(SICI)1096-9063(199905)55:5<610::AID-PS986>3.0.CO;2-3

[87] Unger TA. Pesticide Synthesis Handbook. William Andrew; 1996

[88] Khorramnejad A, Perdomo HD, Palatini U, Bonizzoni M, Gasmi L. Cross talk between viruses and insect cells cytoskeleton. Viruses. 2021;**13**. DOI: 10.3390/v13081658

[89] Pollard TD, Cooper JA. Actin, a central player in cell shape and movement. Science. 2009;**326**(5957):1208-1212. DOI: 10.1126/science.1175862

[90] Neher R, Biophysical properties of the cytoskeleton, neherlab@biozentrum, 2017. Available from: https://neherlab.org/20171031_theoretical_biophysics.html [Accessed: March 22, 2022]

[91] Van Damme J-C, De Ryckel B, Galoux M. Analyse par chromatographie liquide du diamino-2,3 phénazine et de l'hydroxy-2 amino-3 phénazine dans les carbendazimes techniques et formulés. Journal of Chromatography. A. 1990;**518**:375-384. DOI: 10.1016/S0021-9673(01)93197-8

[92] Wang K et al. Carbendazim exposure during the larval stage suppresses major royal jelly protein expression in nurse bees (Apis mellifera). Chemosphere. 2021;**266**:129011. DOI: 10.1016/j.chemosphere.2020.129011

[93] Weidner JP, Block SS. Alkyl and aryl thiolsulfonates. Journal of Medicinal Chemistry. 1964;**7**(5):671-673. DOI: 10.1021/jm00335a028

[94] Zefirov NS, Zyk NV, Beloglazkina EK, Kutateladze AG. Thiosulfonates: Synthesis, reactions and practical applications. Sulfur Reports. 1993;**14**(1):223-240. DOI: 10.1080/01961779308055018

[95] Gallardo-Godoy A, Torres-Altoro MI, White KJ, Barker EL, Nichols DE. 1-Methylpyridinium-4-(4-phenylmethanethiosulfonate) iodide, MTS-MPP+, a novel scanning cysteine accessibility method (SCAM) reagent for monoamine transporter studies. Bioorganic & Medicinal Chemistry. 2007;**15**(1):305-311. DOI: 10.1016/j.bmc.2006.09.058

[96] Sugata K, Song L, Nakamura M, Ueki S, Fajer PG, Arata T. Nucleotide-induced flexibility change in neck linkers of dimeric kinesin as detected by distance measurements using spin-labeling EPR. Journal of Molecular Biology. 2009;**386**(3):626-636. DOI: 10.1016/j.jmb.2008.12.079

[97] Liu T, Ren X, Cao G, Zhou X, Jin L. Transcriptome analysis on the mechanism of ethylicin inhibiting *Pseudomonas syringae* pv. actinidiae on Kiwifruit. Microorganisms. 2021;**9**(4):4

[98] Livingstone TL et al. Plant bioactives and the prevention of prostate cancer: Evidence from human studies. Nutrients. 2019;**11**(9):2245. DOI: 10.3390/nu11092245

[99] Determination and Dynamics of Ethylicin Residues in Cotton-Field Ecosystem. SpringerLink." Available from: https://link.springer.com/article/10.1007/s00128-012-0754-0; [Accessed: March 22, 2022]

[100] Zheng Y, Qing F-L, Huang Y, Xu X-H. Tunable and practical synthesis of thiosulfonates and disulfides from sulfonyl chlorides in the presence of tetrabutylammonium iodide. Advanced Synthesis and Catalysis. 2016;**358**(21):3477-3481. DOI: 10.1002/adsc.201600633

[101] "Digestive System," *ENT 425 - General Entomology*. Available from: https://genent.cals.ncsu.edu/bug-bytes/digestive-system/ [Accessed: March 22, 2022]

[102] De Loof A, Schoofs L. Intraluminal Farnesol and Farnesal in the Mealworm's Alimentary Canal: An Unusual Storage Site Uncovering Hidden Eukaryote Ca2+-Homeostasis-Dependent 'Golgicrine' Activities. Frontiers in Endocrinology. 2019;**10**

[103] Mazoir N, Benharref A, Vaca L, Reina M, González-Coloma A. Optimization of insecticidal triterpene derivatives by biomimetic oxidations with hydrogen peroxide and iodosobenzene catalyzed by MNIII and FEIII porphyrin complexes. Chemistry & Biodiversity. 2020;**17**(9):e2000287. DOI: 10.1002/cbdv.202000287

[104] Gallon ME, Gobbo-Neto L. Plant metabolites involved in the differential development of a heliantheae-specialist insect. Metabolites. 2021;**11**(3):134. DOI: 10.3390/metabo11030134

[105] McBride HM, Neuspiel M, Wasiak S. Mitochondria: More than just a powerhouse. Current Biology. 2006;**16**(14):R551-R560. DOI: 10.1016/j.cub.2006.06.054

[106] Van den Bergh SG. [22] Insect mitochondria. In: Methods in Enzymology. Vol. 10. Academic Press; 1967. pp. 117-122. DOI: 10.1016/0076-6879(67)10025-6

[107] Kumari A. Chapter 3 - Electron Transport Chain. In: Kumari A, editor. Sweet Biochemistry. Academic Press; 2018. pp. 13-16. DOI: 10.1016/B978-0-12-814453-4.00003-0

[108] Sapkota A. Electron transport chain- definition, components, steps, FAQs. Microbe Notes. 2021;**8**

Chapter 5

# Side Effects of Pesticides on Population Growth Parameters, Life Table Parameters, and Predation of the Subsequent Generation of Phytoseiid Mites

*Nayereh Hamedi*


## Abstract

Simply estimating pesticide effects on natural enemies of pests by measuring only lethal effects, or sublethal effects on the only treated natural enemies, may underestimate the total negative effects on them. So sublethal effects on subsequent generations should be assessed to estimate the total effects of their applications. Sublethal effects of commonly used acaricides on population growth parameters, life table parameters, and predation of the predatory mites of the family phytoseiidae were investigated. For this reason, offspring of treated females were used. Gross reproductive rate (*GRR*), the intrinsic rate of birth (*b*), the intrinsic rate of death (*d*), mean generation time (*T*), survivorship ($L_x$), life expectancy ($e_x$), and prey consumption were affected in comparison with control. It could be concluded that sublethal concentrations of most applied pesticides can significantly reduce population growth and life table parameters, and this should be considered in integrated pest management (IPM) programs.


**Keywords:** sublethal concentrations, pesticides, phytoseiidae, population growth parameters, life table parameters, predation

## 1. Introduction

Despite various control methods such as chemical, cultural, and biological control, the common control method of many insect pests is pesticide application, and chemical controls are often the dominant tactic used in integrated pest management (IPM) programs [1–3]. On the other hand, biological control has been a valuable tactic in pest management programs around the world for many years. Integration of biological control with chemical control within an IPM system could reduce pesticide applications and environmental hazards. For this reason, compatibility evaluation of pesticides with naturally existing or augmented biological control agents seems necessary. So, knowledge of the lethal and sublethal effects of pesticides on biological control agents is necessary for the successful implementation of IPM programs.

## 2. Importance of *Tetranychus urticae*

Mites of the family Tetranychidae (commonly known as spider mites) are important pests in agricultural and forestry ecosystems and can be found on many field crops, fruit trees, vegetables, and ornamental plants. Many spider mites naturally inhabit ephemeral and patchily distributed resources such as weeds. The most notorious and important tetranychid mite is the globally-distributed two-spotted spider mite, *Tetranychus urticae* Koch, 1836 [4]. It is one of the important pests on many crops, greenhouse, and garden products [5, 6]. It can create multiple generations (12–25 generations) and adapt to new climates quickly. It also has a broad host range, short life cycle, haploid-diploid sex-determination, and high fecundity lead to the rapid development of resistance to pesticides [7]. So, pesticide resistance, the high cost of pesticides, and loss of production time have raised interest by growers to introduce predatory phytoseiid mites to manage two-spotted spider mites and reduce their need for acaricide applications [8].

## 3. Chemical control of *Tetranychus urticae*

The rapid developmental rate of spider mites and their high fecundity allows them to attain destructive population levels very quickly. In addition, they became resistant to several extensively used acaricides. Consequently, the extensive use of pesticides led to the outbreaks of *Thrips urticae* during the last few decades [9]. Due to the environmental and health hazards resulting from the chemical pesticides as well as their side effects on the nontarget organisms (e.g., soil microorganisms) [10] and predators [11–14], their use has been regulated firmly [15]. Many chemical-based insecticides and acaricides have been registered to control *T. urticae* all over the world such as in Iran, including abamectin and fenpyroximate [16].

Abamectin is a macrocyclic lactone derived from the soil microorganism, *Streptomyces avermitilis*, and acts on gamma-aminobutyric acid (GABA) and glutamate-gated chloride channels [17–19]. Researchers reported that abamectin potentiates the effect of neurotransmitters and increases the influx of chloride ions into nerve cells, disrupting nerve impulses and nerve functions. Abamectin as an insecticide, miticide, and nematicide is widely used in different parts of the world, including America, Europe, and Asia [20–22] and was found to be one of the most toxic chemicals to *T. urticae* [23].

Fenpyroximate is a pyrazole acaricide and insecticide with selective activity against important phytophagous mites in the families Tetranychidae, Eriophyiidae, and Tarsonemidae [17, 24, 25]. After spraying this acaricide, oxygen consumption and ATP production in the pest decline, causing knockdown and paralysis [24]. It is active against all stages of agriculturally important mites, showing higher efficacy against larvae than against other life stages [17].

## 4. Biological control of *Tetranychus urticae* with emphasis on family Phytoseiidae

Natural enemies have been utilized in the management of agricultural pests for centuries. However, the last 100 years have seen a dramatic increase in their use [26].

Biological control, or biocontrol, is the use of an organism to reduce the population density of another organism and it is the core component of IPM that is growing in popularity, especially among organic growers [27]. It is one of the most economical and environmentally harmless methods of pest control for farmers [28]. Two types of biocontrol, natural biocontrol and applied biocontrol, are often distinguished. Natural biocontrol is the reduction of native pest organisms by their indigenous natural enemies. In contrast, applied biocontrol is achieved through human efforts or intervention and consists of three main approaches: conservation, inoculative (classical), and augmentative biocontrol [27]. In some agricultural systems, the natural enemies can suppress the spider mite populations below levels of economic damage [29]. Mite predators play an important role in the IPM of phytophagous mites, particularly in complex cropping systems where they may remove the need for any chemical intervention. Further information on IPM definitions and history can be found [27].

Predatory mites from families Phytoseiidae, Ameroseiidae, Parasitidae, Stigmaeidae, Anystidae, and Bdellidae as natural enemies of Tetranychidae were founded during sampling from Northwestern Iran and Varamin province (2007–2008). Among predator insects, *Stethorus gilvifrons* Mulsant (Coleoptera: Coccinellidae), *Oenopia conglobata* (Linnaeus) (Coleoptera: Coccinellidae), *Exochomus quadripustulatus* (Linnaeus) (Coleoptera: Coccinellidae), *Chrysoperla carnea* (Stephens) (Neuroptera: Chrysopidae), *Scolothrips* sp. (Thysanoptera: Thripidae), and *Orius horvathi* Reuter (Heteroptera: Anthocoridae) were found [30]. Among the predatory mites that have been found, we worked on family Phytoseiidae. Because predaceous mites of the family Phytoseiidae are considered one of the most important groups of natural enemies used in biological control [31]. Indeed, they are considered the most effective natural enemies of tetranychid mites and other microarthropods of economic importance such as thrips [6, 32]. In different countries, phytoseiid mites are successfully used in the management of *T. urticae* in protected environments and open fields [33–36]. Certain phytoseiids consume large numbers of prey and maintain plant-feeding mites at low densities. They have a high reproductive rate, a rapid developmental rate comparable to their prey, a female-biased sex ratio equivalent to their prey allowing them to respond numerically to increased prey density, and can easily be mass-reared [6, 32]. Furthermore, several species within the family may utilize pollen as a food source and can develop and reproduce on pollen as well. Phytoseiids may persist or even maintain themselves at relatively high densities in the crop at times when their main prey is scarce or absent. Therefore, phytoseiids can prevent prey resurgence, without the normal time lag usually associated with a numerical response [37]. Among the predatory mites of the family Phytoseiidae, we worked on *Phytoseius plumifer* (Canestrini and Fanzago) (Acari: Phytoseiidae) because it is an effective predator of phytophagous mites distributed in several countries such as Iran, Egypt, France, Italy, and Israel [38–40], but there is a little available data about acaricides side effects on this predator's performance [13]. One of the projects was the assessment of lethal and sublethal effects of two commonly used acaricides on *Plumozetes plumifer* [12–14].

## 5. Side effects of pesticides on phytoseiid mites

Recently, a plant protection strategy has been recommended, minimizing the use of chemical pesticides. Every crop is infested by various pests; some but not all of them may be controlled by biological means using pathogens, predators, parasitoids, and spiders. But to achieve satisfactory control of complexes of pests, selective pesticides are also indispensable. In fact, they are a prerequisite of IPM [41]. Therefore, studying the side effect of insecticides on natural enemies is highly required to exclude the detrimental effects on the natural enemies.

Pesticide use can be modified to favor natural enemies in a variety of ways, including treating only when economic thresholds dictate, use of active ingredients and formulations that are selectively less toxic to natural enemies, use of the lowest effective rates of pesticides, and temporal and spatial separation of natural enemies and pesticides. Decisions regarding pesticide use for insect pests in IPM programs are typically based on sampling pest populations to determine if they have reached economic threshold levels, although some work has been done to incorporate natural enemy sampling into these pesticide use decisions [26]. IPM also endeavors to use chemicals that act selectively against pests but not against their enemies. However, living organisms are finely tuned systems; a chemical does not have to be lethal in order to threaten the fitness (physical as well as reproductive) of the animal, with unpredictable results on the structure of the biological community [42–44]. However, pesticides may affect the predatory and reproductive behavior of beneficial arthropods short of having direct effects on their survival, few studies investigate the sublethal effects of insecticides other than their direct toxicity (usually LD50) on nontarget animals. Thus, to show that a pesticide is relatively harmless, or indeed has no measurable effect at all, behavioral studies on the effects of sublethal concentration are necessary [41].

Several studies showed that integrating biological control with chemicals in the IPM program for spider mites is particularly attractive. In different countries, phytoseiid mites are successfully used in the management of *T. urticae* in protected environments and open fields [33–36]. Therefore, it is essential to acquire information on the toxicity of commonly used acaricides to these predators [13]. Based upon the study of the effects of two acaricides (abamectin and hexythiazox) on six phytoseiid mites, it is recommended that the frequent use of acaricides against phytophagous mites should be avoided and the feasibility of biological control programs should be promoted to protect the environment, health of living individuals, and the nontarget organisms [45]. Our studies of the effects of two acaricides (abamectin and fenpyroximate) on a phytoseiid mite revealed a similar result. Currently, great efforts are directed toward reducing the use of traditional pesticides and increasing the use of IPM techniques. Therefore, finding the pesticides that are compatible with phytoseiid mites in IPM programs is an interesting and logical approach [46].

Our studies on side effects of acaricides on phytoseiid mites illustrated that evaluating the toxicity of acaricides and insecticides to phytoseiid mites by measuring only female mortality underestimates the real effects of residual exposure, and assessment of sublethal effects is important to determine the total impacts of acaricides and insecticides on the performance of predatory mites. Our studies also demonstrated that the evaluation of pesticide effects based solely on treated mites would have incomplete endpoints. Therefore, to evaluate the total effects of the pesticides on predators, determining these effects on subsequent generations is necessary [12–14]. For example, some studies on the relative toxicity of abamectin to the treated predatory mite of Phytoseiidae family without assessing offsprings reported that the intrinsic selectivity of abamectin makes it a promising candidate for use in integrated mite management (IPM) [47–49]. In contrast, our study in 2 generations of treated predatory mites illustrated this acaricide decrease the biological performance of *P. plumifer*; therefore, it is not a proper candidate in the IPM program [14]. Some other studies in consistence with our studies [13, 14] reported that abamectin and fenpyroximate are harmful to these species and did not recommend them in the IPM program [50, 51]. They evaluated the toxic effects of hexythiazox (Nisorun®, EC 10%), fenpyroximate (Ortus®, SC 5%), and abamectin (Vertimec®, EC 1.8%) on *Phytoseiulus persimilis*. The results showed that the total effect of all concentrations of fenpyroximate and field and one-half the field concentration of abamectin, were found to be toxic to this predatory mite and above the upper threshold. But the total effect values of all concentrations of hexythiazox were below the lower threshold thus it could be considered a harmless acaricide to this predatory mite [51].

In our studies to assess the toxicity and sublethal effects of acaricides on the predatory mites, a modified leaf-dip technique was used [13, 14]. Concentration-response bioassay was conducted to determine the sublethal concentrations of acaricides. $LC_5$, $LC_{10}$, $LC_{20}$, and $LC_{30}$ values were selected for fenpyroximate [13]. For abamectin $LC_{10}$, $LC_{20}$, and $LC_{30}$ were used [14]. The eggs laid by treated females were collected and transferred separately in a leaf disc on a petri dish. Methods were comprehensively described [13, 14]. All reproductive, survival, and voracity parameters of offspring of treated females were recorded from egg to death of the last female.

### 5.1 Side effects of pesticides on life table and population growth parameters of the subsequent generation of treated phytoseiid mites

Demographic toxicology has been considered as a better measure of response to toxicants than individual life-history traits [52]. Life table parameters influence the population growth rates of a mite in the current and next generations. In the female life table, the number of female progeny, the survival rate of immature and female adult stages, daily fecundity, and sex ratio were used for the estimation of different life table parameters. Some estimated parameters were the age-specific survival rate ($l_x$), life expectancy ($e_x$), age-specific fecundity ($m_x$), gross reproductive rate (*GRR*), mean generation time (*T*), the intrinsic rate of birth (*b*), and the intrinsic rate of death (*d*) [27]. The equations and life table construction were adopted from Birch (1948) and Carey (1993) [53, 54]. In the construction of a female age-specific life table, it is necessary to calculate age-specific survival rate ($l_x$) and age-specific fecundity ($m_x$) based on female individuals, where $l_x$ shows the probability that a newborn individual will survive to age $x$, and $m_x$ is the mean number of female eggs laid per female adult at age $x$. *GRR* is total lifetime reproduction in the absence of mortality. This is the average lifetime reproduction of an individual that lives to senescence, useful in considering potential population growth if all ecological limits (predation, competitors, disease, and starvation) were removed for a population. *GRR* is rarely if ever attained in nature, but useful to consider how far below this a population is held by ecological limits. The jackknife method was used to estimate the pseudo-values of the above-described parameters and compare them statistically [55].

Several researchers have reported that life-table parameters of predatory mites of family Phytoseiidae were affected by sublethal concentrations of pesticides [10, 11, 56–63]. In the above-mentioned studies, the population parameters of the subsequent generation of a lot of phytoseiid mites were decreased when exposed to sublethal concentrations of pesticides. Such as offspring of *Neoseiulus longispinosus* exposed to abamectin; *P. plumifer* exposed to abamectin and fenpyroximate; *Amblyseius swirski* exposed to bifenazate and fenazaquin; *Neoseiulus californicus* exposed to pyridaben, spirodiclofen, spiromesifen, and imidacloprid; and *P. persimilis* exposed to fenpyroximate, Propamocarb-Hydrochloride, imidaclopride, and abamectin. And they reported that the mentioned pesticides cannot be considered compatible acaricides with the exposed phytoseiid mite and should not be used with those predatory mites in integrated pest management programs.

In our studies, population growth and life table parameters of offspring of treated predatory mite *P. plumifer* were significantly and in some parameters severely affected by sublethal concentrations of two acaricides abamectin and fenpyroximate [10, 11]. The gross reproductive rate (*GRR*) was 35.66 females per female in the subsequent generation of untreated predators. It was significantly decreased in offspring of the treated female with all sublethal concentrations of fenpyroximate. It was reached to 5.4 females in offspring of $LC_{30}$ treatment. In offspring of treated females with the sublethal concentration of abamectin, *GRR* was decreased significantly too. It was 10.30 in offspring of treated females with $LC_{20}$ (treated females with $LC_{30}$ of abamectin laid no egg, so it was not subsequent generation in this concentration) (**Figure 1**). However, abamectin affected the reproductive of treated females more than fenpyroximate, *GRR* was decreased less in the subsequent generation. It was because of the severe effect of fenpyroximate on the sex ratio of treated females. The sex ratio was 16:8 (female: male) in the subsequent generation of untreated females that changed to 10:26 (female: male) in the subsequent generation of treated females with $LC_{30}$ of fenpyroximate. So, decreasing the number of females in the subsequent generation of the treated female with fenpyroximate can be the reason for lower GRR in offspring of treated females with fenpyroximate in comparison with offspring of treated females with abamectin [10, 11]. Other studies have investigated the sublethal effects of fenpyroximate and pyridaben on two predatory mites from family Phytoseiidae, *Neoseiulus womersleyi* and *P. persimilis*, and reported similar data [59, 61]. The other studies reported a similar decrease in this parameter due to abamectin application on phytoseiid mites, *Notoplites longispinosus* and *P. persimilis* [56, 64]. *GRR* was decreased by other insecticides in the subsequent generation of treated phytoseiid mites [60–62, 65]. In contrast, in the other study spirodiclofen did not affect the *GRR* of the

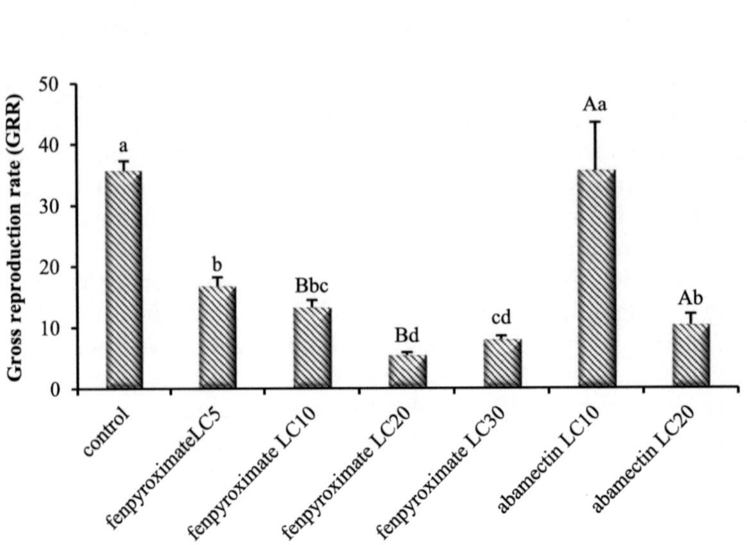


**Figure 1.**
*The gross reproductive rate (GRR) of offspring of the treated and untreated females of Phytoseius plumifer.*

subsequent generation of treated predatory mite, *A. swirskii* [65]. But fenazaquin was affected on *GRR* of the subsequent generation of this species [60].

The intrinsic rate of birth (*b*) was significantly decreased and the intrinsic rate of death (*d*) was significantly increased in offspring of treated females of *P. plumifer* with fenpyroximate and abamectin. The ratio of birth to death (*b*/*d*), which is the number of births per death, was 6.55 in control, which decreased to 0.56 in offspring of treated females with $LC_{30}$ of fenpyroximate. It was 4.11 in offspring of treated females with $LC_{20}$ of abamectin (as mentioned earlier, treated females with $LC_{30}$ of abamectin laid no egg, so it was not subsequent generation in this concentration). Mean generation time (*T*) in offspring of treated females with fenpyroximate was decreased from 17.07 days in control to 13.55 days in $LC_{30}$ treatment [13]. This parameter does not change significantly in offspring of treated females with abamectin [14]. This is in consistence the other study of the effect of spirodiclofen, spiromesifen on predatory mite, *A. swirskii*, *N. colifornicus*, respectively [62, 65]. *T* was decreased in the subsequent generation of *N. longispinosus* treated females with abamectin [56]. The age-specific survival rate of the subsequent generation of the treated and untreated females of *P. plumifer* are given in **Figure 2**. Life expectancy (*ex*) on the first day of adult emergence showed a noticeable reduction in offspring of treated individuals in comparison with control (**Figure 3**).

In contrast to our findings, the other study suggests that sublethal concentrations of spirodiclofen may not affect the population parameters of offspring from treated females of *A. swirskii* [65]. This difference may be due to different predatory mite species or acaricides mode of action. Indeed, fenpyroximate functions as mitochondrial electron transport inhibitors (METI) at Complex I [66], and abamectin acts on gamma-aminobutyric acid (GABA) and glutamate-gated chloride channels [19] while spirodiclofen inhibits the acetyl-CoA carboxylase [67]. In another study, however, reproductive and total fecundity of the subsequent generation of *A. swirskii* were affected by sublethal concentrations of propargite, researchers suggested that the usage of spirodiclofen and propargite as a selective acaricide and at sublethal dosage against spider mites may not affect the life table parameters. However, it is necessary to pay attention to the direct toxicity of spirodiclofen on *A. swirskii* for considering this acaricide in IPM programs [65, 68]. In contrast, fenazaquin and bifenazate are not compatible acaricide with *A. swirskii* and should not be used with this predatory mite in the integrated management of *T. urticae* [60, 69].

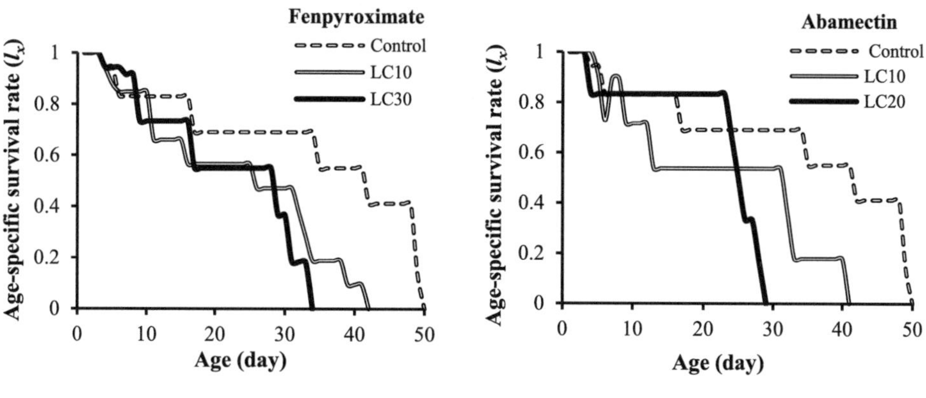


**Figure 2.**
*Survival rate (lx) of offspring of the treated and untreated females of Phytoseius plumifer with fenpyroximate and abamectin.*

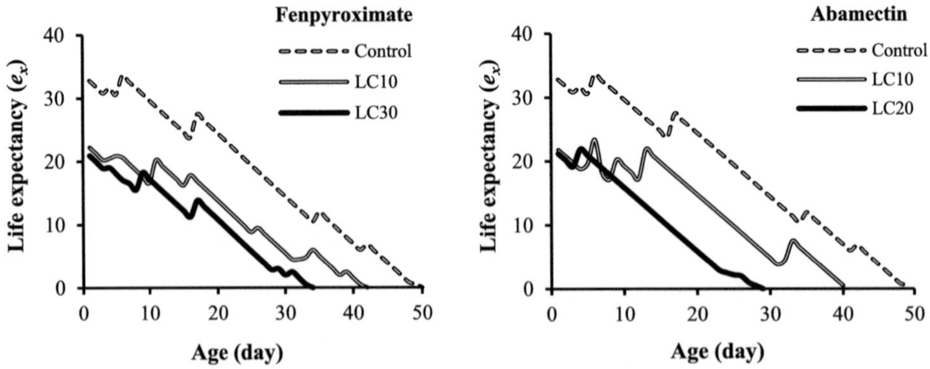


**Figure 3.**
*Life expectancy (ex) of offspring of the treated and untreated females of Phytoseius plumifer with fenpyroximate and abamectin.*

Different small letters above each bar indicate a statistically significant difference between concentrations. Different capital letters above each bar indicate a statistically significant difference between acaricides ($P < 0.05$) (LSD).

## 5.2 Side effects of pesticides on predation of the subsequent generation of treated phytoseiid mites

Besides demographic and life table parameters, the predation rate is an important factor in the biological performance of predatory mites. A direct effect of predation rate on biological performance is suppressing the pest population. The indirect effect of predation rate in biological control success is maintaining egg production and developmental success of predator. Predation rate is potentially affected by sublethal concentrations of pesticides and ignoring this effect may lead to underestimating the negative effect of pesticides on the population of predators [12]. A few studies have evaluated the sublethal effect of pesticides on predation of treated phytoseiid mites [70], but to date, apart from our study [12], no data is available on the side effects of acaricides on prey consumption of subsequent generations of treated phytoseiid mites. For example, a study evaluated the effects of four selective pesticides on predation of treated females of a phytoseiid mite, *Neoseiulus cucumeris*. They reported that flubendiamide, spirotetramat, and metaflumizone had significant impacts on the predation of immature stages; spirotetramat had the greatest effect. The four selective pesticides significantly reduced prey consumption of treated females [71]. In the other study, evaluation of the sublethal effect of abamectin on the functional response of *P. persimilis*, a significant decrease in attack rate and an increase in the handling time (Th) observed and indicating a negative effect of abamectin on the predation. Therefore, they reported that the predator requires more time to identify, persecute, consume, and digest the prey when it is under the influence of acaricide compared to control [70]. About it and as a result of predator biological behavior, Reddy (2013) mentions that the decrease in feeding is reflected by the exposure of thin layer residuals abamectin, when it enters in contact with the mite, it affects the capacity of the neurotransmitters GABA and glutamate stimulating the flow of chlorine ions into the nerve cells resulting in the loss of the function, these ions that flow inside the channel to an opening result in the loss of the cellular function and

| Treatment | μg a.i./ml | | Total prey consumption (Protonymph) | Total prey consumption (Deutonymph) | Total prey consumption (nymph stage) |
|---|---|---|---|---|---|
| Control | 0 | 0 | 6.28 ± 0.14[a] | 9.24 ± 0.59[a] | 14.40 ± 0.63[a] |
| Fenpyroximate | 3.899 | $LC_5$ | 5.80 ± 0.44[a] | 7.00 ± 0.37[b] | 12.95 ± 0.62[a] |
| | 5.607 | $LC_{10}$ | 5.46 ± 0.58[Aa] | 6.74 ± 0.56[Ab] | 12.60 ± 0.56[Aa] |
| | 10.290 | $LC_{20}$ | 5.46 ± 0.45[Aa] | 6.41 ± 0.73[Ab] | 12.07 ± 0.7[Aa] |
| | 11.956 | $LC_{30}$ | 2.65 ± 0.31[b] | 3.19 ± 0.48[c] | 5.96 ± 0.57[b] |
| Abamectin | 0.021 | $LC_{10}$ | 7.00 ± 1.06[Aa] | 6.00 ± 0.93[Aa] | 13.00 ± 0.89[Aab] |
| | 0.033 | $LC_{20}$ | 5.12 ± 0.66[Ab] | 7.75 ± 1.82[Aa] | 10.71 ± 1.47[Ab] |
| | 0.046 | $LC_{30}$ | — | — | — |

*Means followed by different small letters in each column are significantly different in each acaricide. Means followed by different capital letters in each column are significantly different between acaricides in each concentration ($P < 0.05$) (LSD).*

**Table 1.**
*Effect of sublethal concentrations on prey consumption (mean ± SE) of immature stages of the subsequent generation of exposed females of Phytoseius plumifer.*

interruption of the nervous impulses and consequently, the mites stop their feeding, concluding a negative affect for the predatory mites [72]. The other study reported that the effect of pesticides on predation may be due to a repellent effect of the pesticide. Their findings after evaluating four selective pesticides on development, fecundity, and predation of phytoseiid mite, *N. cucumeris*, showed that chlorantraniliprole could be used in fields with *N. cucumeris*, whereas flubendiamide and metaflumizone had poor compatibility with this predatory mite. It would be counterproductive to combine the use of this biological control agent with spirotetramat [71].

Prey consumption of nymphs in subsequent generation of treated females with abamectin and fenpyroximate was significantly decreased in comparison with control. Total prey consumption of nymph was 14.40 in control and decreased to 5.96 in the subsequent generation of treated females with $LC_{30}$ (**Table 1**). Daily prey consumption of females of the subsequent generation of exposed females was not significantly decreased. But total prey consumption of them was decreased. That was because of the decrease in longevity.

## 6. Conclusion

Due to the considerable effects of abamectin and fenpyroximate, in lower than the recommended field rate for *T. urticae* control, on population growth and life table parameters and predation of *P. plumifer* resulted in our studies and a lot of phytoseiid mites resulted in other studies quoted in this chapter, they are not compatible with a lot of species of phytoseiid mites so could not be recommended to use in two-spotted spider mites IPM programs.

## Author details

Nayereh Hamedi
Department of Plant Protection, Agricultural Research, Education and Extension Organization of Khuzestan, Khuzestan, Iran

*Address all correspondence to: nayereh.hamedi81@gmail.com

## References

[1] Mohammadpour K, Namvar P, Hamedi N. Efficacy of proud3 (EC 5.6%) as an organic insecticide against melon aphid, *Aphis gossypii*. Biopesticides International. 2020;**16**(1):1-4

[2] Jones VP. Parrella MP, Development of integrated pest management strategies in floricultural crops. Bulletin of the ESA. 1987;**33**:28-34

[3] Brunner JF. Integrated pest management in tree fruit crops. Encyclopedia of Agriculture and Food Systems. 2014;**10**:15-30. DOI: 10.1016/b978-0-444-52512-3.00175-3

[4] Sedaratian A, Fathipour Y, Moharramipour S. Comparative life table analysis of *Tetranychus urticae* (Acari: Tetranychidae) on 14 soybean genotypes. Journal of Insect Science. 2011;**18**:541-553

[5] Jeppson LR, Keifer HH, Baker EW. Mites Injurious to Economic Plants. Berkeley. California: University of California Press; 1975. p. 614

[6] Hoy MA. Agricultural Acarology: Introduction to Integrated Mite Management. Boca Raton, Florida: CRC Press; 2011. p. 430. DOI: 10.1201/b10909

[7] Van Leeuwen T, Vontas J, Tsagkarakou A, Dermauw W, Tirry L. Acaricide resistance mechanisms in the two-spotted spider mite, Tetranychus urticae, and other important Acari: A review. Insect Biochemistry and Molecular Biology. 2010;**40**:563-572. DOI: 10.1016/j.ibmb.2010.05.008

[8] Cote KW. Using Selected Acaricides to Manipulate Tetranychus Urticae Koch Populations in Order to Enhance Biological Control Provided by Phytoseiid Mites. MSc Thesis. Blacksburg: Virginia Polytechnical Institute and State University; 2001. p. 107

[9] El-Saiedy ESM, Fahim SF. Evaluation of two predatory mites and acaricide to suppress Tetranychus urticae (Acari: Tetranychidae) on strawberry. Bulletin of the National Research Centre. 2021;**45**:97-106. DOI: 10.1186/s42269-021-00558-2

[10] Scheepmaker JWA, van de Kassteele J. Effects of chemical control agents and microbial biocontrol agents on several non-target microbial soil organisms: A meta-analysis. Biocontrol Science and Technology. 2011;**21**:1225-1242

[11] Bergeron PE, Schmidt-Jefris RA. Not all predators are equal: Miti cide non-target effects and differential selectivity. Pest Management Science. 2020;**76**:2170-2179

[12] Hamedi N, Fathipour Y, Saber M, Sheikhi GA. Sublethal effects of two common acaricides on the consumption of *Tetranychus urticae* (Prostigmata: Tetranychidae) by *Phytoseius plumifer* (Mesostigmata: Phytoseiidae). Systematic and Applied Acarology. 2009;**14**:197-205

[13] Hamedi N, Fathipour Y, Saber M. Sublethal effects of fenpyroximate on life table parameters of the predatory mite *Phytoseius plumifer*. BioControl. 2010;**55**:271-278

[14] Hamedi N, Fathipour Y, Saber M. Sublethal effects of abamectin on the biological performance of the predatory mite, *Phytoseius plumifer* (Acari: Phytoseiidae). Experimental and Applied Acarology. 2011;**53**:29-40

[15] Horikoshi R, Goto K, Mitomi M, Oyama K, Sunazuka T, Omura S. Identifcation of pyripyropene a as a promising insecticidal compound in a microbial metabolite screening. The Journal of Antibiotics. 2017;**70**:1-5

[16] Fatemi M, Torabi E, Olyaie-Torshiz A, Taherian M. The efficacy of some chemical and botanical pesticides against *Tetranychus urticae* (Acari: Tetranychidae) on *Platanus orientalis* (Platanaceae) in urban areas. Persian Journal of Acarology. 2021;**10**(3):309-319. DOI: 10.22073/pja.v10i3.67629

[17] Dekeyser MA. Acaricide mode of action. Pest Management Science. 2005;**61**:103-110

[18] Van Leeuwen T, Witters J, Nauen R, Duso C, Tirry L. The control of eriophyoid mites: state of the art and future challenges. Experimental and Applied Acarology. 2010;**51**:205-224. DOI: 10.1007/s10493-009-9312-9

[19] Pitterna T. Avermectin insecticides and acaricides. In: Lamberth C, Dinges J, editors. Bioactive Heterocyclic Compound Classes: Agrochemicals. Weinheim, Germany: Wiley-VCH Verlag GmbH & Co, KGaA; 2012. pp. 195-207. DOI: 10.1002/9783527664412.CH16

[20] Clark JM, Scott JG, Campos F, Bloomquist JR. Resistance to avermectins extent, mechanisms, and management implications. Annual Review of Entomology. 1995;**40**:1-30

[21] Sato ME, Marcos Z, Da Silva A, Dalton R, De Souza MF. Abamectin resistance in *Tetranychus urticae* Koch (Acari: Tetranychidae): Selection, cross-resistance and stability of resistance. Neotropical Entomology. 2005;**34**:991-998

[22] Ikeda H, Ishikawa J, Hanamoto A, Shinose M, Kikuchi H, Shiba T, et al. Complete genome sequence and comparative analysis of the industrial microorganism Streptomyces avermitilis. Nature Biotechnology. 2003;**21**(5):526-531

[23] Martin DE, Latheef MA, López JD. Evaluation of selected acaricides against two-spotted spider mite (Acari: Tetranychidae) on greenhouse cotton using multispectral data. Experimental and Applied Acarology. 2015;**66**(2):227-245. DOI: 10.1007/s10493-015-9903-6

[24] Lummen P. Complex I inhibitors as insecticides and acaricides. Biochimica et Biophysica acta. 1998;**1364**(2):287-296. DOI: 10.1016/s0005-2728(98)00034-6

[25] Cloyd RA, Galle CL, Keith SR, Kemp KE. Effect of fungicides and miticides with mitochondria electron transport inhibiting activity on the two-spotted spider mite, *Tetranychus urticae* (Acari: Tetranychidae). HortScience. 2010;**45**:687-689

[26] Orr D. Biological control and integrated pest management. In: Peshin R, Dhawan AK, editors. Integrated Pest Management: Innovation-Development Process. Dordrecht: Springer; 2009. pp. 207-239. DOI: 10.1007/978-1-4020-8992-3_9

[27] Fathipour Y, Maleknia B. Mite predators. In: Omkar, editor. Ecofriendly Pest Management for Food Security. 1st ed. San Diego: Academic Press; 2016. pp. 329-366. DOI: 10.1016/b978-0-12-803265-7.00011-7

[28] Cock MJW, van Lenteren JC, Brodeur J, Barratt BIP, Bigler F, Bolckmans K, et al. Do new access and benefit sharing procedures under the convention on biological diversity threaten the future of biological control? BioControl. 2010;**55**:199-218

[29] Nyrop J, English-Loeb G, Roda A. Conservation biological control of spider mites in perennial cropping systems. In: Barbosa P, editor. Conservation Biological Control. San Diego: Academic Press; 1998. pp. 307-333. DOI: 10.1016/B978-012078147-8/50063-3

[30] Khodayari S, Hamedi N. Biological control of Tetranychidae by considering the effect of insecticides. In: Ranz RER, editor. Insecticides. 1st ed. London: IntechOpen; 2021. DOI: 10.5772/intechopen.100296

[31] Pérez-Sayas C, Aguilar-Fenollosa E, Hurtado MA, Jaques JA, Pina T. When do predatory mites (Phytoseiidae) attack? Understanding their diel and seasonal predation patterns. Journal of Insect Science. 2018;**25**(6):1056-1064. DOI: 10.1111/1744-7917.12495

[32] Hagen KS, Mills NJ, Gordh G, Mcmurtry JA. Terrestrial arthropod predators of insect and mite pests. In: Bellows TS, Fisher TW, editors. Handbook of Biological Control. San Diego: Academic Press; 1999. pp. 383-503. DOI: 10.1016/b978-012257305-7/50063-1

[33] Strong WB, Croft BA. Inoculative release of phytoseiid mites into the rapidly expanding canopy of hop for control of *Tetranychus urticae* Koch. Environmental Entomology. 1995;**24**:446-453

[34] Gerson U, Smiley RL, Ochoa R. Mites (Acari) for Pest Control. Oxford, UK: Blackwell; 2003. p. 539

[35] Zhang Z-Q. Mites of Greenhouses. Identification, Biology and Control. UK: CABI; 2003. p. 244

[36] Croft BA, Pratt DA, Luh HK. Low-density release of *Neoseiulus fallacis*, provide for rapid dispersal and control of *Tetranychus urticae* (Acari: Phytoseiidae, Tetranychidae) on apple seedlings. Experimental and Applied Acarology. 2004;**33**:327-339

[37] Aguilar-Fenollosa E, Ibáñez-Gual MV, Pascual-Ruiz S, Hurtado M, Jacas JA. Effect of ground-cover management on spider mites and their phytoseiid natural enemies in clementine mandarin orchards (II): Top-down regulation mechanisms. Biological Control. 2011;**59**(2):171-179. DOI: 10.1016/j

[38] Gomaa EA. Reda AS efficiency of *Phytoseius finitimus* Ribaga. Bulletin of Zoological Society of Egypt. 1985;**35**:30-33

[39] Kreiter S, Sentenac G, Weber M, Valentin G. Les Phytoseiidae des vignobles français. Synthe'se de 8 anne'es de recensement. In: Leclant F, Reboulet JN, editors. Proceedings of the Third International Conference on Pests in Agriculture. Paris: ANPP Publishing; 1993. pp. 597-609

[40] Nadimi A, Kamali K, Arbabi M, Abdoli F. selectivity of three miticides to spider mite predator, Phytoseius plumifer (Acari: Phytoseiidae) under laboratory conditions. Agricultural Sciences in China. 2009;**8**:326-331

[41] El-Wakeil N, Gaafar N, Sallam A, Volkmar C. Side effects of insecticides on natural enemies and possibility of their integration in plant protection strategies. In: Trdan S, editor. Insecticides—Development of Safer and more Effective Technologies. London: IntechOpen; 2013. pp. 1-56. DOI: 10.5772/54199

[42] Culin JD, Yeargan KV. The effects of selected insecticides on spiders in alfalfa. Journal of the Kansas Entomological Society. 1983;**56**:151-158

[43] Volkmar C, Schützel A. Spinnengemeinschaften auf einem typischen ackerbaustandort mitteldeutschlands und deren beeinflussung durch unterschiedliche pflanzenschutzintensitäten. Archives of Phytopathology and Plant Protection. 1997;**30**(6):533-546. DOI: 10.1080/03235409709383206

[44] Volkmar C, Schier A. Effekte von Maisanbauregime auf epigäische Spinnen. Effects of reduced soil tillage on spider communities. Phytomedizin. 2005;**35**:17-18

[45] Alhewairini SS, Al-Azzazy MM. Side effects of abamectin and hexythiazox on seven predatory mites. Brazilian Journal of Biology. 2022;**83**:1-8. DOI: 10.1590/1519-6984.251442

[46] Sa'enz-de-Cabezo'n Irigaray FJ, Zalom FG, Thompson PB. Residual toxicity of acaricides to Galendromus occidentalis and *Phytoseiulus persimilis* reproductive potential. Biological Control. 2007;**40**:153-159

[47] Zhang ZQ, Sanderson JP. Relative toxicity of abamectin to the predatory mite *Phytoseiulus persimilis* (Acari: Phytoseiidae) and two-spotted spider mite (Acari: Tetranychidae). Journal of Economic Entomology. 1990;**83**(5):1783-1790. DOI: 10.1093/jee/83.5.1783

[48] Fiedler Ż, Sosnowska D. Side effects of fungicides and insecticides on predatory mites, in laboratory conditions. Journal of Plant Protection Research. 2014;**54**(4):349-353. DOI: 10.2478/jppr-2014-0052

[49] Cote KW, Lewis EE, Schultz PB. Compatibility of acaricide residues with *Phytoseiulus persimilis* and their effects on *Tetranychus urticae*. HortScience. 2002;**37**:906-909

[50] Kaplan P, Yorulmaz S, Ay R. Toxicity of insecticides and acaricides to the predatory mite *Neoseiulus californicus* (McGregor) (Acari: Phytoseiidae). International Journal of Acarology. 2012;**38**(8):699-705. DOI: 10.1080/01647954.2012.719031

[51] Nadimi A, Kamali K, Arbabi M, Abdoli F. Side-effects of three acaricides on the predatory mite, *Phytoseiulus persimilis* Athias-Henriot (Acari: Phytoseiidae) under laboratory conditions. Munis Entomology and Zoology. 2008;**3**:556-567

[52] Forbes VE, Calow P. Is the per capita rate of increase a good measurement of population level-effect of in ecotoxicology? Environmental Toxicology and Chemistry. 1999;**18**:1544-1556

[53] Birch LC. The intrinsic rate of natural increase of an insect population. Journal of Animal Ecology. 1948;**17**:15-26

[54] Carey JR. Applied Demography for Biologists with Special Emphasis on Insects. New York: Oxford University Press; 1993. p. 206

[55] Maia AHN, Luiz AJB, Camponhola C. Statistical inference on associated fertility life table parameters using jackknife technique: Computational aspects. Journal of Economic Entomology. 2000;**93**:511-518

[56] Ibrahim YB, Yee TS. Influence of sublethal exposure to abamectin on the biological performance of *Neoseiulus longispinosus* (Acari: Phytoseiidae). Journal of Economic Entomology. 2000;**93**:1085-1089

[57] Marcic D. Sublethal effects of tebufenpyrad on the eggs and immatures of two-spotted spider mite *Tetranychus*

*urticae*. Experimental and Applied Acarology. 2005;**36**:177-185

[58] Marcic D. Sublethal efects of spirodiclofen on life history and life-table parameters of two-spotted spider mite (*Tetranychus urticae*). Experimental and Applied Acarology. 2007;**42**:121-129

[59] Park JJ, Kim M, Lee JH, Shin K, Lee SE, Kim J, et al. Sublethal effects of fenpyroximate and pyridaben on two predatory mite species, *Neoseiulus womersleyi* and *Phytoseiulus persimilis* (Acari, Phytoseiidae). Experimental and Applied Acarology. 2011;**54**:243-259. DOI: 10.1007/s10493-011-9435-7

[60] Alinejad M, Kheradmand K, Fathipour Y. Sublethal effects of fenazaquin on life table parameters of the predatory mite *Amblyseius swirskii* (Acari: Phytoseiidae). Experimental and Applied Acarology. 2014;**64**(3):361-373. DOI: 10.1007/s10493-014-9830-y

[61] Ghaderi S, Minaei K, Kavousi A, Akrami MA, Aleosfoor M, Ghadamyari M. Demographic analysis of the effect of fenpyroximate on *Phytoseiulus persimilis* Athias-Henriot (Acari: Phytoseiidae). Entomologia Generalis. 2013;**34**:225-233

[62] Ghadim-Mollaloo M, Kheradmand K, Sadeghi BR, Talebi AA. Demographic analysis of sublethal effects of spiromesifen on *Neoseiulus californicus* (Acari: Phytoseiidae). Acarologia. 2017a;**57**(3):571-580. DOI: 10.24349/acarologia/20174173

[63] Ghadim-Mollaloo M, Kheradmand K, Talebi AA. Sublethal effects of pyridaben on life table parameters of the predatory mite *Neoseiulus californicus* (McGregor) (Acari: Phytoseiidae). Zoology and Ecology. 2017b;**28**(1):56-63. DOI: 10.1080/21658005.2017.1408939

[64] Ahmed MM, Abdel-Rahman HR, Abdelwines MA. Application of demographic analysis for assessing effects of pesticides on the predatory mite, *Phytoseiulus persimilis* (Acari: Phytoseiidae). Persian Journal of Acarology. 2021;**10**(3):281-298

[65] Alinejad M, Kheradmand K, Fathipour Y. Assessment of sublethal effects of spirodiclofen on biological performance of the predatory mite, *Amblyseius swirskii*. Systematic and Applied Acarology. 2016;**21**(3):375-384. DOI: 10.11158/saa.21.3.12

[66] Hirata K, Kawamura Y, Kuno M, Igarasgi H. Development of a new acaricide, pyridaben. Journal of Pesticide Science. 1995;**20**:177-179

[67] Nauen R. Spirodiclofen: Mode of action and resistance risk assessment in tetranychid pest mites. Journal of Pesticide Science. 2005;**30**:272-274

[68] Alinejad M, Kheradmand K, Fathipour Y. Demographic analysis of sublethal effects of propargite on *Amblyseius swirskii* (Acari: Phytoseiidae): Advantages of using age-stage, two sex life table in ecotoxicological studies. Systematic and Applied Acarology. 2020;**25**(5):906-917

[69] Havasi M, Alsendi A, Sangak Sani Bozhgani N, Kheradmand K, Sadeghi R. The effects of bifenazate on life history traits and population growth of *Amblyseius swirskii* Athias-Henriot (Acari: Phytoseiidae). Systematic and Applied Acarology. 2021;**26**(3):599-610. DOI: 10.11158/saa.26.3.10

[70] Monjarás-Barrera JI, Chacón-Hernández JC, Cerna-Chávez E, Ochoa-Fuentes YM, Aguirre-Uribe LA, Landeros-Flores J. Sublethal effect of abamectin in the functional response

of the predator *Phytoseiulus persimilis* (Athias-Henriot) on *Tetranychus urticae* (Koch) (Acari: Phytoseiidae, Tetranychidae). Brazilian Journal of Biology. 2018;**79**(2):273-277. DOI: 10.1590/1519-6984.180184

[71] You Y, Lin T, Wei H, Zeng Z, Fu J, Liu X, et al. Laboratory evaluation of the sublethal effects of four selective pesticides on the predatory mite *Neoseiulus cucumeris* (Oudemans) (Acari: Phytoseiidae). Systematic and Applied Acarology. 2016;**21**(10):1506-1514. DOI: 10.11158/saa.21.11.6

[72] Reddy PP. Recent Advances in Crop Protection: Avermectins. New Delhi, India: Springer; 2013. p. 268. DOI: 10.1007/978-81-322-0723-8

Chapter 6

# Benefits and Risks of Pesticide Usage in Pets

*Motunrayo Ganiyat Akande, Solomon Usman Abraham and Johnson Caleb Ogunnubi*

## Abstract

The purpose of this chapter was to highlight the advantages of applying pesticides for the optimum care of pet animals, while also outlining the adverse effects that may be associated with their use. Pesticides can be defined as substances that can be applied for the prevention, control or eradication of unwanted organisms in living systems or in the environment. Companion animals, fondly called "pets" include dogs, cats, ferrets, pet birds and some laboratory animals like albino rats, rabbits, guinea pigs, etc. Pesticides are usually applied on pets to control ectoparasites like ticks, fleas, mites, among others. However, pets may be poisoned by pesticides if their dosages and appropriate routes of administration are not strictly adhered to. Pesticides should be administered to pets by Veterinarians and other suitably qualified personnel. Subsequently, the pets should be monitored for signs of toxicity and be treated promptly if such develop.

**Keywords:** pesticides, pets, benefits, risks, toxicity, ectoparasites

## 1. Introduction

Pesticides are substances or mixtures of substances that possess unique chemical properties for the control of detrimental pests and insect vectors [1, 2]. Pests are living organisms that pose health risks such as biting and sucking, transmission of allergy-inducing constituents, diseases, as well as parasites, thereby causing harm to humans, animals and various components of the ecosystem [3]. Pesticides can be classified as algicides, insecticides, fungicides, herbicides, rodenticides, pyrethroids, fumigants, miticides, molluscicides, etc. with discrete chemical characteristics that decrease economic, health, and environmental risks elicited by pests [4, 5]. The inappropriate application of pesticides can evoke deleterious outcomes in several organisms and the environment. Notably, pesticides do not usually differentiate between pests and other living things, consequently they may cause injury to the organisms they encounter [1].

It has been observed that pesticides may gain access into biological systems through diverse routes. For instance, organophosphate and carbamate insecticides are quickly absorbed after dermal, oral, and inhalation exposures [6]. Damalas and Koutroubas [7] reported that pesticide applicators are commonly exposed to pesticides through the dermal route. Besides, pesticides may be absorbed dermally through a splash, spill, or spray device, when being mixed, loaded or disposed of [8]. Liquid preparations of pesticides are more readily absorbed through the dermal route and other body tissues compared to powders, dusts and granular types [7]. According to [8], oral exposure to a pesticide may occur by accident or intentionally. Moreover, marked damages to the nasal, throat and pulmonary tissues have been observed after inhalation of appreciable quantities of pesticides [7].

Furthermore, exposure of populations to pesticides have been associated with negative health conditions including cancers, congenital disorders, immunological aberrations, respiratory, neurobehavioral and reproductive deficits [9]. These undesirable effects of pesticides may be evoked in several tissues and organs through genetic impairments, epigenetic alterations, mitochondrial dysfunction, oxidative damage, endoplasmic reticulum stress, endocrine disruption, among others [10]. Some of the clinical manifestations of pesticide toxicosis are confusion, agitation, lacrimation, salivation, emesis, bronchospasm, respiratory failure, micturition, diarrhoea, muscle weakness, paralysis, fasciculations, etc. [11].

Pets are animals that are domesticated and catered for by human beings for companionship, pleasure, provision of services and assistance, among others. They include dogs, cats, ferrets, pet birds, rodents, rabbits, guinea pigs, as well as exotic species like cubs, reptiles, etc. Pets are an essential part of human lives and they have been existing with human beings for thousands of years [12]. They are continually exposed to fleas and ticks. These ectoparasites may cause distress, itching, anaemia and systemic infections in the pets [13]. It is crucial to control ectoparasites in companion animals to prevent vector-borne diseases that may eventually result in high morbidity and mortality [14]. Moreover, the presence of fleas and ticks on pets may make their owners vulnerable to parasitism and zoonosis [15, 16].

The purpose of this chapter was to highlight the advantages of using pesticides for the optimum care of pet animals, while also outlining the adverse effects that may be associated with their applications.

## 2. Benefits of pesticide usage in pets

Insecticides such as organophosphates (e.g., malathion, diazinon, phosmet, fenthion, chlorfenvinphos, and cythioate) and carbamates (e.g., carbaryl and propoxur) are used to control insect and nematode infestations in animals [17]. They are formulated as sprays, pour-ons, baits, collars, etc. [17]. Carbamates are used more frequently because they are considered safer than organophosphates. However, some signs of intoxication linked to the application of carbamates are abdominal cramping, emesis, diarrhoea, dyspnoea, seizures, among others [18]. Organophosphate and carbamate insecticides competitively impede acetylcholinesterase by binding to its esteric site [19]. The excessive acetylcholine that ensues brings about unwarranted stimulation of smooth muscles and glandular secretions [17]. However, the inhibition of acetylcholinesterase by organophosphates is irreversible, while the inhibition by carbamates is reversible [20]. The classification, examples, routes of administration and mechanisms of toxicity of some insecticides applied to pets are shown in **Table 1**.

Pyrethroids are synthetic derivatives of natural pyrethrins derived from the plant, *Chrysanthemum cinerariaefolium*, and they contain esters of chrysanthemum acid [21]. They are 2250 times more poisonous to insects compared to higher organisms [30]. This is because insects possess additional sensitive sodium channels, a reduced

| Classification of insecticides | Examples | Mode of administration | Mechanisms of toxicity |
|---|---|---|---|
| Organophosphates | Diazinon, phosmet, cythioate | Sprays, pour-ons, collars | Inhibition of acetylcholinesterase [19] |
| Pyrethroids | Cypermethrin, permethrin | Shampoos, dips, spot-ons and sprays | Interruption of sodium channels in neurons [21, 22] |
| Carbamates | Carbaryl and propoxur | Sprays, pour-ons, collars | Inhibition of acetylcholinesterase [19] |
| Neonicotinoids | Imidacloprid, dinotefuran, nitenpyram | Imidacloprid and dinotefuran are applied as spot-on topical products. Nitenpyram is administered per os. | Act as agonists on the postsynaptic acetylcholine receptors in insects [23, 24] |
| Isoxazolines | Fluralaner, afoxolaner, sarolaner | Oral administration | Blockage of arthropod ligand-gated chloride channels [17] |
| Benzoylphenylurea derivative | Lufenuron | Oral suspension and injectable preparation for cats. Oral tablets for dogs. | Chitin (exoskeleton) synthesis inhibitor [25] |
| Insect growth regulators | Methoprene, fenoxycarb, pyriproxyfen | Oral suspensions, sprays and spot-ons | Mimic insect hormones, thereby interfering with the growth and development of insects [26] |
| Oxadiazine insecticide | Indoxacarb | Administered topically in a spot-on formulation | Bioactivation to an active metabolite that blocks the voltage-gated sodium ion conduits in insects [18] |
| Phenylpyrazole insecticide | Fipronil | Topical administration | Binds to gamma-aminobutyric acid receptors and the glutamate-gated chloride channels in the central nervous systems of invertebrates [13, 27–29] |
| Macrocyclic lactones | Selamectin, aprinomectin, milbemycin | Topical administration | Bind to glutamate-gated chloride channels in the nervous systems of parasites [18] |
| Formamidines | Amitraz | Available as a dip. Also formulated as impregnated collars for dogs | Binds to octopamine receptors for its insecticidal effects [18] |
| Spinosyns | Spinosad | Formulated as edible tablets for dogs and cats | Targets the binding sites on nicotinic acetylcholine receptors [18] |

**Table 1.**
*Classification, examples, route of administration and mechanisms of toxicity of some insecticides applied to pets.*

conformation and lower body temperature [30]. Permethrin, a type I pyrethroid, exists in the form of a liquid, yellow-brown and brown crystals, and it is soluble in organic solvents [31]. It may enter the body through the dermal, oral and inhalational routes [32, 33]. It is found in shampoos, dips, spot-ons, and sprays for the control of ectoparasites in companion animals [33]. Also, it is used for the treatment of scabies and lice [31, 34, 35]. Permethrin evokes injury to insect neurons by elevating the impulse conduction, thereby causing paralysis and death of insects [21]. It is broken down in the body by hydrolysis, esterification, oxidation and conjugation [30, 36].

Its metabolites include cis-3-(2,2 dichlorovinyl) 2,2 dimethylcyclopropane-1-carboxylic acid, trans-3-(2,2-dichlorovinyl)-2,2 dimethylcyclopropane-1-carboxylic acid) and (3 phenoxybenzoic acid) [31]. The metabolites of permethrin are principally excreted in the urine and faeces [21].

Furthermore, cypermethrin, a type II pyrethroid insecticide, is used for the control of pests in agricultural, public and animal health programmes [37]. It evokes toxicity through the interruption of sodium channels in neurons, thereby disrupting neuronal transmission [22]. Also, it produces oxidative stress in living organisms [38–40]. Type II pyrethroids are more neurotoxic relative to type I pyrethroids because of their α-cyano constituents [41].

Another class of insecticides administered for pest control in pets are neonicotinoid insecticides such as imidacloprid, nitenpyram and dinotefuran (stated in **Table 1**). Imidacloprid is structurally similar to nicotine, and is endorsed as a topical spot-on for dogs, as well as for agricultural purposes [14, 23]. It exerts its insecticidal activities by binding to the acetylcholine receptor on the postsynaptic region of insect neurons, thereby averting acetylcholine binding [23, 24]. Besides, imidacloprid has been reported to elicit oxidative stress and cause injury to crucial biological molecules such as deoxyribonucleic acid, proteins and lipids [42]. Moreover, nitenpyram is administered per os to eliminate fleas in dogs and cats [18]. It undergoes fast absorption with utmost blood concentrations attained within one and a half hours, and thirty-six minutes in dogs and cats respectively [18]. Dinotefuran is applied as a topical spot-on with different formulations for dogs and cats against external parasites like fleas, flies, lice, etc. [43].

Fluralaner (an isoxazoline) is a systemically administered insecticidal and acaricidal formulation that elicits long-acting efficacy after oral administration to dogs [44]. Another isooxazoline, afoxolaner, has been reported to be efficacious in dogs and cats against fleas [45–47], ticks [46], and mites [47–50]. It is detected in plasma 20–30 minutes following administration through the oral route and it attains its uppermost level in 2–4 hours [51]. Sarolaner is a broad spectrum isooxazoline with efficacy against fleas, ticks and mites in dogs [52, 53]. Isoxazolines bind to the ligand-gated chloride channels in insects and acarines [17]. Consequently, the presynaptic and postsynaptic transmission of chloride ions across the cell membranes ensue, thereby causing hyperexcitation and uninhibited activity of the central nervous system, ultimately resulting in the death of ectoparasites [17].

Lufenuron, a benzoylphenylurea derivative, is a chitin synthesis inhibitor [25]. It is available as an oral suspension and injectable formulation for cats, and an oral tablet for dogs [17]. It eliminates emerging larvae within the egg or after hatching, and female fleas feeding on treated animals are hindered from producing viable eggs or larvae [25].

Methoprene is an insect growth regulator that mimics insect hormones, thereby interfering with the growth and development of insects [26]. It is formulated as suspensions, emulsifiable and soluble concentrates, sprays and spot-ons, etc. [17]. It is used for flea control in dogs and cats, marine mosquito control, as well as agricultural and domestic pest control [54].

Fipronil is a phenylpyrazole insecticide that is approved for agricultural usage, pest control, as well as topical flea and tick treatment for companion animals [55]. It dissolves in sebum because of its high lipid solubility and it is disseminated throughout the body for the manifestation of its insecticidal effect [13]. It has been shown that fipronil binds non-competitively to γ-aminobutyric acid (GABA) receptors and the glutamate-gated chloride channels in the central nervous systems of invertebrates (e.g., fleas and ticks), thereby eliciting excessive excitation [13]. Additionally, fipronil also binds to mammalian GABA receptors, [27], and engenders oxidative stress through the production of reactive oxygen species [28, 29]. Some investigators have asserted that the foremost metabolite of fipronil, fipronil sulfone, exerts a more robust inhibitory effect on $GABA_A$ receptors and brings about cell impairment at lesser concentrations compared to fipronil [27–29].

Selamectin, aprinomectin and milbemycin are macrocyclic lactones that are used for the control of endoparasites and ectoparasites in dogs and cats [18]. They are widely administered for the prevention of heartworm disease in dogs [56]. Selamectin and aprinomectin are semisynthetic avermectins, while moxidectin is semisynthetic. These substances bind to glutamate-gated chloride channels in the nervous systems of parasites, and this culminates in a speedy and sustained entry of chloride ions into neurons [18]. As a result of this, the activity of the neurons is impeded and paralysis of the parasites occurs. The macrocyclic lactones are administered topically, and are swiftly absorbed through the dermal route. Selamectin exhibits effective control against the flea, *Ctenocephalides felis* [57, 58], biting lice (*Felicola subrostratus*) and ear mites (*Otodectes cynotis*), among others in cats [59].

Indoxacarb is an oxadiazine insecticide that is administered topically in a spot-on formulation for the control of fleas on companion animals [18]. It is found in insect baits for home use and granules, as well as liquids for agricultural applications [60]. Moreover, it is bioactivated to an active metabolite that blocks the voltage-gated sodium ion conduits in insects [18].

Furthermore, formamidines are acaricidal compounds that exert their effects through binding to octopamine receptors [18]. Amitraz is the only approved formamidine for use in veterinary medical practice, and it is applied primarily as an acaricide to control ticks and mites [18]. It is available as a dip for the control of demodicosis in dogs, as well as the control of scabies. An amitraz-impregnated collar is also marketed for the control of ticks on dogs.

Spinosyns are a family of insecticides obtained from the fermentation of an actinomycete, *Saccharopolyspora spinosa* [18]. Spinosyns A and D are the main products of the fermentation procedure, as well as the principal components of Spinosad [61]. Spinosyns mostly target the binding sites on nicotinic acetylcholine receptors, and they also influence GABA receptor function [18]. This ensues in spontaneous muscle contractions, prostration, tremors, and paralysis of insects. Spinosad is used to control numerous insects and it is formulated as edible tablets for dogs and cats [61].

## 3. Risks of pesticide usage in pets

There is a predominant exposure of human and animal populations to pesticides and this may be associated with detrimental effects on their health status [4, 62]. According to [17] , clinical signs of pesticide intoxication can occur within a short or long duration of exposure, depending on the dose, route, and noxiousness of the pesticide administered. It has been documented that those pesticides have severe effects on non-target organisms, including various components of the ecosystem [63].

Various pesticides, especially, insecticides applied to pets for the prevention and control of ectoparasites may be associated with some adverse effects. For instance, permethrin poisoning may produce symptoms including epidermal lesions, pharyngitis, salivation, nausea, emesis, abdominal pain, gastrointestinal mucosal irritation and dyspnoea in animals [32, 63, 64]. Cats are more likely than dogs to develop pyrethroid toxicosis because the feline liver cannot conjugate glucuronide efficiently, and conjugation with glucuronide is essential for permethrin metabolism [33]. Permethrins are regarded as the commonest aetiology of poisoning in cats in the United States of America [65]. Cats may be exposed to permethrin from dermal application of topical formulations, oral intake, and direct contact with dogs administered with it topically [66]. The commonest clinical signs of permethrin intoxication in cats are muscle tremors and seizures, but hypersalivation, depression, emesis, anorexia and even death may ensue [33].

Moreover, alpha-cypermethrin (a synthetic pyrethroid like permethrin) intoxication can cause lacrimation, salivation, nausea, emesis, diarrhoea, mucosal irritations, motor coordination dysfunction, chorea, inactivity, tremors and clonic seizures [30, 36]. It has been observed that dogs usually exhibit signs of intoxication such as shaking of their limbs, slight muscle fasciculation, rubbing of the application site, distress and uneasiness after dermal administration of pyrethrins/pyrethroids [67–69].

Cats are more susceptible to insecticides that inhibit acetylcholinesterase such as organophosphates and carbamates compared to dogs [70]. Also, neonate, geriatric and incapacitated animals are more vulnerable to these groups of pesticides. Organophosphates and carbamates elicit muscarinic, nicotinic, and central nervous system signs of toxicity in biological systems. The muscarinic signs are salivation, lacrimation, urination, defecation, respiratory distress, vomiting, pupillary constriction and reduced heart rate [70]. The nicotinic symptoms include muscle tremors, fasciculations, feebleness, incoordination, and paresis that may culminate in paralysis [71], while the central nervous system signs of toxicity comprise hyperactivity, incoordination, convulsion and unconsciousness [71].

The predominant clinical signs linked to isoxazoline toxicity are emesis, anorexia, diarrhoea and exhaustion in dogs and cats [17]. The administration of lufenuron to cats causes pain at the site of injection and oedema [17]. Additionally, dogs treated with the parenteral formulation of lufenuron developed a marked local reaction [25].

Some investigators asserted that young animals are more likely to exhibit exhaustion and incoordination after oral dosing with methoprene (an insect growth regulator) [71], while the commonest clinical signs of toxicity seen in companion animals exposed to indoxacarb are anorexia, emesis, diarrhoea and lethargy [17]. Moreover, amitraz (a formamidine insecticide, mentioned in **Table 1**) can cause temporary pruritus, urticaria and oedema after the initial administration to pets [18]. In addition, a brief sedation has been recorded in dogs after an amitraz bath that may last for one day or three days in puppies.

## 4. Conclusion

This chapter review presented information on the benefits and risks of the applications of pesticides, mainly insecticides, to pets. Even though pests are harmful to companion animals and their owners, they should be controlled cautiously with the use of appropriate pesticides approved by Veterinarians and relevant regulatory agencies in different countries. This will ensure that the hazards inherent in the pesticides are adequately mitigated. Also, there is a need for researchers, Veterinarians, related health care professionals and pesticide manufacturers to collaborate and find out innocuous methods for the prevention and control of pests in pets. This effort can improve human, animal and ecosystem health and integrity.

## Acknowledgements

The authors are thankful to the staff of the Faculty of Veterinary Medicine and the Veterinary Teaching Hospital, University of Abuja, Nigeria, for their support.

## Conflict of interest

The authors declare that there is no conflict of interest.

## Author details

Motunrayo Ganiyat Akande*, Solomon Usman Abraham
and Johnson Caleb Ogunnubi
Faculty of Veterinary Medicine, Department of Veterinary Pharmacology and Toxicology, University of Abuja, Nigeria

*Address all correspondence to: akande.ganiyat@uniabuja.edu.ng

## References


[1] Yadav IS, Devi NL. Pesticides classification and its impact on human and environment. In: Kumar A, Singhal JC, Techato K, Molina LT, Singh N, Kumar P, Kumar P, Chandra R, Caprio S, Upadhye S, Yonemura S, Rao SY, Zhang TC, Sharma UC, Abrol YP, editors. Environmental Science and Engineering. USA: Studium Press LLC; 2017. pp. 140-158

[2] Rani L, Thapa K, Kanojia N, Sharma N, Singh S, Grewal AS, et al. An extensive review on the consequences of chemical pesticides on human health and environment. Journal of Cleaner Production. 2021;**283**:124657. DOI: 10.1016/j.jclepro.2020.124657

[3] Jeger M, Bragard C, Caffier D, Candresse T, Chatzivassiliou E, Dehnen-Schmutz K, et al. Guidance on quantitative pest risk assessment. EFSA Journal. 2018;**16**:5350. DOI: 10.2903/j.efsa.2018.5350

[4] Aktar W, Sengupta D, Chowdhury A. Impact of pesticides use in agriculture: Their benefits and hazards. Interdisciplinary Toxicology. 2009;**2**(1):1-12. DOI: 10.2478/v10102-009-0001-7

[5] de Souza RM, Seibert D, Quesada HB, de Jesus BF, Fagundes-Klen MR, Bergamasco R. Occurrence, impacts and general aspects of pesticides in surface water: A review. Process Safety and Environmental Protection. 2020;**135**:22-37. DOI: 10.1016/j.psep.2019.12.035

[6] Meerdink GL. Anticholinesterase insecticides. In: Plumlee KH, editor. Clinical Veterinary Toxicology. St. Louis (MO): Mosby; 2004. pp. 178-180

[7] Damalas C, Koutroubas S. Farmers' exposure to pesticides: Toxicity types and ways of prevention. Toxics. 2016;**4**(1):1. DOI: 10.3390/toxics4010001

[8] Kim KH, Kabir E, Jahan SA. Exposure to pesticides and the associated human health effects. Science Total Environment. 2017;**575**:525-535. DOI: 10.1016/j.scitotenv.2016.09.009

[9] Hernandez AF, Parron T, Tsatsakis AM, Requena M, Alarcon R, Lopez-Guarnido O. Toxic effects of pesticide mixtures at a molecular level: Their relevance to human health. Toxicology. 2013;**307**:136-145. DOI: 10.1016/j.tox.2012.06.009

[10] Mostafalou S, Abdollahi M. Pesticides and human chronic diseases: Evidences, mechanisms, and perspectives. Toxicology and Applied Pharmacology. 2013;**268**:157-177. DOI: 10.1016/j.taap.2013.01.025

[11] Eddleston M, Buckley NA, Eyer P, Dawson AH. Management of acute organophosphorus pesticide poisoning. Lancet. 2008;**371**:597-607. DOI: 10.1016/S0140-6736(07)61202-1

[12] Bauer EC, Ogg CL, Carlson MP, Hygnstrom JR. Protecting your cats and dogs from pesticide poisoning. Nebraska Extension Publications. Research-based information that you can use Nebraska Extension G2260. Index: Pesticides, General Safety; 2021

[13] Suzuki T, Hirai A, Khidkhan K, Nimako C, Ichise T, Takeda K, et al. The effects of fipronil on emotional and cognitive behaviors in mammals. Pesticide Biochemistry and Physiology. 2021;**175**:104847

[14] Khalil SR, Awad A, Hesham HM, Nassan MA. Imidacloprid insecticide exposure induces stress and disrupts glucose homeostasis in male rats. Environmental Toxicology Pharmacy. 2017;**55**:165-174

[15] Pérez-Osorio CE, Zavala-Velázquez JE, Arias León JJ, Castro Z. *Rickettsia felis* as emergent global threat for humans. Emerging Infectious Diseases. 2008;**14**:1019-1023

[16] Xhaxhiu D, Kusi I, Rapti D, Visser M, Knaus M, Lindner T, et al. Ectoparasites of dogs and cats in Albania. Parasitology Research. 2009;**105**:1577-1587

[17] Wismer T, Means C. Toxicology of newer insecticides in small animals. Veterinary Clinical Small Animal. 2018;**48**:1013-1026. DOI: 10.1016/j.cvsm.2018.06.005

[18] Dryden MW. Ectoparasiticides Used in Small Animals. 2015. Available from: https://www.merckvetmanual.com/pharmacology/ectoparasiticides/ectoparasiticides-used-in-small-animals [Accessed: March 2, 2022]

[19] Hayes WJ Jr, editor. Pesticides Studied in Man. Baltimore, MD: Williams & Wilkins; 1982. pp. 284-435

[20] Osweiler GD. Organophosphorus and carbamate insecticides. In: Toxicology. Philadelphia: Lippincott Williams & Wilkins; 1996. pp. 231-236

[21] Chrustek A, Hołyńska-Iwan I, Dziembowska I, Bogusiewicz J, Wróblewski M, Cwynar A, et al. Current research on the safety of pyrethroids used as insecticides. Medicina. 2018;**54**:61. DOI: 10.3390/medicina54040061

[22] Singh AK, Tiwari MN, Prakash O, Singh MP. A current review of cypermethrin-induced neurotoxicity and nigrostriatal dopaminergic neurodegeneration. Current Neuropharmacology. 2012;**10**:64-71

[23] Bai D, Lummis SCR, Leicht W, Breer H, Sattelle DB. Actions of imidacloprid and a related nitromethylene on cholinergic receptors of an identified insect motor neurone. Pesticide Science. 1991;**33**:197-204

[24] Lui MY, Cassida JE. High affinity of [3H] imidacloprid in the insect acetylcholine receptor. Pesticide Biochemical Physiology. 1993;**46**:40-46

[25] Stansfield DG. A review of safety and efficacy of lufenuron in dogs and cats. Canine Pract. 1997;**22**:34-48

[26] Wick K, Bond C, Buhl K, Stone D. Methoprene General Fact Sheet; National Pesticide Information Center, Oregon State University Extension Services. 2012. Available from: http://npic.orst.edu/factsheets/methogen.html [Accessed: February 24, 2022]

[27] Li P, Akk G. The insecticide fipronil and its metabolite fipronil sulphone inhibit the rat alpha1beta2gamma2L GABA(A) receptor. British Journal of Pharmacology. 2008;**155**:783-794. DOI: 10.1038/bjp.2008.309

[28] Badgujar PC, Pawar NN, Chandratre GA, Telang AG, Sharma AK. Fipronil induced oxidative stress in kidney and brain of mice: Protective effect of vitamin E and vitamin C. Pesticide Biochemistry and Physiology. 2015;**118**:10-18

[29] Ki YW, Lee JE, Park JH, Shin IC, Koh HC. Reactive oxygen species and mitogen-activated protein kinase induce apoptotic death of SH-SY5Y cells in response to fipronil. Toxicology Letters. 2012;**211**:18-28. DOI: 10.1016/j.toxlet.2012.02.022

[30] Bradberry SM, Cage SA, Proudfoot AT, Vale JA. Poisoning due to pyrethroids. Toxicological Reviews. 2005;**24**:93-106. DOI: 10.2165/00139709-200524020-00003

[31] Toynton K, Luukinen B, Buhl K, Stone D. Permethirn Technical Fact Sheet; National Pesticide Information Center, Oregon State University Extension Services. 2009. Available from: http://npic.orst.edu/factsheets/archive/Permtech.html [Accessed: February 28, 2022]

[32] Wylie BJ, Hauptman M, Woolf AD, Goldman RH. Insect repellants during pregnancy in the era of the Zika Virus. Obstetrics and Gynecology. 2016;**128**:1111-1115. DOI: 10.1097/AOG.0000000000001685

[33] Wismer T. Small Animal Toxicoses-Insecticides. Davis, CA: Veterinary Support Personnel Network; 2003

[34] Del Prado-Lu JL. Insecticide residues in soil, water, and eggplant fruits and farmers' health effects due to exposure to pesticides. Environmental Health and Preventive Medicine. 2015;**20**:53-62. DOI: 10.1007/s12199-014-0425-3

[35] Sheets LP. Imidacloprid: A neonicotinoid insecticide. In: Krieger R, editor. Handbook of Pesticide Toxicology. San Diego (CA): Academic Press; 2001. pp. 1123-1130

[36] Costa LG. The neurotoxicity of organochlorine and pyrethroid pesticides. Handbook of Clinical Neurology. 2015;**131**:135-148

[37] Sharma P, Huq AU, Singh R. Cypermethrin-induced reproductive toxicity in the rat is prevented by resveratrol. Journal of Human Reproduction Science. 2014;7:99-106

[38] Mossa ATH, Heikal TM, Belaiba M, Raoelison EG, Ferhout H, Bouajila J. Antioxidant activity and hepatoprotective potential of *Cedrelopsis grevei* on cypermethrin induced oxidative stress and liver damage in male mice. BMC Complementary and Alternative Medicine. 2015;**15**:251. DOI: 10.1186/s12906-015-0740-2

[39] Das T, Pradhan A, Paramanik A, Choudhury SM. Ameliorative role of zinc on cypermethrin-induced changes in haematological parameters and oxidative stress biomarkers in rat erythrocytes. Toxicology and Environmental Health Sciences. 2016;**8**:234-246. DOI: 10.1590/s1984-29612019064

[40] Akande MG, Okoronkwo SM, Abalaka SE, Idoko SI, Ubah SA, Abah KO, et al. Evaluation of the impacts of taurine on oxidative stress indices in sera and brain of rats exposed to cypermethrin. Journal of Applied Sciences and Environmental Management. 2021;**2**:213-217. DOI: 10.4314/jasem.v25i2.12

[41] Soderlund DM, Clark JM, Sheets LP, Mullin LS, Piccirillo VJ, Sargent D, et al. Mechanisms of pyrethroid neurotoxicity: Implications for cumulative risk assessment. Toxicology. 2002;**171**:3-59. DOI: 10.1016/s0300-483x(01)00569-8

[42] El-Gendy KS, Aly NM, Mahmoud FH, Kenawy A, El-Sebae AKH. The role of vitamin C as antioxidant in protection of oxidative stress induced by imidacloprid. Food and Chemical Toxicology. 2010;**48**:215-221. DOI: 10.1016/j.fct.2009.10.003

[43] Junquera P. Dinotefuran for veterinary use against fleas in dogs and cats, and against houseflies. 2021a. Available from: https://parasitipedia.net/index.php?option=com_content&view=article&id=2470&itemid=2737 [Accessed: March 3, 2022]

[44] Walther FM, Allan MJ, Roepke RKA, Nuernberger MC. Safety of fluralaner chewable tablets (Bravecto TM), a novel systemic antiparasitic drug, in dogs after oral administration. Parasites & Vectors. 2014;**7**:87. DOI: 10.1186/1756-3305-7-87

[45] Cvejic D, Schneider C, Neethling W, Hellmann K, Liebenberg J, Navarro C. The sustained speed of kill of ticks (*Rhipicephalus sanguineus*) and fleas (*Ctenocephalides felis felis*) on dogs by a spot-on combination of fipronil and permethrin (Effitix®) compared with oral afoxolaner (NexGard®). Veterinary Parasitology. 2017;**243**:52-57. DOI: 10.1016/j.vetpar.2017.06.011

[46] Machado MA, Campos DR, Lopes NL, Bastos IPB, Alves MSR, Correia TR, et al. Efficacy of afoxolaner in the flea control in experimentally infested cats. Revista Brasileira de Parasitologia Veterinária. 2019;**28**:760-763. DOI: 10.1590/S1984-29612019064

[47] Beugnet F, de Vos C, Liebenberg J, Halos L, Larsen D, Fourie J. Efficacy of afoxolaner in a clinical field study in dogs naturally infested with *Sarcoptes scabiei*. Parasite. 2016a;**23**:26. DOI: 10.1051/parasite/2016026

[48] Beugnet F, Halos L, Larsen D, de Vos C. Efficacy of oral afoxolaner for the treatment of canine generalised demodicosis. Parasite. 2016b;**23**:14. DOI: 10.1051/parasite/2016014

[49] Lebon W, Beccati M, Bourdeau P, Brement T, Bruet V, Cekiera A, et al. Efficacy of two formulations of afoxolaner (NexGard® and NexGard Spectra®) for the treatment of generalised demodicosis in dogs, in veterinary dermatology referral centers in Europe. Parasites & Vectors. 2018;**11**:506. DOI: 10.1186/s13071-018-3083-2

[50] Machado MA, Campos DR, Lopes NL, Barbieri Bastos IP, Botelho CB, Correia TR, et al. Efficacy of afoxolaner in the treatment of otodectic mange in naturally infested cats. Veterinary Parasitology. 2018;**256**:29-31. DOI: 10.1016/j.vetpar.2018.04.013

[51] Beugnet F, de Vos C, Liebenberg J, Halos L, Fourie J. Afoxolaner against fleas: Immediate efficacy and resultant mortality after short exposure on dogs. Parasite. 2014;**21**:42. DOI: 10.1051/parasite/2014045

[52] Mctier TL, Chubb N, Curtis MP, Hedges L, Inskeep GA, Knauer CS, et al. Discovery of sarolaner: a novel, orally administered, broad-spectrum, isoxazoline ectoparasiticide for dogs. Veterinary Parasitology. 2016;**222**:3-11. DOI: 10.1016/j.vetpar.2016.02.019

[53] Packianathan R, Colgan S, Hodge A, Davis K, Six RH, Maeder S. Efficacy and safety of sarolaner (Simparica®) in the treatment and control of naturally occurring flea infestations in dogs presented as veterinary patients in Australia. Parasite Vectors. 2017;**10**:387

[54] United States Environmental Protection Agency (USEPA). R.E.D. Facts: Methoprene. Washington, DC: Office. 1991. Available from: https://archive.epa.gov/pesticides/reregistration/web/pdf/0030fact.pdf [Accessed: February 27, 2022]

[55] Teerlink J, Hernandez J, Budd R. Fipronil washoff to municipal wastewater from dogs treated with spot-on products. Science Total Environment. 2017;**599-600**:960-966. DOI: 10.1016/j.scitotenv.2017.04.219

[56] McTier TL, Six RH, Pullins A, Chapin S, Kryda K, Mahabir SP, et al. Preventive efficacy of oral moxidectin at various doses and dosage regimens

against macrocyclic lactone-resistant heartworm (*Dirofilaria immitis*) strains in dogs. Parasites & Vectors. 2019;**12**:444. DOI: 10.1186/s13071-019-3685-3

[57] Ritzhaupt LK, Rowan TG, Jones RL, Cracknell VC, Murphy MG, Shanks DJ. Evaluation of the comparative efficacy of selamectin against flea (*Ctenocephalides felis felis*) infestations on dogs and cats in simulated home environments. Veterinary Parasitology. 2002;**106**:165-175. DOI: 10.1016/s0304-4017(02)00051-1

[58] Mctier TL, Evans NA, Martin-Short M, Gration K. Comparison of the activity of selamectin, fipronil, and imidacloprid against flea larvae (*Ctenocephalides felis felis*) *in vitro*. Veterinary Parasitology. 2003;**116**:45-50. DOI: 10.1016/s0304-4017(03)00163-8

[59] Packianathan R, Pittorino M, Hodge A, Bruellke N, Graham K. Safety and efficacy of a new spot-on formulation of selamectin plus sarolaner in the treatment and control of naturally occurring flea infestations in cats presented as veterinary patients in Australia. Parasite Vectors. 2020;**13**:227

[60] Meister RT, Sine C. Crop Protection Handbook. Willoughby (Ohio): Meister Publishing; 2014. p. 266

[61] Junquera P. Spinosyns for veterinary use in dogs, cats and livestock. 2021b. Available from: https://parasitipedia.net/index.php?option=com_content&view=article&id=2415&Itemid=2680 [Accessed: March 3, 2022]

[62] Damalas CA, Eleftherohorinos IG. Pesticide exposure, safety issues, and risk assessment indicators. International Journal of Environmental Research and Public Health. 2011;**8**:1402-1419. DOI: 10.3390/ijerph8051402

[63] Kaur R, Mavi GK, Raghav S. Pesticides classification and its impact on environment. International Journal of Current Microbiology and Applied Science. 2019;**8**:1889-1897

[64] DeGroot WD. Intravenous lipid emulsion for treating permethrin toxicosis in a cat. Canadian Veterinary Journal. 2014;**55**:1253-1254

[65] Meyer EK. Toxicosis in cats erroneously treated with 45 to 65% permethrin products. Journal of the American Veterinary Medical Association. 1999;**215**:198-203

[66] Kuo K, Odunayo A. Adjunctive therapy with intravenous lipid emulsion and methocarbamol for permethrin toxicity in 2 cats. Journal of Veterinary Emergency and Critical Care. 2013;**23**:436-441

[67] Malik R, Ward MP, Seavers A, Fawcett A, Bell E, Govendir M, et al. Permethrin spot-on intoxication of cats: Literature review and survey of veterinary practitioners in Australia. Journal of Feline Medicine and Surgery. 2010;**12**:5-14. DOI: 10.1016/j.jfms.2009.12.002

[68] Haworth MD, Smart L. Use of intravenous lipid therapy in three cases of feline permethrin toxicosis. Journal of Veterinary Emergency and Critical Care. 2012;**22**:697-702. DOI: 10.1111/j.1476-4431.2012.00804.x

[69] Ceccherini G, Perond F, Lippi I, Grazia G, Marchetti V. Intravenous lipid emulsion and dexmedetomidine for treatment of feline permethrin intoxication: A report from 4 cases. Open Veterinary Journal. 2015;**5**:113-121

[70] Nafe LA. Selected neurotoxins. The Veterinary Clinics of North America.

Small Animal Practice. 1988;**18**:593-604. DOI: 10.1016/s0195-5616(88)50057-8

[71] Humphreys DJ. Veterinary Toxicology. 3rd ed. Philadelphia: WB Saunders; 1988

Chapter 7

# Effects of Noise Associated with Pesticides in the Hearing and Vestibular Systems of Endemic Disease Combat Agents

*Adriana Bender Moreira de Lacerda,*
*Patrícia Arruda de Souza Alcarás,*
*Maria Cristina Alves Corazza,*
*Adrian Fuente and Bianca Simone Zeigelboim*

## Abstract

The current study aimed to assess the effect of the concomitant exposure to noise and pesticides on the auditory and vestibular systems of endemic disease combat agents. The sample comprised 58 participants, males, divided into two groups. The exposed group (EG) comprised 40 agents, adults, exposed to the noise and pesticides. The control group (CG) comprised 18 participants, without exposure, paired according to age range and gender. The participants from both groups underwent conventional pure-tone audiometry and high-frequency audiometry, evoked otoacoustic emissions and suppression of the emissions, immittance testing, brainstem evoked response audiometry, and dichotic digits test. The vestibular assessment was only carried out in the experimental group. Results showed no difference between the groups in the findings of the pure-tone audiometry and suppression effect of the evoked otoacoustic emissions. Difference was evidenced between the groups in the acoustic reflex testing, the tympanometry, the brainstem evoked response audiometry, and the dichotic digits test, with worse results among the EG. In the vestibular assessment, there was the prevalence of altered tests among EG in 36.4% of the cases, more evidence for the peripheral vestibular dysfunction. In conclusion, noise and pesticide exposure impaired the auditory and vestibular systems of endemic disease control agents.

**Keywords:** noise, pesticides, community health agents, hearing, hearing loss

## 1. Introduction

Organophosphate pesticides can change the efferent auditory system's mechanism of action. Such a change is caused by the inhibition of acetylcholinesterase, which in turn leads to an accumulation of acetylcholine in the peripheral and central auditory pathways [1–4], affecting the action potential of the efferent system from the superior olivary nucleus to the cochlea [3, 5, 6]. This way, in humans the exposure to pesticides including organophosphate and/or pyrethroid, either alone or in combination with noise, damages the peripheral auditory system [7–10] and the central auditory functions [3, 11–22] as well as vestibular function [8, 10, 21].

There is evidence that hearing loss can be considered an early manifestation of chronic intoxication by pesticide [23]. For this reason, both basic and complementary audiological assessments contribute to early identifying the intoxication and determine the causal nexus in the pesticide-exposed populations [6, 24, 25].

Endemic disease combat agents (EDCA) are not only the most exposed healthcare professionals to pesticides [26] but also to the noise from the motorized knapsack sprayers and vehicle-mounted ultra-low volume (ULV) aerosol generators. Even though noise levels have been reported in few studies on pesticide exposure, their assessment is recommended [1–6, 27].

The EDCA, Public Health professionals who prevent and fight environmental diseases, like dengue, Chagas, Leishmaniosis, and malaria [28]. Their job entails surveillance of houses, waste land, warehouses, and commercial facilities. In addition, they guide the population in the prevention and treatment of infectious diseases, and they manipulate (liquid preparation) and apply larvicides and insecticides to fight vectors [29]. The contamination may occur by skin absorption and inhalation, mainly among agents who make use of the mist-spraying system [30].

Thus, the current study aimed to assess the effect of the concomitant exposure to noise and pesticides on the hearing system and vestibular system of endemic disease combat agents.

## 2. Methods

It is a cross-sectional, field study, quantitative and descripture, approved by the Research Ethics Committee (REC) of the Worker's Hospital/SES/PR by *Plataforma Brasil*, protocol number 1.242.014. Please be informed that all ethical precepts have been respected, including the Informed Consent Forms (ICF).

It was developed at the speech-language-hearing university clinic of a private university, with civil servants of the state of Paraná, Brazil. It partnered with the Syndicate of Federal Civil Servants in Health, Labor, Social Security, and Welfare of the State of Paraná (SindPrevs/PR), Federal University of Paraná, Paraná State Department of Health, and the Public Ministry of Labor.

The inclusion criteria for the exposed group encompassed being an endemic disease control agent, being a civil servant of the State of Paraná, and being over 18 years old. The exclusion criterion was having conductive hearing loss. Recruitment was an oral invitation by the person responsible for the SindPrevs/PR; to those who were interested, the union offered transportation to the site of the field study.

The inclusion criterion for the control group was being of the same age group and gender as the exposed group. The exclusion criteria encompassed not having an occupational history of exposure to physical and chemical agents and the presence of conductive and/or mixed hearing loss. Recruitment was an invitation letter from the researchers.

The study sample comprised 58 professionals, divided into two groups. The Exposed Group (EG) entailing 40 EDCA, all males, ages ranging from 48 to 72 years, occupationally exposed to noise and pesticides such as organophosphates and pyrethroids insecticides (as well as a history of past exposure to other types of pesticides, such as organochlorines, carbamates, and larvicides), generated by automatic pesticide sprayers on average for 31.33 years (range of exposure from 20 to 42 years. Tasks performed by the EDCA included pesticide preparation, application, and material cleaning after application. Pesticides were applied by spraying the poison via backpack pump, hand pump, and tracked vehicles.

Usually, the EDCA is exposed to noise and pesticides for 6–10 h a day on average, besides the time they take maintaining the equipment preparing the substances. For example, the exposure time of the backpack Ultra Low Volume (ULV) can be up to eight hours a day, avoiding the hours of intense sunlight.

According to information provided by the Union of Federal Health, Labor, Social Security, and Social Project Civil Servants, Parana State/Brazil, noise levels generated by motorized knapsack sprayers are 107 dBA/4h (Leq decibel in weighting A for four hours), while the vehicle-coupled heavy ULV generates a 75 dBA/4h noise inside the vehicle with closed windows and 110 dBA/4h outside the vehicle. Regarding the use of personal protective equipment, 27 (82%) EDCA reported using hearing protectors, also wearing a breathing mask, disposable clothing, hats, boots, waterproof gloves, and goggles. The Control Group (CG), comprised of 18 workers, males, aged over 48 years (mean = 56 years old; SD-5.6), from several occupational areas, with no history of exposure to chemical or physical agents.

During the hearing screening, two participants were excluded from the EG featuring mixed hearing loss, thus totaling a sample of 38 endemic disease control agents. No participants were excluded from the CG.

Data collection was performed on a single day, from 7:30 a.m. to 11:00, with groups of three to four workers per day/evaluated. In total, there were six months for data collection, according to the following steps: (1) external acoustic meatus screening, (2) conventional pure-tone audiometry and high-frequency audiometry, (3) immittance testing, (4) transient evoked and distortion-product otoacoustic emissions (TEOAE, and DPOAE), (5) suppression effect otoacoustic emissions, (6) brainstem evoked response audiometry (BERA), (7) dichotic digits test (DDT) in the step integrating bilateral and (8) the vestibular function evaluation is composed of many labyrinthine function and ocular tests. The first part of the evaluation was clinical and consisted of Brandt & Daroff's maneuver.

Data collection was carried out using descriptive statistics. Non-parametric tests were used to compare results between the groups (EG and CG). The results of the study's groups were compared through the *t*-test, Fisher's exact, chi-square, and Pearson correlation, according to each appropriate situation, with a significance level of 0.05 (5%).

## 3. Results

In the conventional audiometry, at frequencies between 250 Hz and 8000 Hz, there was no difference in the means of the tone thresholds between the groups, and there was no difference between the EG and CG in the means of the tone thresholds at high frequencies (9000–16,000 Hz), once p-value, measured by means of the Mann-Whitney statistical test, was greater than 0.05 (5%) for each analyzed frequency.

By verifying the occurrence of hearing loss in the conventional audiometry, in the EG, 15 (39.5%) right ears and 13 (34.2%) left ears were considered normal at all frequencies, while neurosensory hearing loss was evidenced in 23 (60.5%) right ears and 25 (65.8%) left ears. In the CG, 9 (50%) right ears and 9 (50%) left ears were considered normal at all frequencies, while neurosensory hearing loss was evidenced in 9 (50%) right ears and 9 (50%) left ears. Binaurally, greater occurrence of hearing loss in the EG than in the CG was observed. However, such differences were not considered significant by applying the Chi-Square Test (RE = p-value 0.46, and LE-p-value 0.26).

In the tympanometric findings for the EG, 31 right ears and 33 left ears were considered normal, that is, no alteration in the tympanus-bone system, while in the CG, 16 right and left ears were considered normal. Tympanometric alterations were AD, AS, and C-type curves. By comparing the findings in the tympanometry, G-Test showed differences between the EG and CG in the right ear (p = 0.0374), that is, there was greater number of tympanometric anomalies in the EG than in the CG. Between the groups, that difference was not evidenced in the left ear (p = 0.8232).

Regarding the acoustic reflex (stapedius) testing, ipsilateral and contralateral pathways, greater number of reflex absences in the EG was observed, as shown in **Table 1**.

By comparing the present/absent results of the acoustic reflex, by means of the Fisher's Exact Test, difference between the groups was observed in the right ear at 1000 Hz and 4000 Hz, ipsilateral pathway, and at 4000 Hz, contralateral pathway. In the left ear, that difference was observed at 2000 Hz, contralateral pathway.

By taking into consideration the normality of the tympanus-bone system, and the normality of the tone hearing thresholds between the frequencies of 500 Hz and

| Hertz | | RE (afference) | | *P* | LE (afference) | | *P* |
|---|---|---|---|---|---|---|---|
| | | Present | Absent | | Present | Absent | |
| 500 C | EG | 32 | 5 | 0.3508 | 29 | 7 | 0.1735 |
| | CG | 17 | 1 | | 17 | 1 | |
| 1000 C | EG | 30 | 7 | 0.1839 | 26 | 10 | 0.1486 |
| | CG | 17 | 1 | | 16 | 2 | |
| 2000 C | EG | 26 | 11 | 0.1156 | 25 | 11 | 0.0352* |
| | CG | 16 | 2 | | 17 | 1 | |
| 4000 C | EG | 22 | 15 | 0.0246* | 20 | 16 | 0.0962 |
| | CG | 16 | 2 | | 14 | 4 | |
| 500 I | EG | 30 | 6 | 0.0754 | 32 | 5 | 0.1253 |
| | CG | 18 | 0 | | 18 | 0 | |
| 1000 I | EG | 29 | 7 | 0.0471* | 30 | 7 | 0.0507 |
| | CG | 18 | 0 | | 18 | 0 | |
| 2000 I | EG | 30 | 6 | 0.0754 | 32 | 5 | 0.1253 |
| | CG | 18 | 0 | | 18 | 0 | |
| 4000 I | EG | 27 | 9 | 0.0177* | 26 | 11 | 0.1156 |
| | CG | 18 | 0 | | 16 | 2 | |

*Legend: C = contralateral; I = ipsilateral; RE = right ear; LE = left ear.*
*Fisher's Exact Test at the level of significance of 0,05 (significant p-value*).*

**Table 1.**
*Occurrence of presence/absence of the acoustic reflex, contralateral and ipsilateral pathways, among the participants of the exposed group (EG) and control group (CG), separated by right and left ears.*

4000 Hz, the TEOAE testing was held. In the EG, passing rates were 88.89% in the right ear, and 82.35% in the left ear, while in the CG, passing rates evidenced 92.31% in the right ear and 93.33% in the left ear. Despite greater passing rates in the CG in the TEOAE, when comparing the findings between the groups, there was no statistical difference, by means of the Fisher's Exact Test, p-value for the right ear = 0.64 and p-value for the left ear = 0.35. Similar finding was verified in the DPOAE testing, no statistical difference, by means of the Wilcoxon Test, between the groups, in the signal/noise relation at frequency bands of 1501, 2002, 3003, 4004, 6006, and general response ($p \geq 0.05$).

In the suppression effect of the TEOAE testing, greater occurrence of the suppression effect was verified in the participants of the CG than in the EG (**Figure 1**), expressed in percentages. However, by comparing the results of the suppression effect, present and absent, between the groups (EG and CG), by means of the Fisher's Exact Test (RE), and the Chi-Square Test (LE), no difference was evidenced between the groups, in the right ear ($p = 0.2478$), as well as in the left ear ($p = 0.5466$).

Regarding the findings of the brainstem evoked response audiometry (BERA), the statistical difference can be observed, using the student's t-test, between the groups (EG and CG) in the absolute latencies of waves III (right and left ears), and V (right ear), as well as in the latency interpeak I-III (right and left ears), and I-V (right ear) (**Table 2**).

In relation to the scores of the Dichotic Digits Test (DDT)—binaural integration stage, only participants with mean hearing thresholds up to 25 dB HL at frequencies from 500 to 4000 Hz were included in the DDT analyses. Hence, the exposed group comprised 30 participants, and the control group, 14 participants. The boxplot (mean, standard deviation, minimum, and maximum) for the right ear, left ear, and binaural DDT results are shown in **Figure 2**. There is a great variation in DDT results among those exposed, which did not happen in the nonexposed group.

Regarding the findings of the vestibular assessment, only carried out with the participants of the EG, 63.7% of the 33 EDCA featured vestibular testing within normal standards, 36.3% of them had alterations in the exams, of which 15.2%

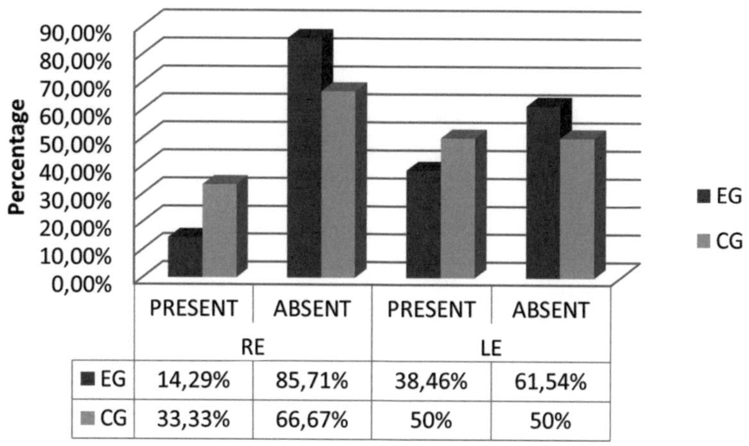


**Figure 1.**
*Percentage of occurrence of the suppression effect in the TOAE testing among the participants, right ears (RE) and left ears (LE). Legend: RE = right ear; LE = left ear; EG = exposed group; CG = control group.*

| BERA | | | Mean | SD | P |
|---|---|---|---|---|---|
| Wave I | RE | EG (N = 32) | 1.65 | 0.12 | 0.1686 |
| | | CG (N = 18) | 1.62 | 0.07 | |
| | LE | EG (N = 31) | 1.64 | 0.12 | 0.1018 |
| | | CG (N = 18) | 1.60 | 0.07 | |
| Wave III | RE | EG (N = 32) | 3.88 | 0.19 | 0.0090* |
| | | CG (N = 18) | 3.76 | 0.11 | |
| | LE | EG (N = 31) | 3.91 | 0.18 | 0.0013* |
| | | CG (N = 18) | 3.76 | 0.12 | |
| Wave V | RE | EG (N = 32) | 5.82 | 0.30 | 0.0185* |
| | | CG (N = 18) | 5.66 | 0.13 | |
| | LE | EG (N = 31) | 5.82 | 0.27 | 0.0843 |
| | | CG (N = 18) | 5.72 | 0.18 | |
| Interpeak I-III | RE | EG (N = 32) | 2.24 | 0.19 | 0.0403* |
| | | CG (N = 18) | 2.15 | 0.13 | |
| | LE | EG (N = 31) | 2.27 | 0.16 | 0.0041* |
| | | CG (N = 18) | 2.15 | 0.12 | |
| Interpeak III-V | RE | EG (N = 32) | 1.94 | 0.19 | 0.1595 |
| | | CG (N = 18) | 1.89 | 0.12 | |
| | LE | EG (N = 31) | 1.91 | 0.20 | 0.1739 |
| | | CG (N = 18) | 1.96 | 0.13 | |
| Interpeak I-V | RE | EG (N = 32) | 4.17 | 0.30 | 0.0450* |
| | | CG (N = 18) | 4.04 | 0.14 | |
| | LE | EG (N = 31) | 4.18 | 0.22 | 0.1328 |
| | | CG (N = 18) | 4.11 | 0.19 | |
| Amplitude I' | RE | EG (N = 32) | 0.12 | 0.09 | 0.4443 |
| | | CG (N = 18) | 0.09 | 0.04 | |
| | LE | EG (N = 31) | 0.13 | 0.06 | 0.1478 |
| | | CG (N = 18) | 0.11 | 0.07 | |
| AmplitudeV' | RE | EG (N = 32) | 0.25 | 0.12 | 0.5000 |
| | | CG (N = 18) | 0.25 | 0.10 | |
| | LE | EG (N = 31) | 0.27 | 0.11 | 0.1206 |
| | | CG (N = 18) | 0.23 | 0.12 | |

*Student´s t-test at the level of significance of 0.05 (significant p-value*).*
*Legend: RE = right ear; LE = left ear; EG = exposed group; CG = control group; SD = standard deviation.*

**Table 2.**
*Mean and standard deviation of the absolute latencies, interlatencies, and amplitudes of waves I and V, right and left ears of the exposed (EG) and control (CG) groups.*

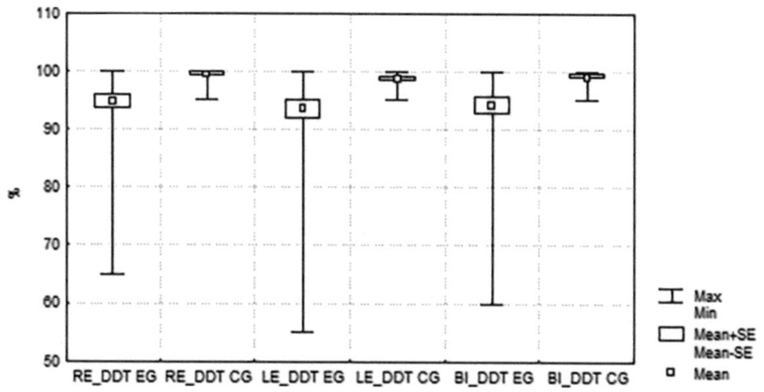


**Figure 2.**
*Boxplot of the exposed group's (EG) and control group's (CG) participants' score in the Dichotic Digits Test (DDT) of the right ear (RE), left ear (LE), and binaural (BI).*

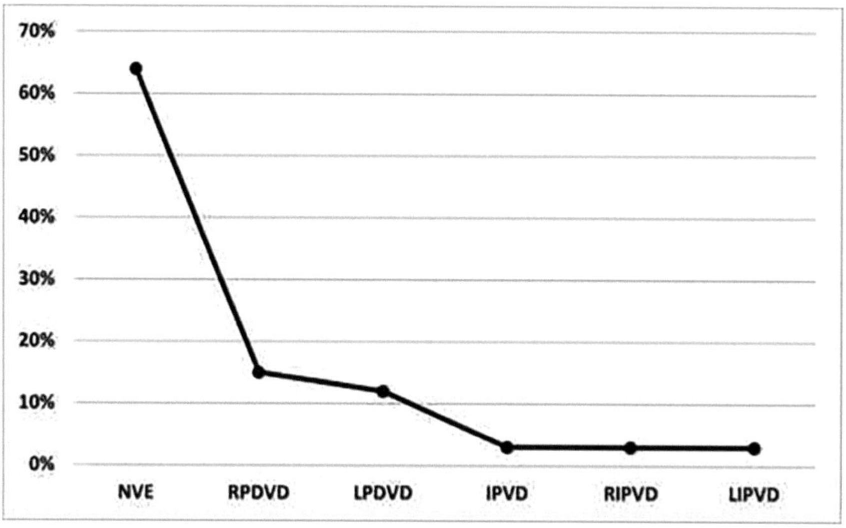


**Figure 3.**
*Vestibular examination result (N=33). Legend: NVE = normal vestibular exam; RPDVD = right deficit peripheral vestibular dysfunction; LPDVD = left deficit peripheral vestibular dysfunction; IPVD = irritative peripheral vestibular dysfunction; RIPVD = right irritative peripheral vestibular dysfunction; LIPVD = left irritative peripheral vestibular dysfunction.*

presented with right deficit peripheral vestibular dysfunction (n = 5), 12.1% with left deficit peripheral vestibular dysfunction (n = 4), and 9% presented irritative peripheral vestibular dysfunction (**Figure 3**). The prevalence rate of altered results was the 12/33 (p = 0.364).

By Fisher's Exact Test, there was no significant statistical correlation between age range (p = 0.1132) and time of exposure to risk agents (p = 0.2825). However, by means of the Mann-Whitney non-parametric test, the participants who evidenced worse auditory thresholds in the right ear at the frequency of 4000 Hz (p = 0.0494), also featured abnormal results in the vestibular assessment.

## 4. Discussion

The results in the current study suggest that the simultaneous exposure to noise and pesticides (used in the Public Health) possibly affected some areas of the peripheral and central auditory system, as well as of the peripheral vestibular system in endemic disease control agents.

However, the impact of that exposure on conventional and high-frequency auditory thresholds on the cochlear physiology, on the efferent medial auditory system, and on the central vestibular system, was not confirmed in the current study, probably due to the size and/or age range of the sample. To confirm the effects on those areas, the use of a similar protocol would be interesting in further studies, with a larger and younger sampling, being held in the country or abroad.

The results of the effects of pesticides associated to noise in the peripheral auditory system showed that there was no difference in the means of the auditory thresholds in the conventional and high-frequency audiometry between the studied groups. Similar results were evidenced in another study [19].

Regarding the tympanometric findings, in the EG, greater number of tympanometric abnormalities were observed in the right ear than in the CG, thus, pesticides may affect the middle ear cavity. Even though this result is observed in other studies with pesticide-exposed populations [31, 32], that finding should be further investigated.

Concerning the findings of the acoustic reflex, EG participants evidenced greater number of absent cases than the CG. Similar findings were observed in other studies with populations exposed to otoagressive agents [19, 31–33]. It can be inferred that, despite the presence of neurosensory hearing loss, there were worse results in the acoustic reflex among the population exposed to pesticides and noise, and the exposure to such harmful agents may lead to damages in the afferent hearing pathways.

In relation to the findings of the evoked otoacoustic emissions, there was no statistical difference regarding signal/noise in the transient stimuli, as well as in the product of distortion. However, it was possible to observe greater levels of responses in the group of participants not exposed to pesticides and noise. Similar observation was verified in the pass/fail of the TEOAE. That finding can be justified by the age factor of the studied population (EG and CG), as all organs undergo organic changes along the years [34].

Regarding the effects of the pesticides associated to noise in the central nervous system, assessed by means of the brainstem evoked response audiometry (BERA), dichotic digits testing (DDT) and suppression effect of the otoacoustic emissions, results evidenced greater impact of pesticides associated to noise on the BERA and DDT, with worse results for the group of participants exposed to pesticides and noise than in the control group. These findings evidence the fact that the central hearing functions of the exposed population have been impaired using pesticides associated to noise.

When assessing the central auditory system of endemic disease combat agents who are exposed to pyrethroid and organophosphate insecticides, the authors identified 56% of central auditory dysfunction, whose relative risk was 7.58. Similar results were observed in other studies with farmworkers exposed to organophosphate pesticides [14] and herbicides, insecticides, and fungicides [17]. Through the long-latency potential (P300), authors verified an increase in the latency of farmworkers exposed to organophosphate insecticides [13]. Such a result suggests that chronic exposure to the pesticide can delay the neurophysiological processes and alter the central auditory system. The same results were observed in a study involving 14 workers responsible for spraying organophosphate insecticides [20].

However, in the suppression effect testing, no difference was observed between the studied groups, which can be attributed to the age range of the sampling in the current study, as age increases, mainly from 60 years and over, there is significant reduction in the suppression effect of the otoacoustic emissions, fundamentally when ipsilateral and contralateral effects are assessed [35].

Concerning the findings of the vestibular screening, 1/3 of the endemic disease control agents were observed to feature peripheral vestibular abnormalities, related to the anterior and posterior labyrinth, once there was statistical difference between the tonal auditory thresholds, at the frequency of 4000 Hz in the right ear, and the abnormal results of the vestibular screening. This finding may be consistent with Cochlear-Vestibular Syndrome. This known fact in the literature justifies the importance of researching the system's integrity through the auditory exams and the vestibular exams [36]. In a study, hearing normality was verified by conventional audiological evaluation, among 61.14% of 18 rural workers exposed to organophosphate insecticides. While 16 workers had irritative peripheral body balance disorder and seven workers had sensorineural hearing loss, thus suggesting that agricultural pesticides cause vestibular alterations through a slow and silent intoxication [37].

## 5. Conclusions

The results presented lead to the conclusion that exposure to pesticides and noise (used in Public Health) possibly affected some areas of the peripheral and central auditory system, as well as of the peripheral vestibular system in endemic disease control agents. And induces harmful effects on the central auditory functions, particularly on the brainstem and figure-ground speech-sound auditory skill, identified through the brainstem auditory evoked potentials and the dichotic digits test. The most common peripheral vestibular effect was of the deficit type, revealing the chronicity of the condition.

## Conflict of interest

The authors declare no conflict of interest.

## Author details

Adriana Bender Moreira de Lacerda[1*], Patrícia Arruda de Souza Alcarás[2], Maria Cristina Alves Corazza[3], Adrian Fuente[1] and Bianca Simone Zeigelboim[4]

1 Université de Montréal, Québec, Canada

2 Tuiuti University of Paraná, Presidente Prudente, Brazil

3 University of Western São Paulo, Presidente Prudente, Brazil

4 Tuiuti University of Paraná, Curitiba, Brazil

*Address all correspondence to: adriana.lacerda@umontreal.com

## References

[1] Koelle GB. Pharmacology of organophosphates. Journal of Applied Toxicology. 1994;**14**(2):105-109. DOI: 10.1002/jat.2550140211

[2] Sidell FR. Clinical effects of organophosphorus cholinesterase inhibitors. Journal of Applied Toxicology. 1994;**14**:111-113. DOI: 10.1002/jat.2550140212

[3] França DMV, Lacerda ABM, Lobato D, Ribas A, Dias KZ, Leroux T, et al. Adverse effects of pesticides on central auditory functions in tobacco growers. International Journal of Audiology. 2017;**56**(4):233-241. DOI: 10.1080/14992027.2016.1255787

[4] Singh M, Minhas RS, Machhan P, Azad RK, Mohindroo S. Audiological assessment in organophosphorus poisoning. International Journal of Otorhinolaryngology and Head and Neck Surgery. 2018;**4**:1-4

[5] Cáceres T, Megharaj M, Venkateswarlu K, Sethunathan Nethunathan N, Naidu R. Fenamiphos and related organophosphorus pesticides: Environmental fate and toxicology. Reviews of Environmental Contamination and Toxicology. 2010;**205**:117-162. DOI: 10.1007/978-1-4419-5623-1_3

[6] Secretaria de Estado da Saúde do Paraná – SESA/PR. Protocolo de avaliação das intoxicações crônicas por agrotóxicos. Superintendência de Vigilância em Saúde; Curitiba. Brazil; 2013; p. 75

[7] Teixeira CF, Augusto LGS, Morata TC. Saúde auditiva de trabalhadores expostos a ruído e inseticidas. Revista de Saúde Pública. 2003;**37**(4):417-423

[8] Körbes D, Silveira AF, Hyppolito MA, Murano G. Alterações no sistema vestibulococlear decorrentes da exposição ao agrotóxico: Uma revisão de literatura. Revista da Sociedade Brasileira de Fonoaudiologia. 2010;**15**(1):146-152

[9] Kós MI, Hoshino AC, Asmus CIF, Mendonça R, Meyer A. Efeitos da exposição a agrotóxicos sobre o sistema auditivo periférico e central: Uma revisão sistemática. Cadernos de Saúde Pública. 2013;**29**(8):1491-1506. DOI: 10.1590/0102-311X00007013

[10] Gatto MP, Fioretti M, Fabrizi G, Gherardi M, Strafella E, Santarelli L. Effect of potential neurotoxic on hearing loss: A review. Neurotoxicology. 2014;**42**:24-32. DOI: 10.1016/j.neuro.2014.03.009

[11] Teixeira CF, Augusto LGS, Morata TC. Occupational exposure to insecticides and their effects on the auditory system. Noise Health. 2002;**4**(14):31-39

[12] Lizardi OS, O'Rourke MK, Morris RJ. The effects of organophosphate pesticide exposure on Hispanic children's cognitive and behavioral functioning. Journal of Pediatric Psychology. 2008;**33**(1):91-101. DOI: 10.1093/jpepsy/jsm047

[13] Dassanayake T, Gawarammana IB, Weerasinghe V, Dissanayake PS, Pragaash S, Dawson A, et al. Auditory event-related potential changes in chronic occupational exposure to organophosphate pesticides. Clinical Neurophysiology. 2009;**120**(9):1693-1698. DOI: 10.1016/j.clinph.2009.07.034

[14] Camarinha CR, Frota SMMC, Pacheco-Ferreira H, Lima MAMT. Avaliação do processamento auditivo

temporal em trabalhadores rurais expostos ocupacionalmente a agrotóxicos organofosforados. Jornal da Sociedade Brasileira de Fonoaudiologia. 2011;**23**(2):102-106

[15] Jayasinghe SS, Pathirana KD. Effects of deliberate ingestion of organophosphate or paraquat on brain stem auditory-evoked potentials. Journal of Medical Toxicology. 2011;**7**(4):277-280. DOI: 10.1007/s13181-011-0173-3

[16] Andrade MIKP. Efeitos da exposição ao agrotóxico no sistema auditivo eferente através das emissões otoacústicas transientes com supressão [thesis]. Rio de Janeiro: Federal University of Rio de Janeiro, Institute of Public Health Studies; 2012

[17] Bazílio MMM, Frota S, Chrisman JR, Meyer A, Asmus CIF, Camara VM. Processamento auditivo temporal de trabalhadores rurais expostos a agrotóxicos. Jornal da Sociedade Brasileira de Fonoaudiologia. 2012;**24**(2):174-180. DOI: 10.1590/S2179-64912012000200015

[18] Murthy VA, Reddy IJV. Audiological assessment in organophosphorus compound poisoning. Indian Journal of Otolaryngology and Head & Neck Surgery. 2012;**66**(1):22-25. DOI: 10.1007/s12070-012-0547-0

[19] França DMV. Efeitos do uso dos agrotóxicos no sistema auditivo central dos fumicultores da Região do Centro-Sul do Paraná [thesis]. Curitiba: Tuiuti University of Paraná; 2013

[20] Delecrode CR. Processamento auditivo em trabalhadores expostos a ruído e inseticidas: testes de ordenação temporal e P300 [dissertation]. Marília: Paulista State University; 2014

[21] Alcarás PAS, Zeigelboim BS, Corazza MCA, Lüders D, Marques JM, Lacerda ABM. Vestibular function evaluation in endemic diseases combat agentes. Health. 2021;**13**(2):144-156. DOI: 10.4236/health.2021.132013

[22] Alcarás PAS, Zeigelboim BS, Corazza MCA, Lüders D, Marques JM, Lacerda ABM. Finding on the central auditory functions of endemic diseases control agents. International Journal of Environmental Research and Public Health. 2021;**18**(13):7051. DOI: 10.3390/ijerph18137051

[23] Sena TRR, Dourado SSF, Lima LV, Antoniolli AR. The hearing of rural workers exposed to noise and pesticides. Noise & Health. 2018;**20**(92):23-26

[24] Murakami Y, Pinto NF, Albuquerque GSC, Perna P, Lacerda A. Chronic pesticide poisoning in the tobacco farming. Saúde Debate. 2017;**41**(113):563-576

[25] França DMVR, Lobato DCB, Moronte EA, Albuquerque GSC, Alcarás PA, Gonçalves CGO, et al. Study on hearing loss and its relationship with work in pesticide-exposed tobacco growers. Revista CEFAC. 2020;**22**(3):1-9. DOI: 10.1590/1982-0216/20202231151

[26] Lima EP, SMB L, MIM A, LHS A, Neves KT, Maia ER. Pesticide exposure and its repercussion in the health of sanitary agents in the State of Ceará, Brazil. Ciência & Saúde Coletiva. 2009;**14**(6):2221-2230

[27] Morata TC, Little MB. Suggested guidelines for studying the combined effects of occupational exposure to noise and chemicals on hearing. Noise & Health. 2002;**4**(14):73-87

[28] Carneiro FF, Rigotto RM, Augusto LGS, Friedrich K, Búrigo AC,

(orgs). Dossiê ABRASCO: Um alerta sobre os impactos dos agrotóxicos na saúde. Expressão Popular: São Paulo. EPSJV: Rio de Janeiro. First edition. 2015:624

[29] Torres R. Agente de combate a endemias. Revista Poli – saúde, educação e trabalho. 2009;**3**:16-17

[30] OSHA. Occupational Safety and Health Administration [Internet]. c2018. Available from: https://www.osha.gov/news/newsreleases/infodate-y/2018 [Accessed: April 20, 2019]

[31] Tomiazzi JS, Pereira DR, Judai MA, Antunes PA, Favareto APA. Performance of machine-learning algorithms to pattern recognition and classification of hearing impairment in Brazilian farmers exposed to pesticide and/or cigarette smoke. Environmental Science and Pollution Research. 2018;**26**(7):6481-6491. DOI: 10.1007/s11356-018-04106-w

[32] Judai MA. Estudo dos limiares auditivos em frequências convencionais e altas frequências de moradores e trabalhadores expostos a praguicidas na região do Pontal do Paranapanema [dissertation]. Presidente Prudente: University of Western São Paulo; 2015

[33] Lobato DCB. Disfunção auditiva induzida por agrotóxicos em trabalhadores agrícolas do Paraná [thesis]. Curitiba: Tuiuti University of Paraná; 2015

[34] Calais LL, Borges ACLC, Baraldi GS, Almeida LC. Efeitos do envelhecimento na função coclear. Distúrb Comum. 2007;**19**(3):365-373

[35] Hood LJ. The role of otoacoustic emissions in identifying carriers of hereditary hearing loss. In: Berlin C.I. Otoacoustic Emissions Basic Science and Clinical Applications. San Diego: Sing. Pub.Group; 1998

[36] Louren EA, Lopes KC, Pontes-Jr A, Oliveira MH, Umemura A, Vargas AL. Distribution of neurotological findings in patients with cochleovestibular dysfunction. Brazilian Journal of Otorhinolaryngology. 2005;**71**:288-296. DOI: 10.1016/S1808-8694(15)31325-2

[37] Hoshino ACH, Pacheco-Ferreira H, Taguchi CK, Tomita S, Miranda MF. Ototoxicity study in workers exposed to organophosphate. Revista Brasileira de Otorrinolaringologia. 2008;**74**(6):912-918

Chapter 8

# Biological and Molecular Effects of Pesticides on Human Health

*Aida Doroudian, Mahdieh Emadi, Reza Hosseinzadeh and Parvaneh Maghami*

## Abstract

Pesticides are widely used in agriculture and are practical and economical to improve the quality of food safety for the permanent population around the world. Even though insecticides are beneficial to cropping views, their extensive use may result in severe consequences due to their biocompatible and permanent nature. Various pesticides can cause serious health risks of direct or indirectly contaminated air, water, soil, and the general ecosystem. The effect of pesticides on blood in the mammalian cell is significant because blood can act as a target and carrier for pesticides. However, the mechanism by which they bind to biopolymers, particularly blood proteins, is not clearly understood yet. This chapter investigates the molecular effects of pesticides on biomacromolecules, especially hemoglobin.

**Keywords:** pesticides, hemoglobin, biomolecule

## 1. Introduction

Pesticides are an extensive range of materials to destroy, control, and protect plants from any pest, due to the wide range of applications of pesticides in agriculture, industry, and households.

They are the most common chemical that people have a risk of exposure to them. There are two groups of pesticides based on their origins: chemical pesticides and biological pesticides. Chemical pesticides act nonspecific and affect many off-target organisms, while biopesticides operate host-specific. Various pesticides are shown based on their origin or target insect and function [1]. Almost 2 million tons of pesticides are used worldwide each year, overgrowing. Pesticides directly or indirectly contaminate air, water, soil, and entire ecosystems, posing a severe threat to the health of living things [2, 3].

At present, just about 2 million tons of pesticides are used worldwide, and 47.5% belong to herbicides, 29.5% to insecticides, 17.5% to fungicides, and 5.5% to other pesticides [4]. Pesticides can enter the body in different ways, such as through direct contact, digestion, or inhalation. As pesticides enter the body, they enter the blood circulation and the entire body.

Pesticides may enter life forms totally in different ways. Due to contrasts in the digestion system and other characteristics, species, strains, and individuals may significantly change their defenselessness to pesticides. Oceanic life forms may retain chemicals specifically from the water over respiratory organs (e.g., gills), the body surface, or utilize admissions of sullied nourishment, suspended particles, or sediments [5].

The overall passing and constant illnesses because by pesticide harm number around 1 million every year. The pesticide's high gamble bunches incorporate creation laborers, formulators, sprayers, blenders, loaders, and rural homestead laborers. During production and formulation, the chance of perils might be higher because the cycles implied are not hazard-free. In modern settings, laborers are at expanded hazard since they handle harmful synthetics, including pesticides, poisonous solvents, and inactive transporters [6].

The human well-being impacts of pesticides can happen through one of three courses: ingestion, inward breath, and skin contact that occur through pesticide items. Microorganisms in water and soil are the primary natural system of pesticides debasement. In contrast, the moment is the pesticide's digestion system when living life forms expend it as the portion of their nourishment taken up [7].

The sum of the chance of pesticide introduction depends on the harmfulness and the opening of the pesticide. Harmfulness may be a degree of how destructive or harmful a pesticide is (causing sickness or other undesirable impacts). In contrast, the introduction may be a degree of contact (length) with a pesticide. The poisonous quality is measured as a deadly dosage (LD50). The LD50 esteem is the factual assessment of a pesticide (mg/kg of body weight) that can murder 50% of the test creatures within an expressed period (24 hours to 7 days). The LD50 esteem moreover depends on the course of section of a pesticide; oral LD50 for oral ingestion, dermal LD50 for skin contact introduction, and deadly concentration (LC50) for the inward breath of fumigants and pesticide vapors [8].

Because blood is the body's central circulatory system that can transport a variety of substances to the organs; therefore, it is essential to understand the effects of pesticides on blood and hemoglobin. Investigating the effects of pesticides on biopolymers can help to understand the molecular mechanisms and the hazardous effects of these compounds. Therefore, this chapter aims to study the molecular impact of pesticides.

## 2. Effects of pesticides on cell and metabolism

Pesticides can affect enzyme actions and metabolic pathways related to the whole-cell function (**Figure 1**) [9].

A few pesticides can essentially diminish the action of NADH-dehydrogenase—the fundamental compound of the mitochondrial electron transport chain. The weakness of NADH-dehydrogenase movement by chlorpyrifos might intervene oxidative stress and neurotoxicity. In addition, pesticides can induce the generation of reactive oxygen species (ROS) and receptive nitrogen species (RNS) in cells, which at last prompts oxidative stress and harm to cell structures. Moreover, increased ROS/RNS in vertebrates during digestion and biotransformation of poisonous substances caused hepatotoxicity [9].

Low degrees of pesticides might create an assortment of biochemical changes, some of which might be answerable for the antagonistic organic impacts announced in people and creatures. The harmfulness of pesticides could influence biological organ capacities and biochemical dysfunctions. It was reported a nephrotoxic change in specialists' occupational exposure to pesticides. Changed liver enzymes, like serum alanine aminotransferase (ALT), and aspartate aminotransferase (AST), have been accounted for among pesticide laborers presented to pesticides [10].

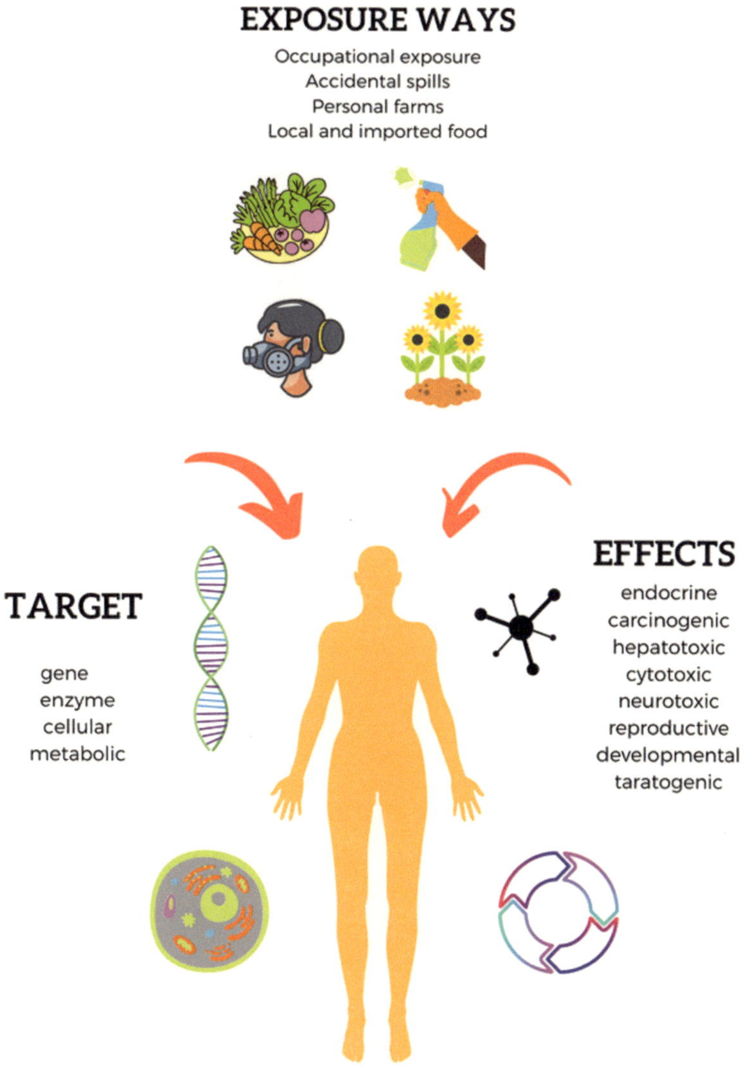


**Figure 1.**
*Pesticides: Formulates, distribution pathways, and impacts on human well-being [9].*

The interaction between different pesticides may result in numerous responses, depending on contrasts within each compound's chemical properties and modes of poisonous activity. For an improved understanding of the toxic quality of pesticide blends, it is essential to have enough knowledge of the chemical reactivity, the toxicokinetics, metabolic pathways, and the components of activity of each compound. An epidemiological view of physiologically based toxicokinetic and toxicodynamic models, factual modeling, and computational (in silico) toxicology approaches can be used to assess toxicological intelligence [11].

Toxicity effects of pesticides can happen on the activation mechanism of the enzymes at low concentrations; for instance, methyl parathion which is broadly utilized in rural fields, inhibits carbonic anhydrase (CA) and bovine erythrocytes demonstrating that fish in natural and cultural environments are powerless to this pesticide and that methyl parathion pollutions would cause fish and bovine deaths [12].

A few pesticides could regulate the action of efflux carriers or compounds engaged with xenobiotic digestion, prompting an adjustment of the bioavailability and toxicity of other xenobiotics.

Pesticides directly impact a few cell processes and essential proteins engaged with general digestion, cell development, differentiation, and endurance [13].

Moreover, specialists showed a positive connection between openness to pesticides and improvement of certain diseases, especially the brain, prostate, kidney malignant growths, and NHL and leukemia. Some of the examinations on kids observed expanded hazards of illness related to primary times of openness, pre-birth and post-pregnancy, and parental openness at work. Many studies showed developed danger and dose-response connections [14].

Different studies show cytotoxicity of the most commonly used pesticides in agriculture on human cell lines. In 2014, Mesnage et al. designed a toxicity test with nine pesticides on three cell lines, including HepG2, HEK293, and JEG3. The results agreed with cytotoxicity after 24 treatments with pesticides by assessing apoptosis and necrosis [15].

Also, similar results were seen by exposing prostate epithelial WPM-Y.1 cell line with imidacloprid and herbicide glyphosate in the study of Abdel-Halim and Osman [16].

Moreover, the study by Abhishek et al. showed the toxicity of parathion methyl (PM) and carbofuran (CN) pesticides on human keratinocyte (HaCaT) cell lines through MTT assay [17].

## 3. Molecular effects of pesticides

Besides the effect of pesticides on cells, metabolic systems, and others, still few studies have been done in vitro and in silico analysis that demonstrates these chemical pesticides have a damaging outcome in genetic and epigenetic alterations, specifically on people who are exposed to them by DNA methylation and miRNA expression [18].

DNA methylation status and miRNAs' over-expression are linked to crucial cell and molecular pathways, leading to different human diseases [19].

Genetic susceptibility has accounted for modulating the degree of genotoxic hazard. Many investigations have shown a relationship between DNA harm and glutathione-S-transferase polymorphisms [20].

It was known that oxidative stresses because of ROS created by pesticides disrupt DNA and its repair instrument, prompting transformation and illnesses. DNA fix components help to correct the DNA harm brought about by pesticides. Hard well-being impacts because of pesticides range from intense to constant sicknesses, for example, malignant growth, birth absconds, neurological imperfections, reproductive deformities, and immunotoxicity [21].

Oxidative stress is a potential mechanism of toxicity that assumes a critical part in the toxicological pathway of various classes of pesticides, most likely because of their digestion or mitochondrial interruption [22].

### 3.1 Effects of different pesticides on hemoglobin

In explaining the destructive effect of pesticides on proteins, extensive experiments have been performed on the interaction of pesticides with hemoglobin as a vital protein.

DOI: http://dx.doi.org/10.5772/intechopen.104811

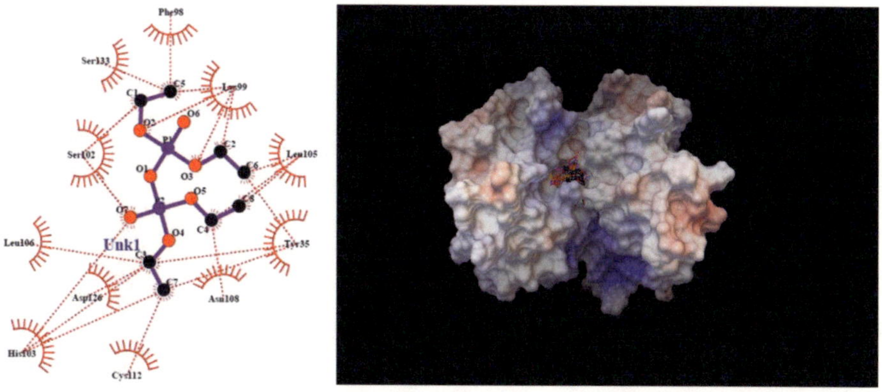

(a) (b)

**Figure 2.**
*(a) Ligplot analysis of the interaction between TEPP and hemoglobin [23]. (b) Penetration of TEPP through hydrophobic pocket-3D model [23].*

### *3.1.1 Interaction of tetraethyl pyrophosphate with hemoglobin*

TEPP[1] can dissolve REC and enter red blood cells; also, it can interact with the heme prosthetic group (internal) and induce heme degradation when interacting with Hb.[2]

LIGPLOT analysis of this interaction (**Figure 2a**) shows that TEPP interacts hydrophobically with aromatic moieties. Docking studies also confirmed the penetration of TEPP into the hydrophobic pocket (**Figure 2b**). The increase in hydrophobicity around aromatic moieties induces a redshift of the globin moiety, as seen in the results of UV-Vis experiments (**Figure 3**).

The blue shift of the Soret (**Figure 4**) band observed is a result of the action of the pesticide on the hydrophobic pocket of Hb. Negative values of Gibbs free energy indicate spontaneous binding of TEPP to Hb. Oxygen affinity measurements and fluorescence studies have shown that this is due to the interaction of TEPP with Hb.

The concentrations of Hb variants (i.e., deoxy-Hb and metHb) increased and [oxyHb] decreased, suggesting that the oxygen transport capacity of Hb was reduced due to the formation of heme degradation products (**Figure 5**). The ATR-FTIR study showed that tetraethyl pyrophosphate could alter the secondary structure of hemoglobin by reducing the alpha-helix content [23].

### *3.1.2 Interaction of Cartap with hemoglobin*

Carbamate insecticide has a lethal effect on the structure and function of human hemoglobin. As shown in (**Figure 6a–c**), changes in Hb absorption were observed in the 200–700 nm range in various concentrations of Cartap.

In addition, the absorption intensity increased at 280 nm, and a bathochromic effect was observed due to the interaction of Cartap and globin through hydrophobic interaction and the change in shape, which was confirmed by molecular docking analysis. In addition, a decrease in the Soret band and Q peak was observed,

[1] Tetraethyl pyrophosphate

[2] Hemoglobin

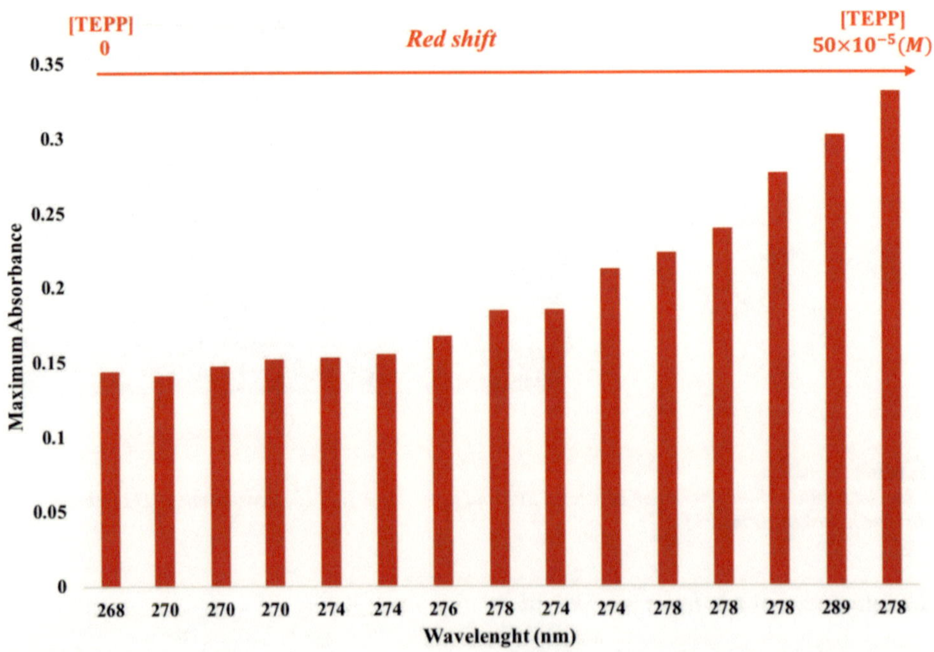


**Figure 3.**
*Red shift in globin region of hemoglobin UV-Vis spectrum due to interaction with TEPP [23].*

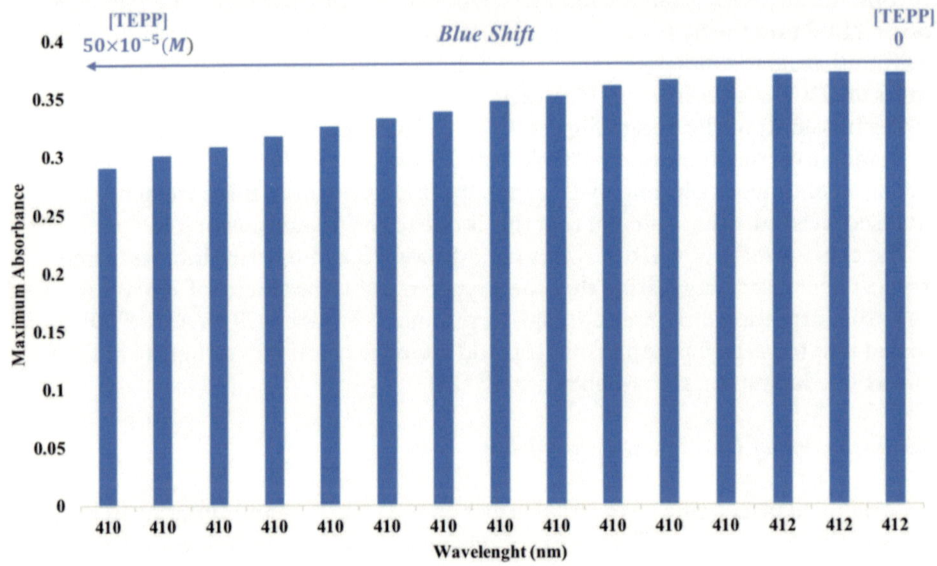


**Figure 4.**
*Blue shift in Soret band of hemoglobin UV-Vis spectrum due to interaction with TEPP [23].*

indicating the affinity of hemoglobin for oxygen in the presence of turbulence in the heme medium and Cartap.

This study shows that the Cartap cause loses standard functionality and negatively affects oxygen affinity and transport. Also, based on thermodynamic analysis (**Figure 7**), HB's stability is reduced in the presence of Cartap. According to the

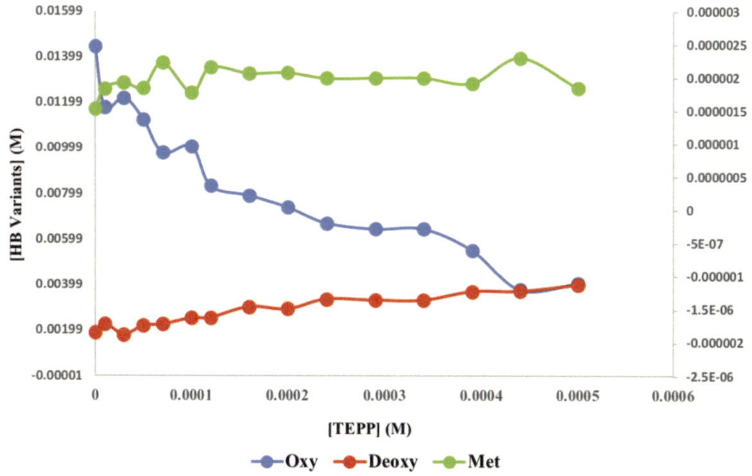


**Figure 5.**
*Formation of different hemoglobin variants due to interaction with Hb [23].*

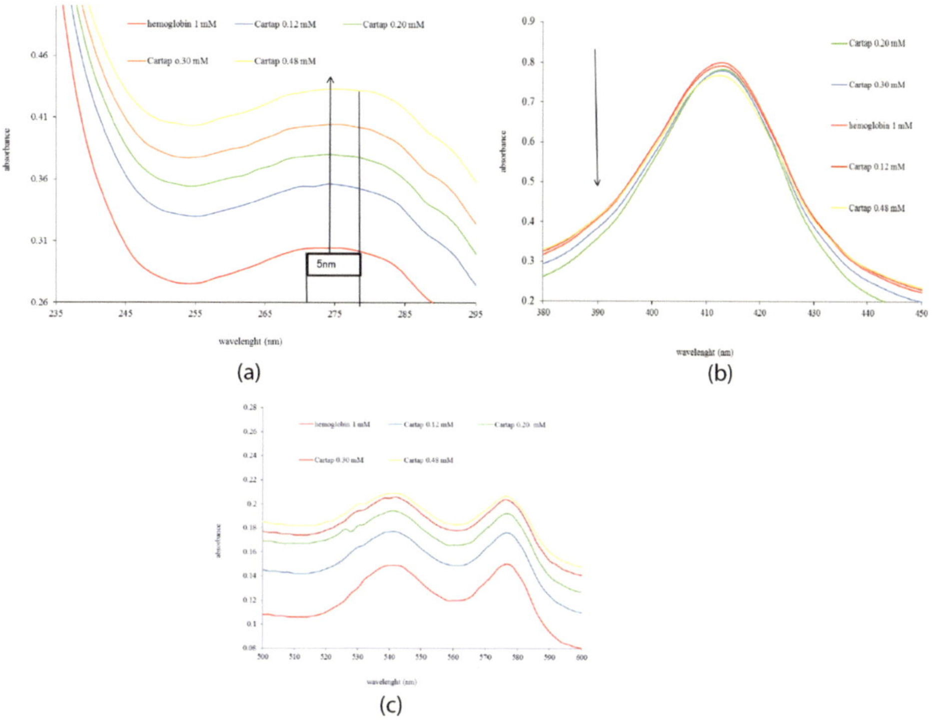


**Figure 6.**
*(a) UV-Vis spectral changes and red shift in globin region due to interaction with different concentrations of Cartap hydrochloride with hemoglobin [24]. (b) UV-Vis spectral changes in the Soret region due to interaction with different concentrations of Cartap hydrochloride with hemoglobin [24]. (c) UV-Vis spectral changes and blue shift in Q bands due to interaction with different concentrations of Cartap hydrochloride with hemoglobin [24].*

molecular docking, Hb has two binding sites Cartap hydrochloride (**Figure 8**), and is effectively related to proteins through hydrogen bonding and pocket residual water keys. According to the results, hydrogen bonding and hydrophobic interactions play

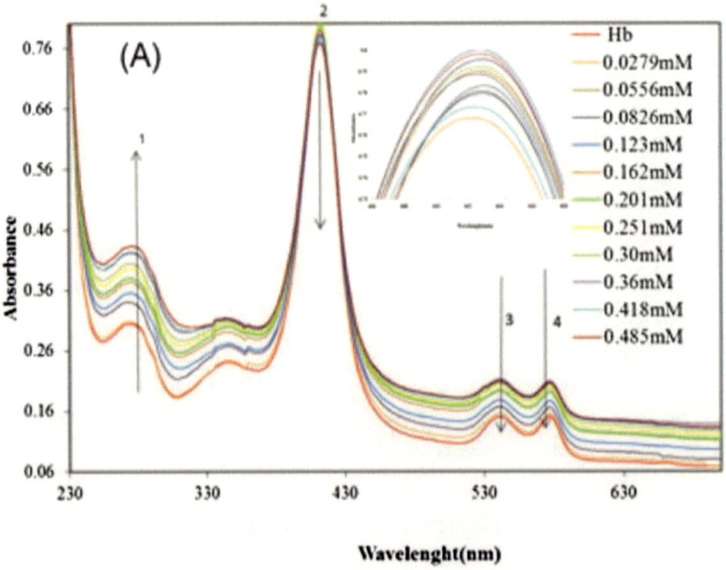


**Figure 7.**
*Thermal denaturation of hemoglobin upon interaction with different concentrations of Cartap hydrochloride [24].*

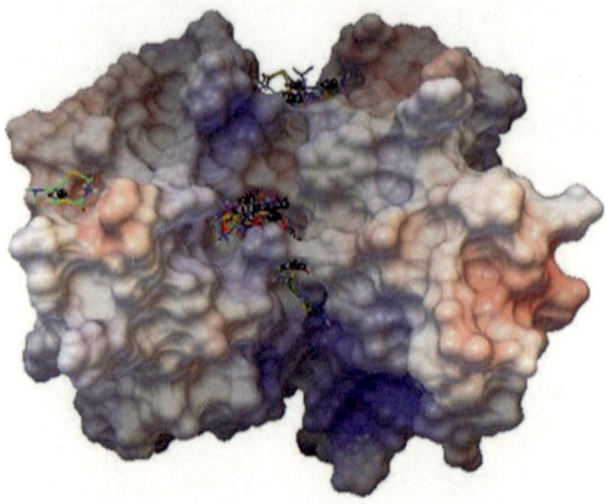

**Figure 8.**
*Docking 3D model of the interaction between Cartap and hemoglobin [24].*

an essential role in the interaction of Cartap with HB, which can denature protein structures.

These results show that the interaction of Cartap and hemoglobin results in structural and functional changes in hemoglobin and porphyrin [24].

### *3.1.3 Interaction of Chlorpyrifos and Cypermethrin with bovine hemoglobin and bovine serum albumin*

Titration experiments showed that the fluorescence intensity of the BSA gradually decreased while the fluorescence intensity of the reaction system containing BHb[3] increased gradually due to interaction with cypermethrin. The maximum emission wavelength was constant at around 340 nm. That is, there was no red or blue shift.

[3] Bovine hemoglobin

Finally, Chlorpyrifos and Cypermethrin were able to bind BSA and Bovin Hb, and both pesticides bind to Albumine much more potent than that hemoglobin [25].

#### *3.1.4 Interaction of Paraquat with bovine hemoglobin*

The reactivity of the heme center with the superoxide anion formed by paraquat is judged by the decrease in the Soret band, and all four heme groups associated with hemoglobin are damaged and eventually destroyed by the superoxide anion formed by the PC.

UV/Vis absorption and synchronous fluorescence spectroscopy revealed that the environmental structure of these Trp (tryptophan) residues was altered and that the results showed that the presence of one class of binding sites on BHb, hydrophobicity, and electrostatic interactions play an essential role in the stabilization of the complex [26].

#### *3.1.5 Interaction of Imidacloprid with hemoglobin*

Ding et al. [27] investigated the binding of Imidacloprid with hemoglobin. They showed that Imidacloprid quenched hemoglobin's intrinsic fluorescence via the static quenching process. The values of enthalpy ($\Delta H = -14.58$ kJ $mol^{-1}$) and entropy ($\Delta S = 32.83$ J $mol^{-1}$ $K^{-1}$) of the reaction indicate that hydrophobic interactions and hydrogen bonding are the dominant intermolecular forces in stabilizing the Imidacloride-Hb complex.

## 4. Insecticide resistance

There are two mechanisms for insecticide resistance: behavioral and physiological. In behavioral resistance, the insect's reaction reduces or prevents exposure that can lead to death. Otherwise, there are different types of modification mechanisms in physiological resistance like decreasing cuticular penetration and target site sensitivity or increasing metabolic detoxification [28]. To explain more, a common feature of insecticide metabolic resistance is the overexpression of detoxification genes at the transcriptional level, leading to high levels of protein and enzymatic activity. Therefore, this detoxification and resistance development level [29]. One of the notable examples of physiological resistance is malaria, which still exists in some African countries like Tanzania, while chemical insecticides are used against them. The straightforward reason is that target-site insensitivity (knockdown resistance' target-site mutations) in malaria vectors, lower penetration, or an enhanced detoxification activity [30].

## 5. Biological control strategies of pesticides

Biopesticides are usually happening compounds or agents acquired from creatures, plants, and microorganisms like microbes, cyanobacteria, and microalgae and are utilized to control farming nuisances and pathogens. There are many kinds of biopesticides, and they are arranged by their extraction sources and the sort of molecule/compound utilized for their readiness. The classifications are microbial pesticides, biochemical pesticides, insect pheromones, plant-based extracts, essential oils, insect growth regulators, and hereditarily adjusted creatures (GMOs) [31].

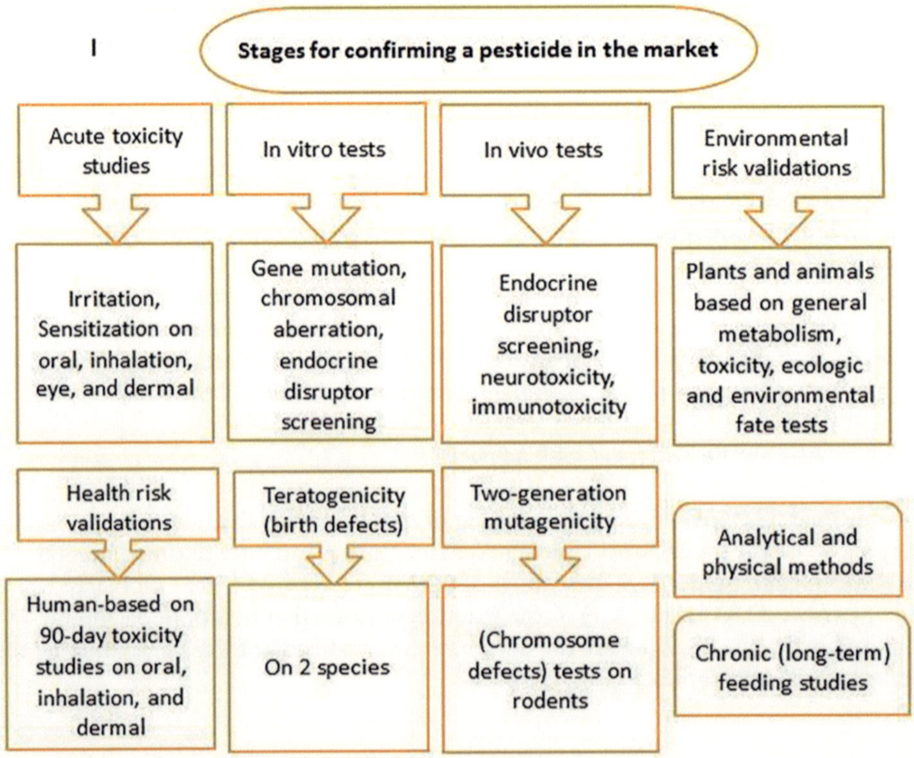


**Figure 9.**
*Stages for confirming a pesticide in the market [33].*

A proficient observing framework that consistently tests food things for pesticide residues, is a solid motivation for framers to utilize synthetic compounds carefully. Except if defiled shipments can be distinguished, ranchers may not know or care whether the products they are selling contain pesticide residues. Notwithstanding, the offices required for compound testing are costly, while there is some debate over the precision of the less expensive bioassay technique. One promising methodology is HACCP—Hazard Analysis at Critical Control Points. This gander at the entire chain of pesticide conveyance and use and chooses the specific places where activity is plausible and will influence [32]. **Figure 9** described the requirements for chemical pesticides to be accepted and used in the market [33].

## 6. Conclusion

In this chapter, the authors tried to review some of the biological and molecular effects of pesticides on the human body in a few critical ways, from cellular to molecular ones. In the past, insufficient information about the biological effects of chemical substances caused an increase in disease and physical damage. Nowadays, by announcements from international organizations and loading, logical papers agro-chemicals more frequently control dangerous bugs and, on a parcel more restricted measure, natural creepy-crawly showers. Despite its viability, the purposeless utilization of chemical pesticides in engaging natural issues causes genuine environmental problems to human well-being, reduces the number of standard adversaries, and gives safe creepy crawlies.

Conversely, biopesticides, utilized for more than a century, are retainers of highlights less significant on the climate and less unsafe to people at any point.

Moreover, biological controlling methods and passing approved and standard processes for manufacturing chemical sprays could be helpful in this way for reducing the consequences of chemical compounds. The study of binding pesticides to proteins is toxicologically essential. This study is expected to provide crucial insights into the interaction of biomacromolecules with pesticides.

There are different molecular assessments of pesticides and their effect on proteins. Still, molecular docking is a well-known program for predicting the interaction of pesticides as ligands and macromolecules like hemoglobin as a target to estimate the penetration of the chemicals on protein pockets. Other analytical assessments, for instance, using spectroscopic methods, could be pretty helpful.

## Author details

Aida Doroudian[1], Mahdieh Emadi[1], Reza Hosseinzadeh[2] and Parvaneh Maghami[1*]

1 Department of Biology, Science and Research Branch, Islamic Azad University, Tehran, Iran

2 Department of Chromatography Research Group, Iranian Academic Center for Education, Culture, and Research(ACECR), Urmia, Iran

*Address all correspondence to: maghami@srbiau.ac.ir

## References

[1] Abubakar Y, Tijjani H, Egbuna C, Adetunji CO, Kala S, Kryeziu TL, et al. Pesticides, history, and classification. In: Natural Remedies for Pest, Disease and Weed control. Amsterdam, Netherlands: Academic Press; 2020. pp. 29-42

[2] Kaloyanova FP, El Batawi MA. Human Toxicology of Pesticides. Boca Raton, Florida, United States: CRC Press; 2019

[3] Zacharia JT. Identity, physical and chemical properties of pesticides. In: Pesticides in the Modern World Trends in Pesticides Analysis. 2011. Vol. 1, pp. 1-8. DOI: 10.5772/17513

[4] Sharma A, Kumar V, Shahzad B, Tanveer M, Sidhu GP, Handa N, et al. Worldwide pesticide usage and its impacts on ecosystem. SN Applied Sciences. 2019;**1**(11):1-6

[5] Lushchak VI, Matviishyn TM, Husak VV, Storey JM, Storey KB. Pesticide toxicity: A mechanistic approach. EXCLI Journal. 2018;**17**:1101. DOI: 10.17179/excli2018-1710

[6] Aktar MW, Sengupta D, Chowdhury A. Impact of pesticides use in agriculture: Their benefits and hazards. Interdisciplinary Toxicology. 2009;**2**(1):1. DOI: 10.2478/v10102-009-0001-7

[7] Hassaan MA, El Nemr A. Pesticides pollution: Classifications, human health impact, extraction and treatment techniques. Egyptian Journal of Aquatic Research. 2020;**46**(3):207-220

[8] Shah R. Pesticides and human health. In: Emerging Contaminants. London, UK: IntechOpen; 2020

[9] Kalyabina VP, Esimbekova EN, Kopylova KV, Kratasyuk VA. Pesticides: Formulants, distribution pathways and effects on human health—A review. Toxicology Reports. 2021 Jan;**1**(8):1179-1192. DOI: 10.1016/j.toxrep.2021.06.004

[10] Araoud M, Neffeti F, Douki W, Hfaiedh HB, Akrout M, Hassine M, et al. Adverse effects of pesticides on biochemicalandhematologicalparameters in Tunisian agricultural workers. Journal of Exposure Science & Environmental Epidemiology. 2012;**22**(3):243-247. DOI: 10.1038/jes.2012.11

[11] Hernández AF, Parrón T, Tsatsakis AM, Requena M, Alarcón R, López-Guarnido O. Toxic effects of pesticide mixtures at a molecular level: Their relevance to human health. Toxicology. 2013;**10**(307):136-145. DOI: 10.1016/j.tox.2012.06.009

[12] Demirdağ R, Yerlikaya E, Aksakal E, Küfrevioğlu ÖI, Ekinci D. Influence of pesticides on the pH regulatory enzyme, carbonic anhydrase, from European Seabass liver and bovine erythrocytes. Environmental Toxicology and Pharmacology. 2012;**34**(2):218-222

[13] Rizzati V, Briand O, Guillou H, Gamet-Payrastre L. Effects of pesticide mixtures in human and animal models: An update of the recent literature. Chemico-Biological Interactions. 2016;**254**:231-246. DOI: 10.1016/j.cbi.2016.06.003

[14] Bassil KL, Vakil C, Sanborn M, Cole DC, Kaur JS, Kerr KJ. Cancer health effects of pesticides: Systematic review. Canadian Family Physician. 2007;**53**(10):1704-1711

[15] Mesnage R, Defarge N, Spiroux de Vendômois J, Séralini GE. Major pesticides are more toxic to human cells than their declared active principles.

BioMed Research International. 2014;**2014**:26. DOI: 10.1155/2014/179691

[16] Abdel-Halim KY, Osman SR. Cytotoxicity and oxidative stress responses of Imidacloprid and glyphosate in human prostate epithelial WPM-Y. 1 cell line. Journal of Toxicology. 2020;**8**: 1-12. DOI:10.1155/2020/4364650

[17] Abhishek A, Ansari NG, Shankhwar SN, Jain A, Singh V. In vitro toxicity evaluation of low doses of pesticides in individual and mixed condition on human keratinocyte cell line. Bioinformation. 2014;**10**(12):716. DOI: 10.6026/97320630010716

[18] Teodoro M, Briguglio G, Fenga C, Costa C. Genetic polymorphisms as determinants of pesticide toxicity: Recent advances. Toxicology Reports. 2019 Jan;**1**(6):564-570. DOI: 10.1016/j.toxrep.2019.06.004

[19] Giambò F, Leone GM, Gattuso G, Rizzo R, Cosentino A, Cinà D, et al. Genetic and epigenetic alterations induced by pesticide exposure: Integrated analysis of gene expression, microRNA expression, and DNA methylation datasets. International Journal of Environmental Research and Public Health. 2021;**18**(16):8697

[20] Kaur R, Kaur S, Lata M. Evaluation of DNA damage in agricultural workers exposed to pesticides using single cell gel electrophoresis (comet) assay. Indian Journal of Human Genetics. 2011;**17**(3):179. DOI: 10.4103/0971-6866.92100

[21] Kaur K, Kaur R. Occupational pesticide exposure, impaired DNA repair, and diseases. Indian Journal of Occupational and Environmental Medicine. 2018;**22**(2):74. DOI: 10.4103/ijoem.IJOEM_45_18

[22] Ledda C, Cannizzaro E, Cinà D, Filetti V, Vitale E, Paravizzini G, et al. Oxidative stress and DNA damage in agricultural workers after exposure to pesticides. Journal of Occupational Medicine and Toxicology. 2021;**16**(1):1-7. DOI: 10.1186/s12995-020-00290-z

[23] Doroudian A, Hosseinzadeh R, Maghami P, Khorsandi K. Spectroscopic investigation on molecular aspects and structural and functional effects of tetraethyl pyrophosphate organophosphorus insecticide interaction with adult human hemoglobin. Journal of Biomolecular Structure & Dynamics. 2021;**16**:1-1. DOI: 10.1080/07391102.2021.1902398

[24] Emadi M, Maghami P, Khorsandi K, Hosseinzadeh R. Biophysical study on the interaction of Cartap hydrochloride and hemoglobin: Heme degradation and functional changes of protein. Journal of Biochemical and Molecular Toxicology. 2019 Jul;**33**(7):e22325. DOI: 10.1002/jbt.22325

[25] Cui Y, Guo J, Xu B, Chen Z. Binding of chlorpyrifos and cypermethrin to blood proteins. Pesticide Biochemistry and Physiology. 2006;**85**(2):110-114. DOI: 10.1016/j.pestbp.2005.11.004

[26] Wang YQ, Zhang HM, Zhang GC, Liu SX, Zhou QH, Fei ZH, et al. Studies of the interaction between paraquat and bovine hemoglobin. International Journal of Biological Macromolecules. 2007;**41**(3):243-250. DOI: 10.1016/j.ijbiomac.2007.02.011

[27] Ding F, Han BY, Liu W, Zhang L, Sun Y. Interaction of Imidacloprid with hemoglobin by fluorescence and circular dichroism. Journal of Fluorescence. 2010;**20**(3):753-762. DOI: 10.1007/s10895-010-0618-0

[28] Dang K, Doggett SL, Veera Singham G, Lee CY. Insecticide resistance and resistance mechanisms in bed bugs,

Cimex spp. (Hemiptera: Cimicidae). Parasites & Vectors. 2017;**10**(1):1-31

[29] Tian F, Mo X, Rizvi SA, Li C, Zeng X. Detection and biochemical characterization of insecticide resistance in field populations of Asian citrus psyllid in Guangdong of China. Scientific Reports. 2018;**8**(1):1-1. DOI: 10.1038/s41598-018-30674-5

[30] Nkya TE, Akhouayri I, Poupardin R, Batengana B, Mosha F, Magesa S, et al. Insecticide resistance mechanisms associated with different environments in the malaria vector Anopheles gambiae: A case study in Tanzania. Malaria Journal. 2014;**13**(1):1-5. DOI: 10.1186/1475-2875-13-28

[31] Kumar J, Ramlal A, Mallick D, Mishra V. An overview of some biopesticides and their importance in plant protection for commercial acceptance. Plants. 2021;**10**(6):1185. DOI: 10.3390/plants10061185

[32] Bhattacharyya A, Barik SR, Ganguly P. New pesticide molecules, formulation technology and uses: Present status and future challenges. Journal of Plant Protection Sciences. 2009;**1**(1):9-15

[33] Reeves WR, McGuire MK, Stokes M, Vicini JL. Assessing the safety of pesticides in food: How current regulations protect human health. Advances in Nutrition. 2019;**10**(1):80-88. DOI: 10.1093/advances/nmy061

Chapter 9

# Deleterious Effects of Banned Chemical Pesticides on Human Health in Developing Countries

*Alaa Eldin Bayoumi*


## Abstract

Nowadays, large quantities of banned chemical pesticides are still in use illegally in various developing countries. The effect of these pesticides on humans, that is, children, adults, including pregnant women, either through chronic residential or occupational exposure, leads to various negative effects. This chapter focuses on the evidence of using the banned pesticides in developing countries and the occurrence of different diseases that affect the quality of life of the affected individuals either at the health, social, and/or labor level. The recorded diseases included obesity, diabetes, Alzheimer's, dementia, Parkinson's, asthma, chronic bronchitis, autism, erectile dysfunction, and psychological disorders as a result of chronic exposure to the banned pesticides. It was highlighted the identification of each disease, some epidemiological studies carried out in developing countries, and the mechanisms of action by which the pesticides are linked to each mentioned disease. In conclusion, it was discussed the major causes behind the incidence of such diseases and suggested suitable solutions that must be presented by the US Environmental Protection Agency, Food and Agriculture Organization among other bodies to the developing countries to avoid and overcome the occurrence of such diseases in the future.


**Keywords:** banned pesticides, human, exposure, developing countries

## 1. Introduction

Chemical pesticides are still considered as an essential tool used in the mass production of agricultural products in developing countries to control a wide range of pests, that is, insects, weeds, plant pathogens among others, and hence maintain high product quantity [1]. The consumption rate of chemical pesticides worldwide reached 2.4 megatons during 2006–2007 [2]. Pesticides have biological activity and toxic action on the targeted pests at the recommended concentration throughout various modes of action mechanisms exerted by the parent compounds and/or their metabolites [3]. If such concentration is exceeded through misuse or incorrect application, these pesticides become toxic to the non-target organisms, including humans [4].

Based on the enormous number of published studies related to the toxic effects of pesticides on non-targeted organisms and humans, some of these pesticides have been banned in developed countries. The most famous banned compounds included organochlorine pesticides (OCPs) in the 1970s [5], some organophosphorus during 2001–2006 [6], certain carbamates insecticides [7], and specific synthetic pyrethroids and neonicotinoids insecticides, among others [8]. Globally, the total number of banned pesticides reached 460 compounds distributed between insecticides, fungicides, and herbicides among other groups of pesticides in 36 and 128 developed and developing countries, respectively [8]. The ban decision is based on various factors, such as a) the high toxicity of such pesticides to the non-targeted organisms (extremely hazardous, i.e., acute oral $LD_{50}$ for the rat <5 mg/kg and highly hazardous, i.e., acute oral $LD_{50}$ for the rat 5–50 mg/kg) [9], b) their carcinogenic effects on humans causing various types of cancer [10], and c) hazardous effects on environmental elements among other reasons such as effects on the endocrine system, that is, hormone disruptors [11].

Practically, the toxic effects of pesticides on mammals, including humans, are the sum of the results obtained by the studies carried out and extrapolation processes through decades from two major categories of studies, that is, *in vivo* and *in vitro*. The *in vivo* studies were carried out using laboratory animals that lead to an understanding of the toxicity mechanisms of the tested pesticides.

Also, it was studied both the toxicokinetics, toxicodynamics of chemical pesticides [12], teratogenicity [13], and carcinogenicity [14] using different mammalian models and affected humans in the epidemiological studies. With carcinogenic pesticides, they were classified into various categories, that is, carcinogenic, probably, possibly be carcinogenic to humans, or not classifiable as carcinogenic to humans [15]. Other studies of pesticides toxicity have been carried out depending on the response of different biomarkers [16], that is, hematological toxicity [17], nephrotoxicity [18], hepatotoxicity [19], neurotoxicity [20], oxidative stress, and DNA damage [21], pulmonary toxicity, reproductive and developmental toxicity and carcinogenesis biomarkers [22].

The *in vitro* toxicity studies have emerged through an approach that has been known as alternative methods, which is based on the application of a principle called (3Rs) designed by Russell and Burch [23] using animal organs, tissues, fertilized eggs, embryos, transplanted organs, such as liver, kidneys, brain, pancreas, and/or tissue pieces [24]. Also, the cell culture techniques, that is, primary cell cultures or cell lines were used as another approach of the alternative methods to understand the cytotoxic effects of xenobiotics including pesticides either at a biological level and/or biochemical level and explaining the genotoxic effects of the tested pesticides [25]. Through such alternative methods in association with the *in vivo* investigation, it was studied the mutagenicity, genotoxicity, and the effects of pesticides at the molecular level [26].

Regarding human health, exposure to different banned chemical pesticides may lead to consequent intoxication symptoms as a result of cellular, biochemical, and genetic effects, that is, congenital malformation, neurochemical and behavioral dysfunctions among others [27]. In developing countries, such as India, it was documented various negative effects resulted from human exposure to pesticides, that is, neurological, respiratory, dermal, and reproductive effects in addition to the impact on general health [28]. In other epidemiological studies related to the usage of banned chemical pesticides in developing countries, it was mentioned the percentages of chemical pesticides that caused a significant number of dead people annually as a result of pesticides intoxication [29]. The mentioned developing countries in such a chapter include India, Egypt, and African countries, Romania, Thailand, Taiwan,

Costa Rica, and Nicaragua among others. In Venezuela, it was recorded various serious problems due to exposure to pesticides either through environmental and/or occupational routes. Such exposure is due to limitations in the regulation acts related to the use of pesticides and lack of health and safety measures [30].

As specific pesticides, it was found that *in utero* exposure to organochlorine insecticides, that is, dichlorodiphenyltrichloroethane (DDT), dieldrin, endosulfan, heptachlor, dicofol, and methoxychlor was associated with neurodevelopmental effects in children [31]. Organophosphorus insecticides, such as parathion, dimethoate, monocrotophos, and chlorpyrifos among others, cause cardiovascular diseases [32]. Moreover, some banned pyrethroids insecticides such as fenvalerate, permethrin, and other compounds including certain metabolites causing DNA damage in human sperm [33] and developmental neurotoxicity [34]. As for banned neonicotinoid pesticides, that is, imidacloprid thiacloprid, they may cause breast cancer by increasing the expression of the aromatase enzyme. Other banned chemical pesticides, such as some triazine herbicides, were associated with breast cancer [35].

As for human exposure to pesticides via oral feeding, it was documented that consumption of contaminated foods by pesticide residues when exceeded the Acceptable Daily Intake (ADI) and/or the Maximum Residue Limits (MRL) values listed by the Codex Alimentarius may lead to various undesirable effects. Based on such type of exposure, high levels of pesticide residues and their metabolites in foodstuff could reach the human blood, including maternal blood, cord blood, placenta, breast milk, and children.

In the case of occupational exposure to pesticides, it was documented that such exposure occurs directly during various processes, such as manufacturing, transporting, storing, retailers, preparation, application by the user, re-entry into treated fields, harvest, and equipment cleaning. The exposure may be due to the misuse during the application of pesticides without protective equipment, which is considered economically expensive in developing countries. In addition, other routes of exposure may occur, such as exposure of children and/or pregnant women, which affects their fetuses.

In recent studies, it was found that all the mentioned negative effects of pesticides exposure are results of bad practices, that is, inadequate, unsafe, and handling applications without wearing the protective instrument. Also, it was reported that the lack of awareness of suitable pesticide use is considered as the main reason behind the occurrence of various diseases in the farmers' works in the Tu Ky district, Vietnam [36]. To overcome such types of problems, the Agricultural Pesticides Committee [APC] in Egypt, has been introduced around 10.000 licensed applicators in the agriculture fields and targets to reach 50.000 during the next years after their training in the suitable pesticides application procedures [37].

Based on the mentioned facts, this chapter has been focused on the most deleterious effects linked to chronic human exposure to various banned chemical pesticides that affect the exposed individuals with various diseases, that is, obesity, Alzheimer's, Parkinson's, asthma, chronic bronchitis, type 2 diabetes, autism, erectile dysfunction, and psychological disorder. Such diseases were selected in this chapter based on the fact that they are affecting the quality of life of the affected individuals either at the health, social, and/or labor level. Based on the lack of scientific and investigation background in most developing countries, the identification of each disease and the different mechanisms followed by the banned pesticides and/or their metabolites to exhibit these diseases either in pregnant women, their fetuses, children, and/or adults who were exposed to the banned pesticides were discussed.

## 2. Evidence of using banned pesticides in the developing countries

Practically, huge amounts of banned pesticides are still in use illegally in various developing and developed countries. Such banned pesticides include organochlorine insecticides, that is, DDT for controlling the public health pests [38] due to their cheap price, ease of production, lack of registration, and control measures systems for pesticides in some of these countries, besides the prevalence of the risk versus benefit theory [39]. Unfortunately, it was documented that the World Health Organization supported the reintroduction of DDT for malaria eradication in 2006 [40].

In agriculture, various published documents have been reported that a list of banned pesticides is still in use in developing countries [41]. For example, it was reported that the banned pesticides, such as chlorpyrifos, endosulfan, and cypermethrin, as insecticides, atrazine and glyphosate, as herbicides, are still in use in various provinces in Argentina in addition to Paraguay [42]. In addition, it was reported that many developing countries are still using the banned chemical pesticides that have been exported from the European Union after their banning in European countries [1].

Based on the export statistics from China Customs, the export volume of pesticides during the period of January to November 2015, African markets containing developing countries represented 13.9% of the total export of pesticides from China to 44 countries. The most exported banned pesticides include paraquat, glyphosate, 2,4-D-dimethylammonium, atrazine, glyphosate-monoammonium, tebuthiuron, as herbicides, lambda-cyhalothrin, imidacloprid, chlorpyrifos, dichlorvos, emamectin, as insecticides, mancozeb, metalaxyl+mancozeb, as fungicides. The top 10 countries by export value were Nigeria, South Africa, Ghana, Ivory Coast, Egypt, Kenya, Cameroon, Tanzania, Ethiopia, and Guinea. The amount of export value for these top 10 countries constitutes 85.9% of the total export value to Africa from China [43]. Moreover, some banned pesticides may be used in form of counterfeit and/or contraband pesticides either in developing or developed countries [44]. The decision toward the herbicide glyphosate has been taken by the Mexican Government to be a period of transition when sustainable alternatives will be promoted [8].

The decision to ban pesticides in developing countries is based on the published information by the USEPA and the European Union, while the executive decision will be in practical form by regulation and legislative acts present in some developing countries. For example, Egypt (Agricultural Pesticides Committee, APC), India, Malaysia, Philippines, Costa Rica, Mexico, Kenya, Bangladesh, and Indonesia, each of them has their regulation and legislative acts [28]. Other low-income countries have not any institutions related to pesticides in general, especially the agricultural ones, and hence their usage of the agricultural chemicals depends on the importing of these products from neighboring countries.

Also, the decision of banning pesticides in other developing countries may be retard to be in practice due to a lack of transparency and also depending on the availability of alternative pesticides in the market. In addition, the decision-makers in the developing countries may offer a period called a period of transition or grace period that may be reached for 6 months or more than 1 year as an expected period to consume the remaining stock of these banned pesticides. Such cases were already offered for the organophosphate insecticide chlorpyrifos, dimethoate, fipronil, alpha-cypermethrin, amitrole, carbendazim, iprodione, diazinon, carbosulfan, diuron, diniconazole among other pesticides, such as the case in Egypt and some other developing countries. During such grace period, the targeted pesticide[s] to be banned is still in use, and hence the exposure processes, that is, farmers, unlicensed applicators, children, and/or pregnant women are continued.

At the governmental level of these countries, the most important question in the mentality of the decision-makers is related to the quantity of such pesticide[s], that is, where and how to treat the remaining amounts of such pesticide[s], that is, there is no possibility to withdrawing the remaining amounts that are still in the pesticide market. However, in a recent action taken by APC, based on the requirements of the European Commission, the MRLs of chlorpyrifos must not exceed 0.01 mg/kg of the exporting crops and such limits must be followed by the Agriculture Export Council [37]. In addition, it was decided to restrict the usage of such insecticides to control insect pests on nonedible crops, such as cotton, to control the desert locust and termites [37]. Unfortunately, from the practical viewpoint, despite such decisions, chlorpyrifos is still illegally available like other banned pesticides in the pesticide market in Egypt at least during the grace period. So, the use of banned pesticides at that moment is considered an inevitable fact either in Egypt or other developing countries.

## 3. Mechanisms of toxic action of some banned chemical pesticides

Depending on the chemical structure of the targeted banned chemical pesticides, it could classify the mechanisms of their action into two main categories, that is, neurotoxic action and hormone disruptors.

### 3.1 Neurotoxicity

Based on the published studies related to human diseases associated with exposure to banned chemical pesticides, it was found that most of these pesticides were neurotoxic compounds. The most famous classes of these pesticides are organochlorine, organophosphorus, carbamates, synthetic pyrethroids, and neonicotinoids insecticides. For organochlorine insecticides, that is, dichlorodiphenyltrichloroethane (DDT), cyclodiens, hexachlorocyclohexane, many studies have shown that its mode of action is based on a reduction in the transport of potassium ions, after blocking the sodium channels, and inhibiting the enzymes ($Na^+$ -$K^+$/$Ca^{2+}$ -$Mg^{2+}$-ATPases), inhibiting the binding between calcium and calmodulin and then affecting the flow of neurotransmitters [45].

Regarding organophosphorus as parent compounds and/or their activated metabolites, such as [−oxon], the mechanisms of their actions depend on the irreversible inhibition of acetylcholinesterase by phosphorylating the amino acid serine in the esteratic site of the AchE enzyme [46], resulting in hyperstimulation of the cholinergic nerves, that is, muscarinic and nicotinic acetylcholine receptor [47]. Carbamate insecticides act as reversible inhibitors to acetylcholinesterase leading to various symptoms of toxicity [48].

As for the synthetic pyrethroids, their mechanisms of toxic action depend on inhibition of ($Ca^{2+}$, $Mg^{2+}$-ATPase) enzymes, binding to gamma-aminobutyric acid receptors in the chloride channels, and inhibition of the calmodulin protein that binds to calcium and thus increase the calcium ions that affect the flow of the neurotransmitters that lead to cause various symptoms of poisoning in humans [49]. In addition, it was documented that neonicotinoids insecticides, that is, imidacloprid act on nicotinic acetylcholine receptors, and hence stimulate these receptors at low doses while blocking such receptors at high doses leading to paralysis and death [50]. Also, some fungicides exhibit neurotoxic action, which affects the peripheral nerves, the motor nerves, and the central nervous system that leads to different symptoms of poisoning [51]. Based on such toxic action observed in humans through different accidents, such fungicide has been banned in the 1970s [52].

### 3.2 Effects on the endocrine system

Based on the published studies, it was documented that various diseases, such as obesity, diabetes, and erectile dysfunction, have occurred as a result of exposure to various banned pesticides. It was reported that such diseases belong to the mechanisms of action of the mentioned pesticides within the endocrine system. Historically, in the 1970s, it had emerged the adverse effects of some chemical pesticides on the endocrine system through various modes of action and it was called for such pesticides the term (Hormone Disruptors) [53]. Such effects may be through the hormone-secreting gland, effects on the composition of the hormone itself, effects on its production and secretion rates, or that it is similar to it in composition or interfere with the hormone in its function by competing with it for binding to the hormonal receptors [54]. Many studies have shown that hormone disruptors cause many adverse health effects on humans, leading to various diseases [55]. Exposure to hormone disruptors maybe not be observable for many years. If the fetus is exposed to any of these disruptors during the pregnancy of the mother, then these substances cause adverse effects on many functions of this fetus that are not observed until after its birth and reaching puberty [56].

## 4. Diseases other than cancer caused by banned chemical pesticides

In the first two decades of this century, several studies that have been published and searched in MEDLINE [through PubMed] revealed the relationship between the exposure of humans, that is, farmworkers, unlicensed pesticide applicators, children, and/or pregnant women to different chemical pesticides and the emergence of various diseases other than cancer [57]. Such studies reported that these diseases do not appear until after reaching adulthood, which affects the quality of life of the affected individual[s] at the health, social, and labor level either in the present and/or the future.

### 4.1 Obesity

Obesity, that is, body mass index [BMI] $\geq 30.0$ has been defined as a chronic disease that affects around 13% of the global population and 62% of people living in developing countries. These huge percentages lead to the death of 2.8 million individuals each year as a result of obesity [58]. Exposure to various classes of banned chemical pesticides and obesity are well documented in many published studies carried out in developed countries. For example, it was found that there was a positive linkage between maternal, prenatal, or postnatal exposure to pesticides and obesity, especially DDT as an organochlorine insecticide [59] and chlorpyrifos as organophosphorus pesticides [60]. Also, other classes of pesticides were found to be associated with the development of obesity, that is, bifenthrin as pyrethroid [61] and imidacloprid as neonicotinoid [62]. In addition, it was reported that exposure of pregnant women to pesticides through agricultural or industrial activities leads to overweight children [63].

Various mechanisms have been followed by pesticides to be associated with obesity. For example, increasing the adipocyte differentiation by quizalofop-*p*-ethyl, diazinon, imidacloprid, fipronil, and permethrin among others [64]. Also, it was documented that proliferation and alteration in the adipose function tissue lead to increasing the lipid uptake and alteration of the neuroendocrine control of feeding that affects the metabolism of nutrients [65]. Also, it was found that some pesticides, that is, organophosphate, carbamate, and organochlorines disrupt hormonal status through oxidative stress, which affects mitochondrial function, especially in the cellular metabolism of lipids, carbohydrates, and proteins [49]. Another theory proved that some pesticides are mimic the natural lipophilic hormone and altering the nuclear receptor transcription factor, which affects the key adipogenic factors, fat depot size, and function [66]. More recently it was reported that some pesticides are linked to obesity by affecting the gust microbiota, metabolic homeostasis by affecting the peroxisome proliferator-activated receptors (PPARs) and the thyroid hormone pathway, altering the fate of mesenchymal stem cells (MSCs), and dysregulation of sex steroid hormone [64].

### 4.2 Diabetes

Diabetes mellitus is a chronic disease with two main types, that is, Type 1 diabetes, which occurs in case of the pancreas does not produce enough insulin, and type 2 (T2D) or insulin resistance, which means that the body cannot effectively use the insulin it produces [67]. Diabetes, especially T2D, is a major cause of various diseases, such as cardiovascular disease [68], endometrial [69], prostate [70], and colon cancer [71] in addition to other diseases, that is, blindness, kidney failure, heart attacks, stroke and lower limb amputation [72].

The relationship between human exposure to the banned chemical pesticides and the occurrence of diabetes has been documented. Two studies in Korea found that low-dose background exposure to 10 OCPs, that is, HCH, HCB, heptachlor epoxide, *p,p′*-DDE, *p,p′*-DDD, *p,p′*-DDT, *o,p′*-DDT, oxychlordane, trans-nonachlor, and mirex were strongly associated with prevalent type 2 diabetes in Koreans people [73].

In Thailand, it was reported that endosulfan as an organochlorine insecticide, mevinphos an organophosphorus, carbaryl/Sevin as carbamate, and benlate as fungicides were positively associated with diabetes, as described in the case–control study carried out by Juntarawijit and Juntarawijit, [74]. In India, various pesticides including herbicides, that is, atrazine, butylate, 2,4,5-T, diazinon, fonofos, phorate, and parathion as organophosphorus insecticides, *p,p′*-DDT, *p,p′*-DDE, β-HCH, and oxychlordane as organochlorine insecticides were positively associated with hyperglycemia and diabetes [75]. In Egypt, it was found that lindane followed by *o,p′*-DDD, and *p,p′*-DDE as DDT metabolites as organochlorine compounds and malathion as organophosphate insecticide was strongly associated with type 1 diabetes in children, as reported in the preliminary study carried out by El-Morsi et al. [76].

As a mechanism of action by which the banned chemical pesticides induces diabetes, various specific studies have been proved that OCPs, as it is well known that these pesticides are lipophilic, hydrophobic, and highly resistant to metabolic degradation, so that, they are bioaccumulated in fatty tissues for many years, and their serum concentration is considered to be a good reflection of lifetime exposures [77].

However, it was documented that OCPs have variable molecular and cellular targets and thus they cannot be considered to have a single mode of action. Inflammation in adipose tissue, ectopic lipid accumulation [lipotoxicity] in liver, muscle, and pancreas, and mitochondrial dysfunction caused by oxidative damage caused by OCPs lead to the development of insulin resistance and T2D [77]. Also, these pesticides may affect pancreatic β cells and trigger insulin resistance, thus impairing both lipid and glucose metabolism [78].

Besides, it is well known that *p,p′*-DDE is antiandrogenic and can bind to the androgen receptor and that DDT has estrogenic properties; both estrogen and androgen receptors are involved in the mediation of insulin sensitivity [79]. Another study showed that certain OCPs exposure can disrupt glucose homeostasis, which could contribute to the development of type 2 diabetes in the future [80]. As for banned organophosphorus pesticides, it was documented that exposure to sufficiently high levels of these compounds would be expected to result in increased accumulation of acetylcholine, potentially leading to overstimulation and eventual downregulation of its receptors and reducing insulin production [81].

### 4.3 Alzheimer's disease

Alzheimer's disease (AD) is the major form of dementia and is considered the fourth leading cause of death in the elderly. AD is the most common progressive neurological disease and results in an irreversible loss of neurons [82]. One of the most symptoms of AD is loss of short-term memory, speech problems, confusion, mood swings, self-care inability, and behavioral issues [83]. Few studies have been carried out in developing countries on the link between exposure to pesticides and AD. In India, it was found that OCPs, that is, β-HCH, dieldrin, and *pp′*-DDE are associated with the risk of AD in the north Indian population [84]. The same finding was reported with organophosphates insecticides [85].

In China, a positive association between pesticide exposure and AD, confirms the hypothesis that pesticide exposure is a risk factor for AD, as shown in the systematic review and meta-analysis carried out by Yan et al. [86]. One internal exposure investigation evaluated the relationship between serum dichlorodiphenyldichloroethylene (DDE) levels and AD, observing a 3.8-fold increase in serum levels of organochlorine metabolites of DDE in patients with AD when compared with control participants [87].

As mechanisms behind the occurrence of AD, it was found that such disease is a progressive neurodegenerative disorder associated with the loss of cholinergic neurons and the presence of excessive neuritic plaques containing amyloid β protein and abnormal tau protein filaments as neurofibrillary tangles [88]. Decreased level of acetylcholine in AD patients appears to be a critical element in producing dementia and memory disorders [89]. It was documented that various chemical pesticides cause uncoupled oxidative phosphorylation, which increases the levels of free radicals [90], which affect the mitochondrial function and hence increased the production of ROS and higher levels of oxidative stress that lead to cellular damage in form of synaptic linked with the development and progression of AD [91]. Baltazar et al. [92] found that various pesticides share many features, such as the ability to induce oxidative stress, mitochondrial dysfunction, α-synuclein fibrillization, and neuronal loss.

At *in vivo* level, various studies revealed that some pesticides may disrupt the metabolic pathways, such as the homeostasis of amyloid-β, causing a significant elevation in amyloid-β levels in the cortex and hippocampus, as well as increasing memory loss and reduced motor activity in experimental animals [93]. Thus, some researchers have documented that pesticide exposure is a potential risk factor for AD, and hence proved such results through several epidemiological studies [94].

Chin-Chan et al. [95] reported that some pesticides have been associated with AD due to their ability to elevate beta-amyloid [Aβ] peptide and the phosphorylation of Tau protein [P-Tau], causing senile/amyloid plaques and neurofibrillary tangles (NFTs) characteristic of AD. Tang et al. [96] showed the proposed neuropathological mechanisms that included oxidative stress through the reactive oxygen species [ROS] generated by pesticides, neuroinflammation enhancement that leads to amyloid-beta Aβ and tau protein expression, promotion of amyloidogenesis, such as amyloid plaque formation, DNA damage, and dysfunction of the brain-Gut axis. Like dementia, it was published in a nationwide population-based cohort study that revealed the relationship between exposure to pesticides and dementia [97].

### 4.4 Parkinson's disease

Parkinson's disease (PD) is the second most common neurodegenerative disorder after Alzheimer's disease (AD), having an overall prevalence ranging from 1 to 2 per 1000 people. PD is characterized by various motor dysfunctions, such as rigidity, bradykinesia, resting tremor, gait freezing, and postural reflex impairment, and neuropsychological dysfunctions, such as cognitive decline, depression, and sleep disturbance, all of which negatively affect patients' quality of life (QOL), as presented through the systematic review and meta-analysis carried out by Zhao et al. [98].

From an epidemiological viewpoint, the association between the use of pesticides and PD was first reported by Barbeau et al. [99]. Pesticides have been implicated as one of the most likely major environmental risk factors for PD [100]. In the case of the relation between pesticides and PD, it was documented that people exposed to pesticides at workplaces have a higher risk of PD than people exposed at home, and exposure at both workplaces and residences has the highest PD risk [101].

Occurrence of PD in developing and developed countries concerning the exposure to pesticides, the results of a study of the meta-analysis carried out by Ahmed et al. [102] showed that both types of countries suffered from such disease in a significant association with pesticide exposure. The pesticides linked with PD included trifluralin and paraquat as herbicides, maned and mancozeb as fungicides, diazinon, chlorpyrifos, parathion, β-HCH, permethrin, and dieldrin as insecticides. In another case–control study, paraquat was closely associated with a higher risk of developing PD [103].

Many mechanisms have been involved in the role of pesticides in PD development. Karen et al. [104] reported a significant reduction in the mitochondrial function in the *in vivo* synaptosome preparations, there was an increased dopamine turnover and decreased motor activity. In addition, dopaminergic neurotransmission was affected by exposure to permethrin. Also, dieldrin as organochlorine insecticide-induced apoptotic cell death alters dopamine levels and induces mitochondrial dysfunction and protein aggregation [105], while endosulfan inhibits proteasomal activity [106].

### 4.5 Respiratory disorder diseases

#### *4.5.1 Asthma*

Asthma is a common and global chronic inflammatory disease of the airways that affects children and adults characterized by variable and recurring respiratory symptoms (wheezing, breathlessness, chest tightness, and dry cough), airflow obstruction, and mucus hypersecretion hyperreactivity (AHR), all of which interfere with breathing [107]. Several factors lead to asthma diseases, that is, genetic, allergic conditions, and multiple lifestyle factors in addition to low birth weight, prematurity, exposure to tobacco and indoor and outdoor air pollutants, and occupational exposure to chemicals, such as pesticides.

Two major types of epidemiological studies have been published concerning exposure to pesticides and linkage to asthma, that is, exposure of children and adults. For example, it was reported that occupational exposure to pesticides was associated with the prevalence of asthma [108]. In addition, children of farmers are at risk of pesticide exposure through various routes, that is, living close to agricultural fields, participating in farm work, and eating fruits and vegetables soon after harvest [109].

As for lower or middle-income developing countries, exposure of children to banned chemical pesticides was studied in Mexico [110], Brasil, [111], Costa Rica [112], Sri Lanka [113], and Lebanon [114]. As for exposure of adults through the occupational route, various studies have been carried out in different developing countries, such as Kenya [115], Ghana through a cross-sectional study [116], and Ethiopia on a large-scale cross-sectional study [117]. In Egypt, a published case–control study of adolescent pesticide applicators showed an association between exposure to OPs pesticides, chlorpyrifos, and reduction of lung function [118].

To investigate the relation between exposure to some specifically banned pesticides and the occurrence of asthma, Hoppin et al. [119] reported that paraquat as herbicide, dieldrin, heptachlor, lindane coumaphos, diazinon, parathion, DDT, and ethylene dibromide as insecticides, and captan as fungicide were associated with allergic and nonallergic asthma. In a recent ecological study carried out in Argentina, it was found a strong association between asthma in agricultural workers and occupational exposure to the herbicide glyphosate [120]. In another recent study carried out in Uganda, it was reported a strong association between organophosphate and carbamate insecticide exposure and disorder in lung function, including asthma among smallholder farmers [121]. As for OPs pesticides, it was demonstrated in the review published by Shaffo et al. [122] that exposure to various OPs pesticides, that is, bromofos, chlorpyrifos, diazinon, fenthion, malathion, and parathion were associated with asthma.

As a causal link between organophosphorus pesticides and asthma, mechanistic studies exhibited a blockage of autoinhibitory in the muscarinic receptors present in the parasympathetic nerves that innervate airway smooth muscle by which OPs induce airway hyperreactivity [122]. OPs disrupt the control of the respiratory function in the brain stem, which leads to central apnea.

#### *4.5.2 Chronic bronchitis*

Chronic bronchitis is long-term inflammation of the lining of the bronchial tubes. The most common symptoms include cough, mucus, wheezing, chest discomfort that leads to disability, severe infection in the airways, narrowing of the breathing tubes, and hence trouble to breathe. Globally, such disease is the third leading cause of death, that is, over 3 million in 2019. It was reported that more than 80% of the documented deaths by such disease are in low and middle-income developing countries [123].

Such disease is one of the most common diseases caused by several factors, such as exposure to pesticides [124]. For example, in a case–control study carried out in Lebanon, it was found that pesticide exposure was associated with chronic bronchitis [125]. In India, it was found a higher occurrence of chronic bronchitis, which was associated with OP and carbamate pesticide exposures in agricultural workers [126]. In Vietnam, it was reported that 1499 Vietnam veterans who applied Agent Orange (the mixture of two equal parts of the herbicides 2,4-D and 2,4,5-T) showed a higher frequency of chronic respiratory diseases, such as chronic bronchitis [127]. Also, in Singapore, it was observed that pesticides were strongly associated with chronic bronchitis [128]. In the agricultural health study project, it was found that 16 pesticides were strongly associated with chronic bronchitis [129]. As mechanisms of action that are followed by the chemical pesticides to cause chronic bronchitis, it was documented that OPs as inhibitors to AChE, increase the acetylcholine quantity on nicotine and muscarinic receptors that lead to cholinergic over-expression on the smooth muscle of the airway hence causing broncho-constriction [130].

### 4.6 Autism

Autism has been recognized as the damage that occurred to many important areas of brain development. Autism has been defined as specific conditions of neurodevelopment that are characterized by specific, repetitive behavior, and difficulty in social communication. Also, autism is a condition in which a patient suffers from specific behavioral symptoms that result from many known and unknown biological factors based on brain dysfunctions that affect the developing brain's ability to handle information [131]. It has been observed that most children with autism suffer from difficulty in their ability to learn as a result of their mental retardation, although few of these children with autism have an average level of intelligence, although they sometimes suffer from epilepsy and audiovisual damage [132].

The statistics showed that there are 62 children with autism out of every 10.000 births [133]. It was found that there is an increase in the incidence of autism among children in many parts of the world and that many of these children were due to the exposure of their mothers during pregnancy to pesticides and other environmental pollutants, whether through direct exposure to pesticides during the application or non-direct exposure, such as the consumption of food contaminated with pesticide residues [134].

A study conducted by Shelton, et al. [135] showed that there is an increase in the risk rates of autism among children whose mothers lived near the fields where pesticides were applied. In some detail, Blatt et al. [136] reported the occurrence of disruption of the nerve conduction system of gamma-aminobutyric acid (GABA) by the action of some chemical pesticides, and it was found that there is a relationship between that and the incidence of autism in studies on the density of receptors in brain tissue. Also, the results of the study conducted by Lyall et al. [137] indicated that exposure to high levels of organochlorine pesticides during pregnancy was associated with autism in newborns. Also, it was documented in a cohort study that some exposure of pregnant women to pyrethroids leads to autism spectrum disorder (ASD) in their children [138]. Other pesticides cause autism, that is, metabolites of diazinon and chlorpyrifos as organophosphorus [139], besides organochlorines, pyrethroids, and carbamates insecticides [140].

Genetic analysis studies indicated that there are 206 genes appeared responsible for showing autism. This set of genes is present at many barriers in the human body, such as the blood–brain barrier, skin, intestine, placenta, and cellular barrier trophoblast. To reach this conclusion on the responsibility of these genes, the response of such genes to various chemical compounds was analyzed by the Comparative Toxicogenomics Database (CTD) [141].

### 4.7 Erectile dysfunction

Erectile dysfunction has been defined as a persistent inability to achieve or maintain an adequate erection for satisfactory complete sexual performance. Physiologically, penile erection is a neurovascular phenomenon that involves the coordination of three hemodynamic events, that is, elevated arterial inflow, relaxation of the sinusoidal smooth muscle, and decreased venous outflow. It also implies the interaction of the brain, nerves, neurotransmitters, and smooth and striated muscles. Any alteration in one or more of these components may affect the erectile tissue and cause erectile dysfunction [142]. Many factors play an important and major role in the pathogenesis of erectile dysfunction, such as exposure to pesticides [142].

It was established that sexual behavior in humans is controlled by hormonal and neural regulatory processes, therefore, some pesticides that act as hormone disruptors negatively affect the nature of the sexual relationship. Various studies showed that erectile dysfunction is responsible for infertility for up to 10% of the male population around the world, as mentioned in the review published by Kaur et al. [142].

In Egypt, Soliman et al. [143] conducted a study in the Damietta governorate. The results showed that there is a close relationship between chronic exposure to pesticides (DDT, and some organophosphorus and carbamates) and erectile dysfunction. Besides, it was possible to prove that acetamiprid as a neonicotinoid insecticide has the most damaging effect on erectile dysfunction due to the effect on several inhibitory pathways [144]. Also, it was demonstrated that some pesticides exert their effect on tunica albuginea, TA tissues (the fibrous envelope of connective tissue that surrounds the corpora cavernosa of the penis, TA composed of elastin and collagen, so, the effect on elastin leads to erectile dysfunction) [145].

It was established that some organophosphorus pesticides cause a decrease in the concentration of the male hormone, testosterone, through various mechanisms. Also, it was reported that the decrease of such hormone was related to inhibition in the release of luteinizing hormone (LH) and follicle-stimulating hormone (FSH) [146]. Such inhibition occurs because organophosphorus pesticides inhibit the enzyme acetylcholine esterase and as a result, the level of the neurotransmitter acetylcholine increases and thus affects the inhibition of these hormones [147]. The pesticides were found to be responsible for the induction of apoptosis in Leydig cells, which were responsible for the secretion of 95% of the testosterone in the blood. Therefore, the death of these cells results in a significant reduction in the concentration of that hormone [148].

### 4.8 Psychological disorder

Various studies showed that there is a relationship between poisoning farmers with pesticides and the psychological problems that they suffer especially depressive disorders [149]. London et al. [150, 151] explained that many pesticides, especially organophosphorus, were associated with an increase in the occurrence of psychological problems, that is, depression, which has sometimes reached the suicide of some workers who were previously exposed to such pesticides. With depression, various studies showed the relationship between pesticides and decreasing serotonin levels in workers exposed to pesticides [152]. It is well established for the human being the relationship between the lowering in serotonin levels and depression [153]. At the experimental animal level, it was reported that some pesticides, that is, deltamethrin (pyrethroid insecticide) and acetamiprid (neonicotinoid insecticide) [154], chlorpyrifos (organophosphorus insecticide), and cypermethrin (pyrethroid insecticide) [155] caused a decrease in serotonin and dopamine levels in rats.

Some studies showed links between exposure to pesticides and suicide rates, which has been reported by Faria et al. [156]. For example, it was reported that exposure to high levels of organophosphorus pesticides was associated with higher rates of suicide among workers exposed to these pesticides [157]. The same phenomenon has been recorded in many countries, such as Brazil [156] and Costa Rica [158].

In the case of the developing countries, as mentioned in the systematic review published by Gunnell et al. [159], it was estimated that the suicidal attempts due to pesticide toxicity ranged between 5200 and 21,910 in African countries. In Central America, especially, Belize, Costa Rica, El Salvador among other countries, it was documented that pesticides account for 31% of suicide cases in this region. In Eastern Mediterranean countries about 16.5% and 5629, in South East Asia, that is, India, Bangladesh, Sri Lanka, and Thailand, the proportion and annual total of pesticide suicides in this region reached 20.7% and 51,050, with a range of 47,720 to 82,680 cases.

## 5. Conclusions

Various developing countries are still using the banned chemical pesticides as a reason to the cheap price, ease of production or importation, and lack of both registration and control measures systems. Based on the aforementioned diseases caused by banned pesticides, exposure to such substances must be prevented as much as possible in developing countries. To continue the pest control process, an alternative method must be followed, that is, integrated pest management (IPM) strategies, using biopesticides to ensure that there are no residues of harmful chemical pesticides, whether they have been canceled or are still allowed to be used. The possibilities of getting rid of the large quantities of banned pesticides are technologically not available in all developing countries. Therefore, the developed countries must cooperate and the international institutions, organizations, and/or agencies, such as Food and Agriculture Organization (FAO) and the United States Environmental Protection Agency (USEPA) among others, must contribute to disbursing the material compensation and making alternative pesticides available to the developing countries. Such agencies must cooperate to overcome the problems related to using pesticides in developing countries that suffer from the import, export of banned pesticides, lack of training on the correct handling of pesticides, low pesticide education, lack of legislation, lack of enforcement of the available legislation, and absence of monitoring for pesticides residues on locally consumed products. The expected assistance from these bodies may be as providing safe equipment, education, training the farmers, and licensed pesticides applicators in the developing countries through various sustainable, not temporary programs.

In addition, the governments of developing countries must bear their responsibility to establish specialized bodies responsible for managing everything related to the use of pesticides, which have the authority to prevent the import of internationally banned pesticides, under the supervision of the relevant international bodies and to combat the counterfeit and contraband pesticides. So that, through following these suggested realistic practical solutions, it could possible to stop and prevent the continuation of using the banned pesticides at the international level. Following these policies, it may contribute positively to reducing the incidence of the aforementioned diseases and others associated with exposure to pesticides.

## Author details

Alaa Eldin Bayoumi
Faculty of Agriculture, Department of Plant Protection, Ain Shams University, Shoubra El-Kheima, Cairo, Egypt

*Address all correspondence to: alaaeb@agr.asu.edu.eg

## References

[1] European Parliament. The Use of Pesticides in Developing Countries and their Impact on Health and the Right to Food. Belgium: Policy Department for External Relations, Directorate-General for External Policies of the Union; 2021. pp. 1-44. Available from: https://www.europarl.europa.eu/cmsdata/219887/Pesticides%20health%20and%20food.pdf [Accessed: August 12, 2021]

[2] Atwood D, Jones PC. Pesticides Industry Sales and Usage 2008-2012 Market Estimates. Washington DC, USA: U.S. Environmental Protection Agency; 2017. pp. 1-24

[3] Stenersen J, editore. Chemical Pesticides Mode of Action and Toxicology. 1st ed. USA: CRC Press; 2004. p. 296

[4] Damalas CA, Eleftherohorinos IG. Pesticide exposure, safety issues, and risk assessment indicators. International Journal of Environmental Research and Public Health. 2018;**5**:1402-1419. DOI: 10.3390/ijerph8051402

[5] Tsai WT. Organochlorine Insecticides. In: PH W, editor. Encyclopedia of Toxicology. 3rd ed. Cambridge, Massachusetts, USA: Academic Press; 2014. pp. 711-713

[6] Brenda, G. 2011 Pesticide Exposure in Womb Linked to Lower IQ, Health & Pregnancy. Atlanta, USA: WebMD; 2011. Available from: https://www.webmd.com/baby/news/20110421/pesticide-exposure-in-womb-linked-to-lower-iq#1 [Accessed: September 3, 2021]

[7] Cone M. Toxic Pesticide Banned after Decades of Use. New York, USA: Environ. Health News; 2010. Available from: https://www.scientificamerican.com/article/toxic-pesticide-banned-after-decades-of-use/ [Accessed: September 3, 2021]

[8] PAN, Pesticide Action Network International PAN International, Consolidated List of Banned Pesticides. New Taipei City, Taiwan: PAN International; 2021. Available from: https://pan-international.org/pan-international-consolidated-list-of-banned-pesticides/ [Accessed: August 15, 2021]

[9] WHO. The WHO Recommended Classification of Pesticides by Hazard and Guidelines to Classification. Geneva: World Health Organization; Licence: CC BY-NC-SA 3.0 IGO. 2019 edition; 2020. Available from: https://apps.who.int/iris/bitstream/handle/10665/332193/9789240005662-eng.pdf [Accessed: August 16, 2021]

[10] Alavanja MCR, Ross MK, Bonner MR. Increased cancer burden among pesticide applicators and others due to pesticide exposure. CA: a Cancer Journal for Clinicians. 2013;**63**(2): 120-142. DOI: 10.3322/caac.21170

[11] USEPA, Overview of Risk Assessment in the Pesticide Program. North Carolina, USA: EPA; 2021. Available from: https://www.epa.gov/pesticide-science-and-assessing-pesticide-risks/overview-risk-assessment-pesticide-program [Accessed: August 27, 2021]

[12] Martin T, Thompson H, Thorbek P, Ashauer R. Toxicokinetic-Toxicodynamic modeling of the effects of pesticides on growth of *Rattus norvegicus*. Chemical Research in Toxicology. 2019;**32**(11):2281-2294. DOI: 10.1021/acs.chemrestox.9b00294

[13] Garcês A, Pires I, Rodrigues P. Teratological effects of pesticides in vertebrates: A review. Journal of Environmental Science and Health, Part B. 2020;**55**(1):75-89. DOI: 10.1080/03601234.2019.1660562

[14] Yadav H, Sankhla MS, Kumar R. Pesticides-induced carcinogenic & neurotoxic effect on human. Forensic Research & Criminology International Journal. 2019;7(5):243-245. DOI: 10.15406/frcij.2019.07.00288

[15] IARC, Monographs on the Identification of Carcinogenic Hazards to Humans. Lyon, France: IARC; 2021. Available from: https://monographs.iarc.who.int/agents-classified-by-the-iarc/ [Accessed: August 26, 2021]

[16] Anwar WA. Biomarkers of human exposure to pesticides. Environmental Health Perspectives. 1997;**105**(4):801-806. DOI: 10.1289/ehp.97105s4801

[17] Wafa T, Nadial K, Amel N, Ikbal C, Insaf T, Asma K, et al. Oxidative stress, hematological and biochemical alterations in farmers exposed to pesticides. Journal of Environmental Science and Health. Part. B. 2013;**48**(12):1058-1069. DOI: 10.1080/03601234.2013.824285

[18] Attia AM. Risk assessment of occupational exposure to pesticides. In: Linkov I, Ramadan AB, editors. Comparative Risk Assessment and Environmental Decision Making. Nato Science Series, IV, Earth and Environmental Sciences. Dordrecht: Springer; 2004. pp. 349-362. DOI: 10.1007/1-4020-2243-3_23

[19] Garcia-Garcia CR, Parron T, Requena M, Alarcon R, Tsatsakis AM, Hernández AF. Occupational pesticide exposure and adverse health effects at the clinical, hematological and biochemical level. Life Sciences. 2016;**15**(145): 274-283. DOI: 10.1016/j.lfs.2015.10.013

[20] Kori RK, Jain AK, Yadav RS. Biomarkers: An essential gizmo in pesticide toxicity. Biomarkers. 2016;**2**:1-5. DOI: 10.21767/2472-1646.100009

[21] Muniz JF, McCauley L, Scherer J, Lasarev M, Koshy M, Kow Y. Biomarkers of oxidative stress and DNA damage in agricultural workers: A pilot study. Toxicology and Applied Pharmacology. 2008;**227**:97-107. DOI: 10.1016/j.taap.2007.10.027

[22] WHO. Biomarkers and Risk Assessment: Concepts and Principles. Geneva: Environmental Health Criteria; 1993. p. 155

[23] Russell WMS, Burch RL, editors. The Principles of Humane Experimental Technique. London: Methuen & Co. Limited; 1959. p. 238

[24] Liebsch M, Grune B, Seiler A, Butzke D, Oelgeschäger M, Pirow R, et al. Alternatives to animal testing: Current status and future perspectives. Archives of Toxicology. 2011;**85**(8):841-858. DOI: 10.1007/s00204-011-0718-x

[25] Rees KR. Cells in culture in toxicity testing: A review. Journal of the Royal Society of Medicine. 1980;**73**(4):261-264

[26] Tebourbi O, Sakly M, Rhouma KB. In: Stoytcheva M, editor. Molecular Mechanisms of Pesticide Toxicity, Pesticides in the Modern World - Pests Control and Pesticides Exposure and Toxicity Assessment. London, UK: InTech; 2011. Available from: https://www.intechopen.com/chapters/20786 [Accessed: August 24, 2021]. DOI: 10.5772/17952. ISBN: 978-953-307-457-3

[27] Lopez SL, Delia Aiassa D, Benitez-Leite S, Lajmanovich R, Mañas F,

Poletta G, et al. Chapter two: Pesticides used in south American GMO-based agriculture: A review of their effects on humans and animal models. Advances in Molecular Toxicology. 2012;**6**:41-75. DOI: 10.1016/B978-0-444-59389-4. 00002-1

[28] Kesavachandran CN, Fareed M, Pathak MK, Bihari V, Mathur N, Srivastava AK. Adverse health effects of pesticides in agrarian populations of developing countries. Reviews of Environmental Contamination and Toxicology. 2009;**200**:33-52. DOI: 10.1007/978-1-4419-0028-9_2

[29] Abdou KA, Abo El-Atta HM. Epidemiology of pesticides in developing countries. Advances in Clinical Toxicology (ACT). 2018;**3**(1):1-8

[30] Rojas M, Weiss P, Rodríguez M, Agreda O. Determinants of pesticide poisonings in the central-western region of Venezuela. Chinese Society of Toxicology. 2003;**7**(4):16-16

[31] Eskenazi B, Marks AR, Bradman A, Fenster L, Johnson C, Barr DB, et al. In utero exposure to dichlorodiphenyltrichloroethane (DDT) and dichlo-rodiphenyldichloroethylene (DDE) and neurodevelopment among young Mexican American children. Pediatrics. 2006;**118**(1):233-241. DOI: 10.1542/peds.2005-3117

[32] Hung DZ, Yang HJ, Li YF, Lin CL, Chang SY, Sung FC, et al. The long-term effects of organophosphates poisoning as a risk factor of CVDs: A nationwide population-based cohort study. PLoS One. 2015;**10**:e0137632. DOI: 10.1371/journal.pone.0137632

[33] Jurewicz J, Radwan M, Wielgomas B, Sobala W, Piskunowicz M, Radwan P, et al. The effect of environmental exposure to pyrethroids and DNA dam-age in human sperm. Systems Biology in Reproductive Medicine. 2015;**61**(1):37-43. DOI: 10.3109/19396368.2014.981886

[34] Shafer TJ, Meyer DA, Crofton KM. Developmental neurotoxicity of pyrethroid insecticides: Critical review and future research needs. Environmental Health Perspectives. 2005;**113**:123-136. DOI: 10.1289/ehp.7254

[35] Kettles MK, Browning SR, Prince TS, Horstman SW. Triazine herbicide exposure and breast cancer incidence: An ecologic study of Kentucky counties. Environmental Health Perspectives. 1997;**105**:1222-1227. DOI: 10.1289/ehp.971051222

[36] Huyen VH, Song NV, Thuy NT, Dung LP, Hoan LK. Effects of of pesticides on farmers' health in Tu Ky district, Hai Duong province, Vietnam. Sustainable Futures. 2020;**2**:1-26

[37] Agricultural Pesticide Committee, APC. Activities. Cairo, Egypt: Ministry of Agriculture; 2021. Available from: www.apc.gov.eg [Accessed: August 26, 2021]

[38] Van den Berg H. Global status of DDT and its alternatives for use in vector control to prevent disease. Environmental Health Perspectives. 2009;**117**:1656-1663. DOI: 10.1289/ehp.0900785

[39] Ecobichon DJ. Pesticide use in developing countries. Toxicology. 2001;**160**(1-3):27-33. DOI: 10.1016/s0300-483x(00)00452-2

[40] Rehwagen C. WHO recommends DDT to control malaria. BMJ. 2006;**333**(7569):622. DOI: 10.1136/bmj.333.7569.622-b

[41] Lucena RA, Allam MF, Jimenez SS, Villarejo ML. A review of environmental

exposure to persistent organochlorine residuals during the last fifty years. Current Drug Safety. 2007;**2**:163-172. DOI: 10.2174/157488607780598313

[42] Avila-Vazquez M, Maturano E, Etchegoyen A, Difilippo FS, Maclean B. Association between Cancer and environmental exposure to glyphosate. International Journal of Clinical Medicine. 2017;**8**:73-85. DOI: 10.4236/ijcm.2017.82007

[43] Erwin X. Briefing on Export of Pesticides to Africa from China in 2015. California, USA: AgNews; 2016. Available from: http://news.agropages.com/News/NewsDetail---17241.htm [Accessed: July 30, 2021]

[44] Bayoumi AE. Counterfeit pesticides. ACS Chemical Health & Safety. 2021;**28**:232-237. DOI: 10.1021/acs.chas.1c00010

[45] Coats JR. Mechanisms of toxic action and structure-activity relationships for organochlorine and synthetic pyrethroid insecticides. Environmental Health Perspectives. 1990;**87**:255-262. DOI: 10.1289/ehp.9087255

[46] Pope CN. Organophosphorus pesticides: Do they all have the same mechanism of toxicity? Journal of Toxicology and Environmental Health. Part B, Critical Reviews. 1999;**2**(2):161-181. DOI: 10.1080/109374099281205

[47] Abou-Donia MB. Organophosphorus ester-induced chronic neurotoxicity. Archives of Environmental Health. 2003;**58**(8):484-497. DOI: 10.3200/AEOH.58.8.484-497

[48] Karami-Mohajeri S, Abdollahi M. Toxic influence of organophosphate, carbamate, and organochlorine pesticides on cellular metabolism of lipids, proteins, and carbohydrates: A systematic review. Human & Experimental Toxicology. 2010;**30**:1119-1140. DOI: 10.1177/0960327110388959

[49] Ecobichon DJ. Toxic effects of pesticides. In: Klaassen CD, editor. Casarett & Doull's Toxicology, the Basic Science of Poisons. 6th ed. New York, USA: McGraw Hill; 1996. pp. 763-810

[50] Gervais JA, Luukinen B, Buhl K, Stone D. Imidacloprid Technical Fact Sheet. Vol. 2012. Oregon State University, USA: National Pesticide Information Center; 2012. Available from: http://npic.orst.edu/factsheets/archive/imidacloprid.html. [Accessed: September 4, 2021]

[51] Bakir F, Rustam H, Tikriti S, Al-Damluji SF, Shihristani H. Clinical and epidemiological aspects of methylmercury poisoning. Postgraduate Medical Journal. 1980;**56**(651):1-10. DOI: 10.1136/pgmj.56.651.1

[52] Edwards R, Ferry DC, Temple WA. Fungicides and related compounds. In: Hayes WJ, Law ER, editors. Handbook of Pesticide Toxicology. Classes of Pesticides. Vol. 3. Cambridge, Massachusetts, USA: Academic Press; 1991. pp. 1409-1470

[53] Chmelíková E, Markéta S, Michal J, David N. Endocrine disruptors: General characteristics, chemical nature and mechanisms of action. A review. Medical Journal of Cell Biology. 2018;**6**(4):135-139. DOI: 10.2478/acb-2018-0022

[54] Mnif W, Hassine AIH, Bouaziz A, Bartegi A, Thomas O, Roig B. Effect of endocrine disruptor pesticides: A review. International Journal of Environmental Research and Public Health. 2011;**8**:2265-2303. DOI: 10.3390/ijerph8062265

[55] Diamanti-Kandarakis E, Bourguignon JP, Giudice LC, Hauser R,

Prins GS, Soto AM, et al. Endocrine-disrupting chemicals: An Endocrine Society scientific statement. Endocrine Reviews. 2009;**30**(4):293-342. DOI: 10.1210/er.2009-0002

[56] Meeker JD. Exposure to environmental endocrine disruptors and child development. Archives of Pediatrics & Adolescent Medicine. 2012;**166**(6):E1-E7. DOI: 10.1001/archpediatrics.2012.241

[57] Shah R. Pesticides and Human Health. London, UK: Licensee IntechOpen; 2020. Available from: https://www.intechopen.com/chapters/73921 [Accessed: August 25, 2021]

[58] Ng M, Flemeng T, Robinson M, Thomson B, Graetz N, Margono C, et al. Global, regional, and national prevalence of overweight and obesity in children and adults during 1980-2013: A systematic analysis for the global burden of disease study 2013. Lancet. 2014;**384**:766-781. DOI: 10.1016/S0140-6736(14)60460-8

[59] Schwartz MW, Seeley RJ, Zeltser LM, Drewnowski A, Ravussin E, Redman LM, et al. Obesity pathogenesis: An Endocrine Society scientific statement. Endocrine Reviews. 2017;**38**:267-296. DOI: 10.1210/er.2017-00111

[60] Picon-Ruiz M, Morata-Tarifa C, Valle-Goffin JJ, Friedman ER, Slingerland JM. Obesity and adverse breast cancer risk and outcome: Mechanistic insights and strategies for intervention. CA: a Cancer Journal for Clinicians. 2017;**67**(5):378-397. DOI: 10.3322/caac.21405

[61] Wei C, Wang X, Yao X, Xi F, He Y, et al. Bifenthrin induces fat deposition by improving fatty acid uptake and inhibiting lipolysis in mice. Journal of Agricultural and Food Chemistry. 2019;**67**(51):14048-14055. DOI: 10.1021/acs.jafc.9b06727

[62] Sun Q, Xiao X, Kim Y, Kim D, et al. Imidacloprid promotes high fat diet-induced adiposity and insulin resistance in male C57BL/6J mice. Journal of Agricultural and Food Chemistry. 2016;**64**:9293-9306. DOI: 10.1021/acs.jafc.6b04322

[63] OECD, Overweight and Obesity among Children and Adolescents. 2020. Available from: https://www.oecd-ilibrary.org/sites/7402dbb2-en/index.html?itemId=/content/component/7402dbb2-en [Accessed: August 9, 2021]

[64] Ren XM, Kuo Y, Blumberg B. Agrochemicals and obesity. Molecular and Cellular Endocrinology. 2020;**515**:110926. DOI: 10.1016/j.mce.2020.110926

[65] Mustieles V, Fernández MF, Martin-Olmedo P, Alzaga BG, et al. Human adipose tissue levels of persistent organic pollutants and metabolic syndrome components: Combining a cross-sectional with a 10-year longitudinal study using a multi-pollutant approach. Environment International. 2017;**104**:48-57. DOI: 10.1016/j.envint.2017.04.002

[66] Grun F, Blumberg B. Environmental obesogens: Organotins and endocrine disruption via nuclear receptor signaling. Endocrinology. 2006;**147**:50-55. DOI: 10.1210/en.2005-1129

[67] Guo S. Insulin signaling, resistance, and metabolic syndrome: Insights from mouse models into disease mechanisms. The Journal of Endocrinology. 2014;**220**:T1-T23. DOI: 10.1530/JOE-13-0327

[68] Adeva-Andany MM, Martínez-Rodríguez J, González-Lucán M, et al. Insulin resistance is a cardiovascular risk factor in humans. Diabetes and Metabolic Syndrome: Clinical Research and Reviews. 2019;**13**:1449-1455. DOI: 10.1016/j.dsx.2019.02.023

[69] Mu N, Zhu Y, Wang Y, Zhang H, Xue F. Insulin resistance: A significant risk factor of endometrial cancer. Gynecologic Oncology. 2012;**125**(3):751-757. DOI: 10.1016/j.ygyno.2012.03.032

[70] Di Sebastiano KM, Pinthus JH, Duivenvoorden WC, Mourtzakis M. Glucose impairments and insulin esistance in prostate cancer: The role of obesity, nutrition and exercise. Obesity Reviews. 2018;**19**(7):1008-1016. DOI: 10.1111/obr.12674

[71] Cirillo F, Catellani C, Sartori C, Lazzeroni P, et al. Obesity, insulin resistance and colorectal Cancer: Could miRNA dysregulation play a role? International Journal of Molecular Sciences. 2019;**20**:2922. DOI: 10.3390/ijms20122922

[72] WHO, Diabetes. Geneva, Switzerland: WHO; 2021. Available from: https://www.who.int/news-room/fact-sheets/detail/diabetes [Accessed: August 15, 2021]

[73] Son HK, Kim SA, Kang JH, Chang YS, et al. Strong associations between low-dose organochlorine pesticides and type 2 diabetes in Korea. Environment International. Amsterdam: Elsevier; 2010;**36**:410-414. DOI: 10.1016/j.envint.2010.02.012

[74] Juntarawijit C, Juntarawijit Y. Association between diabetes and pesticides: A case-control study among Thai farmers. Environmental Health and Preventive Medicine. 2018;**23**(3):1-10. DOI: 10.1186/s12199-018-0692-5

[75] Gupta HPK, Pandey R, Anjum S. Occupational exposure to pesticide: A risk of diabetic neuropathy. Journal of Analytical Toxicology. 2018;**1**(14):1-5

[76] El-Morsi DA, Abdel Rahman RH, Abou-Arab AAK. Pesticides residues in Egyptian diabetic children: A preliminary study. Journal of Clinical Toxicology. 2012;**2**(6):1-5. DOI: 10.21608/mjfmct.2012.47773

[77] Lee SH, Kim Y, Kim HY, Kim YH, Kim MS, et al. Aminostyrylbenzofuran directly reduces oligomeric amyloid-ß and reverses cognitive deficits in Alzheimer transgenic mice. PLoS One. 2014;**9**:57-63. DOI: 10.1371/journal.pone.0095733

[78] Ruzzin J, Petersen R, Meugnier E, Madsen L, Lock EJ, et al. Persistent organic pollutant exposure leads to insulin resistance syndrome. Environmental Health Perspectives. 2010;**118**:465-471. DOI: 10.1289/ehp.0901321

[79] Lind PM, Lind L. Endocrine-disrupting chemicals and risk of diabetes: An evidence-based review. Diabetologia. 2018;**61**:1495-1502. DOI: 10.1007/s00125-018-4621-3

[80] Han X, Zhang F, Meng L, Xu Y, Li Y, et al. Exposure to organochlorine pesticides and the risk of type 2 diabetes in the population of East China. Ecotoxicology and Environmental Safety. 2020;**190**:110125. DOI: 10.1016/j.ecoenv.2019.110125

[81] van Koppen CJ, Kaiser B. Regulation of muscarinic acetylcholine receptor signaling. Pharmacology & Therapeutics. 2003;**98**:197-220. DOI: 10.1016/s0163-7258(03)00032-9

[82] Goedert M, Spillantini MG. A century of Alzheimer's disease. Science. 2006;**314**:777-781. DOI: 10.1126/science.1132814

[83] Burns A, Iliffe S. Alzheimer's disease. BMJ. 2009;**338**:b158. DOI: 10.1136/bmj.b158

[84] Singh NK, Chhillar N, Banerjee BD, et al. Organochlorine pesticide levels and risk of Alzheimer's disease in north Indian population. Human & Experimental Toxicology. 2013;**32**(1): 24-30. DOI: 10.1177/0960327112456315

[85] Yadav SS, Singh MK, Yadav RS. Organophosphates induced Alzheimer's disease: An epigenetic aspect. Journal of Clinical Epigenetics. 2016;**2**(1):1-8. DOI: 10.21767/2472-1158.100010

[86] Yan D, Zhang Y, Liu L, Yan H. Pesticide exposure and risk of Alzheimer's disease: A systematic review and meta-analysis. Scientific Reports. 2016;**6**:32222. DOI: 10.1038/srep32222

[87] Richardson JR, Roy A, Shalat SL, von Stein RT, Hossain MM, et al. Elevated serum pesticide levels and risk for Alzheimer disease. JAMA Neurology. 2014;**71**(3):284-290. DOI: 10.1001/jamaneurol.2013.6030

[88] Breunig JJ, Guillot-Sestier MV, Town T. Brain injury, neuroinflammation and Alzheimer's disease. Frontiers in Aging Neuroscience. 2013;**5**:26. DOI: 10.3389/fnagi.2013.00026

[89] Nagele RG, Andrea MR, Anderson WJ, Wang HY. Intracellular accumulation of beta-amyloid (1-42) in neurons is facilitated by the alpha 7 nicotinic acetylcholine receptor in Alzheimer's disease. Neuroscience. 2002;**110**:199-211. DOI: 10.1016/s0306-4522(01)00460-2

[90] Lee J, Giordano S, Zhang J. Autophagy, mitochondria and oxidative stress: Cross-talk and redox signaling. Journal of Biochemistry. 2012;**441**:523-540. DOI: 10.1042/BJ20111451

[91] Wang X, Wang W, Li L, Perry G, et al. Oxidative stress and mitochondrial dysfunction in Alzheimer's disease. Biochimica et Biophysica Acta. 2014;**1842**(8):1240-1247. DOI: 10.1016/j.bbadis.2013.10.015

[92] Baltazar MT, Dinis-Oliveira RJ, Bastos ML, et al. Pesticides exposure as etiological factors of Parkinson's disease and other neurodegenerative diseases–a mechanistic approach. Toxicology Letters. 2014;**230**:85-103. DOI: 10.1016/j.toxlet.2014.01.039

[93] Salazar JG, Ribes D, Cabré M, Domingo JL, et al. Amyloid beta peptide levels increase in brain of AβPP Swedish mice after exposure to chlorpyrifos. Current Alzheimer Research. 2011;**8**:732-740. DOI: 10.2174/156720511797633197

[94] Hayden KM, Norton MC, Darcey D, et al. Occupational exposure to pesticides increases the risk of incident AD: The Cache County study. Neurology. 2010;**74**:1524-1530. DOI: 10.1212/WNL.0b013e3181dd4423

[95] Chin-Chan M, Yepes JN, Vega BQ. Environmental pollutants as risk factors for neurodegenerative disorders: Alzheimer and Parkinson diseases. Frontiers in Cellular Neuroscience. 2015;**9**(124):1-22. DOI: 10.3389/fncel.2015.00124

[96] Tang BL. Neuropathological mechanisms associated with pesticides in Alzheimer's disease. Toxics. 2020;**8**(21):1-16. DOI: 10.3390/toxics8020021

[97] Lin JN, Lin CL, Lin MC, Lai CH, et al. Increased risk of dementia in patients with acute organophosphate and carbamate poisoning: A nationwide population-based cohort study. Medicine (Baltimore). 2015;**94**(29):e1187. DOI: 10.1097/MD.0000000000001187

[98] Zhao N, Yang Y, Zhang L, Zhang Q, et al. Quality of life in Parkinson's disease: A systematic review and

meta-analysis of comparative studies. CNS Neuroscience & Therapeutics. 2021;**27**:270-279. DOI: 10.1111/cns.13549

[99] Barbeau A, Roy M, Bernier G, Campanella G, Paris S. Ecogenetics of Parkinson's disease: Prevalence and environmental aspects in rural areas. The Canadian Journal of Neurological Sciences. 1987;**14**(1):36-41. DOI: 10.1017/s0317167100026147

[100] Wang A, Costello S, Cockburn M, et al. Parkinson's disease risk from ambient exposure to pesticides. European Journal of Epidemiology. 2011;**26**(7):547-555. DOI: 10.1007/s10654-011-9574-5

[101] Liu C, Liu Z, Zhang Z, Li Y, Fang R, et al. Scientometric analysis and visualization of research on parkinson's disease associated with pesticide exposure. Frontiers in Public Health. 2020;**8**(91):1-14. DOI: 10.3389/fpubh.2020.00091

[102] Ahmed H, Abushouk AI, Gabr M, Negida A, Abdel-Daim AM. Parkinson's disease and pesticides: A meta-analysis of disease connection and genetic alterations. Biomedicine & Pharmacotherapy. 2017;**90**:638-649. DOI: 10.1016/j.biopha.2017.03.100

[103] Shrestha S, Parks CG, Umbach DM, Barber MR, et al. Pesticide use and incident Parkinson's disease in a cohort of farmers and their spouses. Environmental Research. 2020;**191**:110186. DOI: 10.1016/j.envres.2020.110186

[104] Karen DJ, Li W, Harp PR, Gillette JS, Bloomquist JR. Striatal dopaminergic pathways as a target for the insecticides permethrin and chlorpyrifos. Neurotoxicology. 2001;**22**:811-817. DOI: 10.1016/s0161-813x(01)00063-8

[105] Kanthasamy AG, Kitazawa M, Kanthasamy A, Anantharam V. Dieldrin-induced neurotoxicity: Relevance to Parkinson's disease pathogenesis. Neurotoxicology. 2005;**26**:701-719. DOI: 10.1016/j.neuro.2004.07.010

[106] Wang XF, Li S, Chou AP, Bronstein JM. Inhibitory effects of pesticides on proteasome activity: Implication in Parkinson's disease. Neurobiology of Disease. 2006;**23**:198-205. DOI: 10.1016/j.nbd.2006.02.012

[107] Brozek G, Lawson J, Szumilas D, Zejda J. Increasing prevalence of asthma, respiratory symptoms, and allergic diseases: Four repeated surveys from 1993-2014. Respiratory Medicine. 2015;**109**:982-990. DOI: 10.1016/j.rmed.2015.05.010

[108] Baldi I, Robert C, Piantoni F, Tual S, et al. Agricultural exposure and asthma risk in the AGRICAN French cohort. International Journal of Hygiene and Environmental Health. 2014;**217**:435-442. DOI: 10.1016/j.ijheh.2013.08.006

[109] Hyland C, Laribi O. Review of take-home pesticide exposure pathway in children living in agricultural areas. Environmental Research. 2017;**156**:559-570. DOI: 10.1016/j.envres.2017.04.017

[110] Marks AR, Harley K, Bradman A, et al. Organophosphate pesticide exposure and attention in young Mexican-American children: The CHAMACOS study. Environmental Health Perspectives. 2010;**118**:1768-1774. DOI: 10.1289/ehp.1002056

[111] Asmus CIRF, Camara VM, Landrigan PJ, Claudio L. A systematic review of Children's environmental health in Brazil. Annals of Global Health. 2016;**82**:132-148. DOI: 10.1016/j.aogh.2016.02.007

[112] de Joode BVW, Mora AM, Lindh CH, et al. Pesticide exposure and

neurodevelopment in children aged 6-9 years from Talamanca, Costa Rica. Cortex. 2016;**85**:137-150. DOI: 10.1016/j.cortex.2016.09.003

[113] Kudagammana ST, Mohotti K. Environmental exposure to agrochemicals and allergic diseases in preschool children in high grown tea plantations of Sri Lanka. Allergy, Asthma and Clinical Immunology. 2018;**14**:1-5. DOI: 10.1186/s13223-018-0308-z

[114] Salameh PR, Baldi I, Brochard P, Raherison C, et al. Respiratory symptoms in children and exposure to pesticides. The European Respiratory Journal. 2003;**22**:507-512. DOI: 10.1183/09031936.03.00107403a

[115] Ohayo-Mitoko GJ, Kromhout H, Simwa JM, et al. Self reported symptoms and inhibition of acetylcholinesterase activity among Kenyan agricultural workers. Occupational and Environmental Medicine. 2000;**57**:195-200. DOI: 10.1136/oem.57.3.195

[116] Quansah R, Bend JR, Abdul-Rahaman A, et al. Associations between pesticide use and respiratory symptoms: A cross-sectional study in southern Ghana. Environmental Research. 2016;**150**:245-254. DOI: 10.1016/j.envres.2016.06.013

[117] Negatu B, Kromhout H, Mekonnen Y, Vermeulen R. Occupational pesticide exposure and respiratory health: A large-scale cross-sectional study in three commercial farming systems in Ethiopia. Thorax. 2017;**72**:1-8. DOI: 10.1136/thoraxjnl-2016-208924

[118] Callahan CL, Al-Batanony M, Ismail AA, et al. Chlorpyrifos exposure and respiratory health among adolescent agricultural workers. International Journal of Environmental Research and Public Health. 2014;**11**:13117-13129. DOI: 10.3390/ijerph111213117

[119] Hoppin JA, Umbach DM, London SJ, et al. Pesticide use and adult-onset asthma among male farmers in the agricultural health study. The European Respiratory Journal. 2009;**34**:1296-1303. DOI: 10.1183/09031936.00005509

[120] Avila-Vazquez M, Difilippo FS, Lean BM, Maturano E. Environmental exposure to glyphosate and risk of asthma in an ecological study. Global Journal of Medical Research. 2021;**21**(1):15-22. DOI: 10.34257/GJMRFVOL21IS1PG15

[121] Hansen MRH, Jørs E, Sandbæk A, Sekabojja D, et al. Organophosphate and carbamate insecticide exposure is related to lung function change among smallholder farmers: A prospective study. Thorax. 2021;**76**(8):780-789. DOI: 10.1136/thoraxjnl-2020-214609

[122] Shaffo FC, Grodzki AC, Fryer AD, Lein PJ. Mechanisms of organophosphorus pesticide toxicity in the context of airway hyperreactivity and asthma. American Journal of Physiology. Lung Cellular and Molecular Physiology. 2018;**315**:L485-L501. DOI: 10.1152/ajplung.00211.2018

[123] WHO, Chronic Obstructive Pulmonary Disease (COPD). Geneva, Switzerland: WHO; 2021. Available from: https://www.who.int/news-room/fact-sheets/detail/chronic-obstructive-pulmonary-disease-(copd) [Accessed: August 27, 2021]

[124] Sprince NL, Lewis MQ, Whitten PS, et al. Respiratory symptoms: Associations with pesticides, silos and animal confinement in the Iowa farm family health and hazard surveillance project. American Journal of Industrial Medicine. 2000;**38**(4):455-462. DOI: 10.1002/1097-0274(200010)38:4<455::aid-ajim12>3.0.co;2-l

[125] Salameh PR, Waked M, Baldi I, Brochard P, Saleh BA. Chronic bronchitis and pesticide exposure: A case–control study in Lebanon. European Journal of Epidemiology. 2006;**21**(9):681-688. DOI: 10.1007/s10654-006-9058-1

[126] Chakraborty S, Mukherjee S, Roychoudhury S, et al. Chronic exposures to cholinesterase-inhibiting pesticides adversely affect respiratory health of agricultural workers in India. Journal of Occupational Health. 2009;**51**:488-497. DOI: 10.1539/joh.l9070

[127] Kang HK, Dalager NA, Needham LL, et al. Health status of Army chemical corps Vietnam veterans who sprayed defoliant in Vietnam. American Journal of Industrial Medicine. 2006;**49**(11):875-884. DOI: 10.1002/ajim.20385

[128] LeVan TD, Von Essen S, Romberger DJ, et al. Polymorphisms in the CD14 gene associated with pulmonary function in farmers. American Journal of Respiratory and Critical Care Medicine. 2005;**171**:773-779. DOI: 10.1164/rccm.200404-530OC

[129] Hoppin JA, Valcin M, Henneberger PK, Kullman GJ, et al. Pesticide use and chronic bronchitis among farmers in the agricultural health study. American Journal of Industrial Medicine. 2007;**50**(12):969-979. DOI: 10.1002/ajim.20523

[130] Tarmure S, Alexescu TG, Orasan O, Negrean V, et al. Influence of pesticides on respiratory pathology – A literature review. Annals of Agricultural and Environmental Medicine. 2020;**27**(2): 194-200. DOI: 10.26444/aaem/121899

[131] Landa RJ. Diagnosis of autism spectrum disorders in the first 3 years of life. Nature Clinical Practice. Neurology. 2008;**4**(3):138-147. DOI: 10.1038/ncpneuro0731

[132] Nazeer A, Ghaziuddin M. Autism Spectrum disorders: Clinical features and diagnosis. Pediatric Clinics of North America. 2012;**59**(1):19-25. DOI: 10.1016/j.pcl.2011.10.007

[133] Elsabbagh M, Divan G, Koh YJ, Kim YS, et al. Global prevalence of autism and other pervasive developmental disorders. Autism Research. 2012;**5**(3):160-179. DOI: 10.1002/aur.239

[134] Liu J, Schelar E. Pesticide exposure and child neurodevelopment: Summary and implications. Workplace Health & Safety. 2012;**60**(5):235-243. DOI: 10.1177/216507991206000507

[135] Shelton JF, Hertz-Picciotto I, Pessah IN. Tipping the balance of autism risk: Potential mechanisms linking pesticides and autism. Environmental Health Perspectives. 2012;**20**(7):944-951. DOI: 10.1289/ehp.1104553

[136] Blatt GJ, Fitzgerald CM, Guptill JT, et al. Density and distribution of hippocampal neurotransmitter receptors in autism: An autoradiographic study. Journal of Autism and Developmental Disorders. 2001;**31**(6): 537-543. DOI: 10.1023/a:1013238809666

[137] Lyall K, Croen LA, Sjödin A, et al. Polychlorinated biphenyl and organochlorine pesticide concentrations in maternal mid-pregnancy serum samples: Association with autism Spectrum disorder and intellectual disability. Environmental Health Perspectives. 2017;**125**(3):474-480. DOI: 10.1289/EHP277

[138] Barkoski JM, Philippat C, Tancredi D, et al. In utero pyrethroid pesticide exposure in relation to autism spectrum disorder (ASD) and other neurodevelopmental outcomes at 3 years in the MARBLES longitudinal

cohort. Environmental Research. 2021;**194**:110495. DOI: 10.1016/j.envres.2020.110495

[139] Philippat C, Barkoski J, Tancredi DJ, et al. Prenatal exposure to organophosphate pesticides and risk of autism spectrum disorders and other non-typical development at 3 years in a high risk cohort. International Journal of Hygiene and Environmental Health. 2018;**221**(3):548-555. DOI: 10.1016/j.ijheh.2018.02.004

[140] Ongono JS, Béranger R, Baghdadli A, Mortamais M. Pesticides used in Europe and autism spectrum disorder risk: Can novel exposure hypotheses be formulated beyond organophosphates, organochlorines, pyrethroids and carbamates? - a systematic review. Environmental Research. 2020;**187**:109646. DOI: 10.1016/j.envres.2020.109646

[141] Carter CJ, Blizard RA. Autism genes are selectively targeted by environmental pollutants including pesticides, heavy metals, bisphenol a, phthalates and many others in food, cosmetics or household products. Neurochemistry International. 2016;**101**:83-109. DOI: 10.1016/j.neuint.2016.10.011

[142] Kaur RP, Gupta V, Christopher AF, Bansal P. Potential pathways of pesticide action on erectile function a contributory factor in male infertility. Asian Pacific Journal of Reproduction. 2015;**4**(4):322-330. DOI: 10.1016/j.apjr.2015.07.012

[143] Soliman SS, Awad-Allah A, Ebied ZM. Erectile dysfunction in workers chronically exposed to pesticides and organic solvents in Damietta governorate. Mansoura Journal of Forensic Medicine & Clinical Toxicology. 2008;**16**(2):63-76

[144] Zhang JJ, Wang Y, Xiang HY, et al. Oxidative stress: Role in acetamiprid-induced impairment of the male mice reproductive system. Agricultural Sciences in China. 2011;**10**(2):786-796. DOI: 10.1016/S1671-2927(11)60063-1

[145] Najafi G, Razi M, Hoshyar A, et al. The effect of chronic exposure with imidacloprid insecticide on fertility in mature male rats. International Journal of Fertility and Sterility. 2010;**4**(1):9-16

[146] Ngoula F, Watcho P, Dongmo MC, et al. Effects of pirimiphos-methyl (an organophosphate insecticide) on the fertility of adult male rats. African Health Sciences. 2007;**7**(1):3-9. DOI: 10.5555/afhs.2007.7.1.3

[147] Slimani S, Boulakoud MS, Abdennour C. Pesticide exposure and reproductive biomarkers among male farmers from north-East Algeria. Annals of Biological Research. 2011;**2**:290-297

[148] Akingbemi BT, Ge RS, Klinefelter GR, et al. A metabolite of methoxychlor, 2,2-bis(p-hydroxyphenyl)-1,1, 1- trichloroethane, reduces testosterone biosynthesis in rat leydig cells through suppression of steady-state messenger ribonucleic acid levels of the cholesterol side-chain cleavage enzyme. Biology of Reproduction. 2000;**62**(3):571-578. DOI: 10.1095/biolreprod62.3.571

[149] Beard JD, Umbach DM, Hoppin JA, et al. Suicide and pesticide use among pesticide applicators and their spouses in the agricultural health study. Environmental Health Perspectives. 2011;**119**(11):1610-1615. DOI: 10.1289/ehp.1103413

[150] London L, Flisher AJ, Wesseling C, et al. Suicide and exposure to organophosphate insecticides: Cause or effect? American Journal of Industrial Medicine. 2005;**47**(4):308-321. DOI: 10.1002/ajim.20147

[151] London L, Beseler C, Bouchard MF, et al. Neurobehavioral and neurodevelopmental effects of pesticide exposures. Neurotoxicology. 2012;**33**(4):887-896. DOI: 10.1016/j.neuro.2012.01.004

[152] Kori RK, Mandrah K, Hasan W, et al. Identification of markers of depression and neurotoxicity in pesticide exposed agriculture workers. Journal of Biochemical and Molecular Toxicology. 2020;**34**(6):1-9. DOI: 10.1002/jbt.22477

[153] Cowen PJ, Browning M. What has serotonin to do with depression? World Psychiatry. 2015;**14**(2):158-160. DOI: 10.1002/wps.20229

[154] Gasmi S. Neurotransmission dysfunction by mixture of pesticides and preventive effects of quercetin on brain, hippocampus and striatum in rats. Toxicology and Environmental Health Sciences. 2020;**12**:203-212

[155] Ibrahim KA, Abdelrahman SM, Elhakim HKA, Ragab EA. Single or combined exposure to chlorpyrifos and cypermethrin provoke oxidative stress and downregulation in monoamine oxidase and acetylcholinesterase gene expression of the rat's brain. Environmental Science and Pollution Research. 2020;**27**(11):12692-12703. DOI: 10.1007/s11356-020-07864-8

[156] Faria NMX, Fassa AG, Meucci RD. Association between pesticide exposure and suicide rates in Brazil. Neurotoxicology. 2014;**45**:355-362. DOI: 10.1016/j.neuro.2014.05.003

[157] Meyer A, Koifman S, Koifman RJ, et al. Mood disorders hospitalizations, suicide attempts, and suicide mortality among agricultural workers and residents in an area with intensive use of pesticides in Brazil. Journal of Toxicology and Environmental Health. Part A. 2010;**73**(13-14):866-877. DOI: 10.1080/15287391003744781

[158] Wesseling C, de Joode B, Keifer M, London L, et al. Symptoms of psychological distress and suicidal ideation among banana workers with a history of poisoning by organophosphate or n-methyl carbamate pesticides. Occupational and Environmental Medicine. 2010;**67**(11):778-784. DOI: 10.1136/oem.2009.047266

[159] Gunnell D, Eddleston M, Phillips MR, Konradsen F. The global distribution of fatal pesticide self-poisoning: Systematic review. BMC Public Health. 2007;**7**(357):1-15. DOI: 10.1186/1471-2458-7-357

Chapter 10

# Toxicity Status and Risks of Common Active Ingredients in Open Markets

*Dele Omoyele Adeniyi*


## Abstract

Agrochemical stores in selected geographical locations in North Central Nigeria were surveyed, a structured questionnaire administered and students of selected Universities and staff of the Ministry of Agricultural and Rural Development were sampled. Seventeen active ingredients: Paraquat dichloride, Glyphosate, Permethrin + pyriproxyfen, Dimethylamine salt, Cypermethrin, Chlorpyrifos, Dichlorvos, Lambda-cyhalothrin, 2,2-dichlorovinyl Dimethyl phosphate, Hexaconazole, Imidacloprid, Dimethoate, Nicosulfuron, Profenofos + cypermethrin, S-metolachlor, Carbendazim, and Mancozeb were recorded, and dominated by herbicides and insecticides. Toxicity analysis of active ingredients in open markets based on the recommendation of international standard organizations showed that some common active ingredients were not approved and some others were not listed for agricultural use. Many of the active ingredients negatively affect plant pollinators, aquatic animals, are highly toxic to birds, honey bees, and poses risk to wildlife. Some others are a possible carcinogen, fatal if inhaled, highly hazardous with high environmental toxicity posing a serious health risk to humans by disrupting the endocrine system, inducing heritable mutations in germ cells, impair fertility and reproductive toxicity.


**Keywords:** pesticides, active ingredients, residue, toxicity, standards, risks

## 1. Introduction

The rise in the number of chemicals being introduced into agriculture and horticulture has given rise to some concerns over the safety of the food crop and that of the operator. A working party was established in the UK which passed some regulations over the possible risks to consumers of treated crops. This led directly to the formation of the Advisory Committee on Poisonous Substances used in agriculture, which extended concern to effects on the environment. However, new toxic chemicals and their formulations need to be brought to the notice of the Government before being put on the market. The introduction of the Pesticides Safety Precautions Scheme (PSPS) strengthened the requirement in which manufacturers of the new chemical were required to provide data relating to the safety of the product; full description, proposed uses, mode of action, toxicity and persistence, relevant to the user of the product, consumer of treated produce, domestic animals and wildlife.

The outcome of such products was published with the key elements included on product labels; advice on operator safety, target crops, dose rate limitations, harvest interval, and environmental safety. The PSPS was accompanied by the voluntary scheme which evaluates the efficacy of crop protection chemicals prior to the approval of chemical and based on trials efficacy data.

The increasing regulatory requirements are seen over decades, and especially in the past 20 years, have placed much financial pressure on the research-based crop protection companies. Increasing demands for toxicology, metabolism, and environmental data to support registration applications have resulted in a cost of approximately £100 million to discover research, develop and register a new product. Earlier, horticultural and vegetable markets were targeted pesticides markets, today such are far too small to justify the investment in required regulatory studies and can only be considered as "add-on markets" to be considered once success in a major market has been achieved. Markets must also be considered at the international level no single country market would justify the investment in pesticides research and development.

## 2. Pesticides in agriculture

Pesticide is defined as a product that kills or controls various types of pests, plant, or animal that is harmful to man or the environment. Pesticides are used in agriculture to protect crops against insects, fungi, weeds, nematodes, and parasitic plant pests, as well as to protect public health in controlling vectors of tropical diseases. They can also be used to prevent, destroy, repel, or mitigate any pest and can either kill pests or render them ineffective. Pesticides are used on fruits, vegetables, wheat, rice, olives, tree crops, canola pressed into oil, and on non-food crops, such as cotton, grass, and flowers. Pesticides applied to food crops in the field can leave potentially harmful residues after pesticides are applied to the crops, they may interact with the plant surfaces, be exposed to environmental factors, such as wind, sun, and maybe washed off during rainfall. The pesticide may be absorbed by the plant surface (waxy cuticle and root surfaces) and enter the plant transport system (systemic) or stay on the surface of the plant (contact). The pesticides that get into the plant tissues may be transformed (metabolized) or sequestered in the tissues to form the pesticide residue.

Pesticide residues are the deposits of pesticide active ingredients, their metabolites or breakdown products are present in some components of the environment after their application, spillage, or dumping. The presence of pesticide residues is a concern for consumers because pesticides are known to have potentially harmful effects on other nontargeted organisms than pests and diseases. Infants, children, and adults are commonly exposed to pesticides by eating them on and in food and animals equally ingest such through feeds and mills. Pesticides are potentially toxic to humans and have been linked to a wide range of human health hazards, ranging from short-term impacts, such as headaches and nausea to chronic impacts, such as cancer, reproductive harm, and endocrine disruption.

## 3. Benefits and risks of pesticides

The application of any chemical to a crop or food raises the question of risks and benefits. This discussion of risk has shifted from dealing with toxicity to the user in the field and the consumer to a much wider focus that includes the whole environment and the ecosystem in which the crops are growing. As a consequence, more and more studies are required before a fungicide can be used, leading to enormous development costs. This leads the industry to concentrate on the big markets, while smaller markets are increasingly left out and in urgent need of effective fungicides. Overall, most analyses come to the conclusion that the benefits of fungicides far outweigh the risks, if they are used carefully and according to the label recommendations. Currently, more than 80% of the fruit and vegetable crops have been known to receive a fungicide every season.

## 4. International standard and requirements

There are standard organizations of international reputes that certify and license agricultural products for safe consumption and to fulfill the international requirement for the trade. These standard organizations are also functional at regional and national levels and requirements at these levels are often benchmarked with the provision of the international organizations. Such organizations include but are not limited to 4C Association, Bonsucro (Better Sugar Cane Initiative), Better Cotton Initiative (BCI), Fairtrade International, FSC, RSB, SAN (Sustainable Agriculture Network), and UTZ.

Growers, produce buyers and agents, warehouse owners, manufacturers, and even the general public, have perceived the use of chemicals for various purposes as part of everyday life, either for domestic or agricultural. This has led to the indiscriminate use of pesticides for varied reasons and in search of quick action and effect. The uncoordinated system in this sector of agriculture, lacking regulation and enforcement required for best practices and safety measures in the handling of agrochemicals prompted this study. This in a way undermined the associated risks of indiscriminate use of these agrochemicals, their toxicity, and residues on plants, animals, man, and the environment. Agrochemicals commonly sold in open markets were surveyed; the target crops, associated hazards/risks, and their safety statuses were evaluated based on the benchmark by international standard organizations.

## 5. Study geography and analysis

The survey of agrochemical stores and trading facilities was conducted in North Central Nigeria. Agrochemical dealers in three major farm-based stores in central towns were randomly selected and visited with a structured questionnaire. The questionnaire was duly completed with the co-operation and support of the respondent and the interviewee. The identity of the chemical stores in selected locations was kept anonymous. Information was sought on the trade name, type of agrochemicals (herbicide, insecticide or fungicide, etc.), active ingredient(s) present in the pesticides, and the crop(s) in which the pesticides were targeted. However, the trade names of the agrochemicals and locations were kept anonymous but the active ingredients were used as the bases of this report. The active ingredients were benchmarked with the requirements of the international standards organization.

The information obtained from the agrochemical stores on the active ingredients on sale in the open market were subjected to the benchmarks of the international standards organizations, such as 4C Association, BCI, Bonsucro, FSC, Fairtrade, RSB, Rainforest Alliance, SAN, and UTZ, as related to the toxicity, restriction status, and effect of such active ingredients on human, animal, and environment.

## 6. Status of agrochemicals in open market

The survey of farmers–based agrochemicals stores showed the presence of seventeen (17) active ingredients common in the open market. A total of eighteen (18) trade names; Weed Crusher, Parae Force, Weed Cut, Grass Cutter, Touch Down, Clear Weed, Force Up, Drysate, Round-Up, Sunsate, Cyperthrone, Vestamine, Relimine, Amino Force, Amino Force Granular, Guard Force, Gramaxone and Meta Force were herbicides, thirteen (13) trade names; Super Care, Cyper Force, Cyper-DiForce, Flush Out, Termifos, Termiclor, Pest Off, Rid-Off, LaraForce, Magic Force, Knock Off, DD-Force and Iron Force were insecticides and nine (9) trade names; Fungi Care, Confidor, Storm Force, ImiForce, Dime Force, Fungus Force, ForceLet, Z-Force, and Zeb-Care were fungicides. No record of nematicides or any agrochemicals against parasitic pest plants were found in the study geographies. These agrochemicals were also dominated by herbicides which were 42.67% on average, the insecticides were 35.0% of the stocks while only 25.5% of agrochemicals across study geographies were fungicides (**Tables 1–3**). This information implied that pesticides used in the geographies were mostly for weed management and insect pests' control both for agricultural and domestic purposes. The commonly used active ingredients by the indication of sales from the selected geographies showed Paraquat dichloride, Glyphosate, Permethrin + pyriproxyfen, Dimethylamine salt, Cypermethrin, Chlorpyrifos, Dichlorvos, Lambda-cyhalothrin, 2,2-dichlorovinyl Dimethyl phosphate, Hexaconazole, Imidacloprid, Dimethoate, Nicosulfuron, Profenofos + cypermethrin, S-metolachlor, Carbendazim, and Mancozeb. The common active ingredients cut across varied pesticides types across the geographies.

**Table 1** showed that geography I was dominated by herbicides with 45%, 36% insecticides, and only 27% were fungicides. Targeted crops were mostly grains, legumes, vegetables, a few tubers, fruits, and tree crops.

Either one or two of the selected geographies had Paraquat dichloride, Glyphosate, Cypermethrin, Dichlorvos, Lambda-cyhalothrin, Imidacloprid, and Dimethoate common to them while Glyphosate and Cypermethrin are most frequent on sale across all the geographies surveyed. These active ingredients were variedly targeted

| S/N | Status | Active ingredient(s) | Targeted crop(s) |
|---|---|---|---|
| 1 | Herbicide | Paraquat dichloride | Maize, weeds, cowpea, rubber, oil palm |
| 2 | Herbicide | Glyphosate | Grasses, weeds, woody shrubs |
| 3 | Herbicide | Permethrin + pyriproxyfen | Maize, weed |
| 4 | Herbicide | Dimethylamine salt | Maize, tomato, cotton, fruit trees |
| 5 | Insecticide | Cypermethrin | Smaller insects |
| 6 | Insecticide | Chlorpyrifos | Vegetables, rice, soya beans, cocoa |
| 7 | Insecticide | Dichlorvos | Insect of vegetable, rice, yam, cowpea |
| 8 | Insecticide | Lambda-cyhalothrin | Insect pest in maize, vegetables, rice |
| 9 | Fungicide | Hexaconazole | Pepper, vegetable |
| 10 | Fungicide | Imidacloprid | Pepper, watermelon, groundnut, cocoa |
| 11 | Fungicide | Dimethoate | Carrot, beans, groundnut |

**Table 1.**
*Active ingredients in Geography I and Targeted crop(s).*

| S/N | Status | Active ingredient(s) | Targeted crop(s) |
|---|---|---|---|
| 1 | Herbicide | 2,4-dimethylamine salt | Rice, rubber, wheat, sugar cane |
| 2 | Herbicide | Nicosulfuron | Maize |
| 3 | Herbicide | Glyphosate | Sugar cane, weeds |
| 4 | Insecticide | 2,2-dichlorovinyl Dimethyl phosphate | |
| 5 | Insecticide | Lambda-cyhalothrin | Pineapple, carrot, orange, rice, beans |
| 6 | Insecticide | Profenofos + cypermethrin | Maize, cotton, orange |
| 7 | Insecticide | Cypermethrin | Carrot, cocoa, groundnut, onion |
| 8 | Fungicide | Imidacloprid | Pepper, groundnut, cocoa |
| 9 | Fungicide | Carbendazim | Fruit and vegetables |

**Table 2.**
*Active ingredients in Geography II and Targeted crop(s).*

| S/N | Status | Active ingredient(s) | Targeted crop(s) |
|---|---|---|---|
| 1 | Herbicide | Glyphosate | Annual grass, sugar cane, |
| 2 | Herbicide | Paraquat dichloride | Non-selective, grasses, broad-leaved weeds |
| 3 | Herbicide | S-metolachlor | Potato, yam, groundnut |
| 4 | Herbicide | Di-methylamine | Corn, weeds, sugarcane |
| 5 | Insecticide | Cypermethrin | Corn, tomato, cocoa, watermelon |
| 6 | Fungicide | Dimethoate | Beans, groundnut |
| 8 | Fungicide | Mancozeb | Fruits, vegetable |

**Table 3.**
*Active ingredients in Geography III and Targeted crop(s).*

to manage weeds, insects, and pathogens in grains, legumes, nuts, tubers (root and stem), fruits and vegetables, and tree crops (**Tables 1–3**).

The presence of insecticides was higher in geography II showing 44% occurrence, the fungicides were only 22% while herbicides showed 33% of the agrochemicals in the open market and these were targeted against varied crop types, for example, corms, vegetables, fruits, grains, and some tree crops (**Table 2**).

However, the report of geography III as shown in **Table 3** indicated that only 12.5% of agrochemicals were fungicide which was the least across the selected geographies and likewise was the 25% insecticides but herbicide occurrence was highest (50%) of the agrochemical in all the geographies (**Table 3**).

## 7. Toxicity of active ingredients in open market

The toxicity analysis of the active ingredients commonly on sale in the open market was based on recorded cases of pesticide active ingredients and formulations that have shown a high incidence of severe or irreversible adverse effects on human health or the environment, in accordance with the recommendation of the standard organizations and Pesticide Action Network International list of highly hazardous pesticide (PAN-HHP).

| S/N | Active ingredient(s) | Status in EU database | Status in Standard Organizations (BCI/RA/FSC/4C/SAN/UTZ) | Status in PAN-HHP |
|---|---|---|---|---|
| 1 | Paraquat dichloride | Not approved | Prohibited, to be faced out by 2024<br>Fatal if inhaled/may cause severe effects<br>Highly toxic to birds/may cause severe effect | Added to PAN-HHP list in 2011, 2019.<br>Acute toxicity: Fatal if inhaled.<br>Not yet formally listed but agreed by PIC |
| 2 | Glyphosate | Approved | May only be used under specific, defined conditions<br>Probable carcinogenic | Added to PAN-HHP list in 2011,2014, 2019.<br>Long-term health effects: possible carcinogens.<br>Environmental toxicity: very persistent in water/sediment. |
| 3 | Permethrin + pyriproxyfen | Approved | Prohibited, highly restricted/ restricted use/risk-specific mitigation measures are mandatory<br>Identified as hazardous, use with extreme caution<br>Minimization of use<br>Probable carcinogen<br>Highly toxic to honey bees<br>Aquatic risk, pollinator risk, wildlife risk | Added to PAN-HHP list in 2011, 2019.<br>Long-term health effects: probable/likely carcinogen.<br>Environmental toxicity: highly toxic to bees |
| 4 | Dimethylamine salt | Not listed | Not listed | Not listed |
| 5 | Cypermethrin | Approved | Highly restricted/restricted use,<br>Risk specific mitigation measures are mandatory<br>Highly aquatic toxicity<br>Highly toxic to honey bees, aquatic risk, pollinator risk | Added to PAN-HHP list in 2011, 2019.<br>Environmental toxicity: highly toxic to bees |
| 6 | Chlorpyrifos | Not indicated | Potentially to be prohibited<br>Highly restricted/restricted use/risk-specific mitigation measures are mandatory<br>May only be used under specific conditions/minimization of the use<br>Inhalation risk, high aquatic toxicity/ highly toxic to bees, birds, aquatic risk.<br>Pollinator risk, wildlife risk | Added to PAN-HHP list in 2011, 2019<br>Environmental toxicity: highly toxic to bees |

| S/N | Active ingredient(s) | Status in EU database | Status in Standard Organizations (BCI/RA/FSC/4C/SAN/UTZ) | Status in PAN-HHP |
|---|---|---|---|---|
| 7 | Hexaconazole | Not approved | Not listed | Added to PAN-HHP list in 2011.<br>Long-term health effects: possible carcinogens.<br>Environmental toxicity: very persistent in water, highly toxic to bees. |
| 8 | Dichlorvos | Not approved | Highly restricted/ prohibited, to be phased out by 2024<br>May only be used under a specific, defined condition<br>Highly hazardous, fatal if inhaled.<br>Highly aquatic toxicity/ highly toxic to honey bees, birds | Added to PAN-HHP list in 2011, 2019.<br>Acute toxicity: highly hazardous, fatal if inhaled.<br>Long term health effect: possible carcinogen<br>Environmental toxicity: highly toxic to bees |
| 9 | Lambda-cyhalothrin | Approved | Highly restricted/ minimization of use/ may only be used under a specific condition, to be phased out by 2024<br>Fatal if inhaled<br>Endocrine disruptor, highly aquatic toxicity/ highly toxic to honey bees/aquatic risk, pollinator risk | Added to PAN-HHP list in 2011, 2019.<br>Acute toxicity: Fatal if inhaled.<br>Long-term health effects: Endocrine disruptor, reproductive toxicity.<br>Environmental toxicity: highly toxic to bees |
| 10 | Imidacloprid | Approved | Restricted, prohibited with an exception for certain pests in certain crops and regions/ minimization of use.<br>Prohibited without exception/potentially prohibited<br>May cause severe effects<br>Highly toxic to honey bees, birds/ Neonicotinoid/may cause severe effects | Added to PAN-HHP list in 2011, 2019.<br>Environmental toxicity: highly toxic to bees |
| 11 | Dimethoate | Not approved | Restricted, minimization of use/ potentially to be prohibited<br>Inhalation risk<br>Highly toxic to honey bees/highly toxic to birds/aquatic risk, pollinator risk, wildlife risk | Added to PAN-HHP list in 2011, 2019.<br>Long-term health effects: probable carcinogen, Endocrine disruptor, reproductive toxicity.<br>Environmental toxicity: highly toxic to bees |

| S/N | Active ingredient(s) | Status in EU database | Status in Standard Organizations (BCI/RA/FSC/4C/SAN/UTZ) | Status in PAN-HHP |
|---|---|---|---|---|
| 12 | Nicosulfuron | Approved | Not listed | Added to PAN-HHP list in 2019. Very persistent in water /sediments |
| 13 | 2,2-dichlorovinyl Dimethyl phosphate | Not listed | Not listed | Not listed |
| 14 | Profenofos + cypermethrin | Not approved + Approved | Restricted, identified as hazardous. Use with extreme caution High aquatic toxicity/ high toxic to honey bees | Added to PAN-HHP list in 2009, 2011,2019. Environmental toxicity: highly toxic to bees |
| 15 | Carbendazim | Not approved | Prohibited/potential to be prohibited, exceptions may apply for certain pests, in certain crops and regions. May only be used under specific, defined conditions Minimization of the use Mutagenic, Reproductive toxin | Added to PAN-HHP list in 2011, 2019. Long term health effect: induce heritable mutations in germ cells of humans, impair fertility in humans, cause developmental toxicity to humans, probable likely carcinogen, Endocrine disruptor, reproductive toxicity |
| 16 | S-metolachlor | Approved | Restricted use, Risk specific mitigation measures are mandatory Aquatic risk | Not listed |
| 17 | Mancozeb | Approved | Restricted use of pesticides, risk-specific mitigation measures are mandatory. May only be used under specific, defined conditions. Minimization of use, prohibited/potentially to be prohibited Probable carcinogen. Endocrine disruptor, wildlife Risk | Added to PAN-HHP list in 2011, 2019. Long-term health effects: Probable likely carcinogen, Endocrine disruptors, reproductive toxicity. |

**Table 4.**
*Pesticides hazardous nature and toxicity status.*

The hazard criteria of the active ingredients were grouped into—acute toxicity, long-term health effects, environmental toxicity, and international regulations (global pesticide-related conventions). The pesticides grouping, hazard, and toxicity status (**Table 4**) were the recommendations of globally harmonized system of classification and labeling of chemicals (GHS), World Health Organization (WHO) recommended classification of pesticides by hazard, the international agency for research on cancer (IARC), U.S. environmental protection agency (U.S. EPA), and European Union categorization of endocrine disruptors. The recommendation of these organizations was benchmarked by the 4C Association, Bonsucro (Better Sugar Cane Initiative), Better Cotton Initiative (BCI), Fairtrade International, FSC, RSB, SAN (Sustainable Agriculture Network), and UTZ.

Glyphosate, herbicide, and very common active ingredient are used for the management of weeds both in agriculture and domestically. The active ingredient is classified as highly restricted for use, with mandatory risk-specific mitigation measures. The active ingredient is prohibited, identified as hazardous and its use should be extremely cautious and minimized. Di-methylamine (2,4 dimethylamine salt) was found to be commonly used by growers and the public in weed management but no record of this active ingredient was found in the databases of EU, Pesticide Action Network International, and other international standard organizations.

Nicosulfuron is an approved active ingredient for the management of weeds but with the environmental hazard of being very persistent in water/sediment. Profenofos + cypermethrin, an insecticide combination is restricted, to be used with extreme caution, shows high toxicity to honey bees and high aquatic toxicity according to FSC and Fairtrade. Another approved herbicide is S-metolachlor although recommended for restricted use and mandatory risk-specific mitigation measures to be taken and has aquatic risk according to RA, SAN.

## 8. Safety statuses of active ingredients in the open market

The three geographies surveyed were major agrochemical markets in the state, which were purposefully selected for the study. Pesticides poisoning most often comes from swallowing chemicals, after consuming contaminated foods or beverages. Frequently exposed persons are also susceptible to poisoning that can cause organs or systems damage.

Paraquat is a leading cause of fatal poisoning in parts of Asia, the Pacific Islands, and the South and Central Americas. It is rapidly but incompletely absorbed and then largely eliminated unchanged in urine within 12–24 hours, the very high case fatality of paraquat is due to inherent toxicity and lack of effective treatments [1]. Paraquat dichloride was shown to be very immobile in the soil, does not hydrolyze nor photodegrade in aqueous solutions, and is resistant to microbial degradation under aerobic and anaerobic conditions. The primary route of environmental dissipation of paraquat is adsorption to biological materials and soil clay particles [2], Paraquat dichloride is highly toxic to birds/may cause severe effects [3]. It is reported that more than 70% of trusted sources of paraquat poisonings result in death. Ingesting small to medium amounts of it can lead to fatal poisoning, lung scarring, and the failure of multiple organs, heart, respiratory, kidney, and liver failure. Ingesting large amounts of paraquat causes confusion, muscle weakness, seizures, difficulty breathing, fast heart rate, and coma [4]. Paraquat dichloride is not an approved active ingredient by the EU standards on safe pesticides. It has been recently listed in PAN as a highly hazardous pesticide in 2019 [5], with restricted use, it is prohibited from use and to be faced out by the year 2024. The effect on humans includes fatality if inhaled and may cause severe effects (SAN, PIC).

Glyphosate and its formulations may not only be considered as having genotoxic, cytotoxic, or endocrine-disrupting properties but a causative agent of reproduction abnormalities in both wildlife and humans. Furthermore, the extensive use of glyphosate-based herbicides in genetically modified glyphosate-resistant plants grown for food and feed should be of grave concern since they can be sources of genotoxicity, cytotoxicity, and reproductive toxicity in wildlife and humans [6]. Although glyphosate is approved for use by the EU, other standards organizations have listed it as a highly hazardous pesticide in 2011, 2014, and 2019 [5]. This active ingredient has been restricted, only be used under specific and defined conditions. It is also a probable carcinogenic substance to humans and has environmental toxicity by being very persistent in water and sediments [7]. Glyphosate provokes oxidative damage in the liver and kidneys of mammals by disrupting mitochondrial metabolism, disrupting endocrine-signaling systems and residues from glyphosate may pose higher risks to the kidneys and liver, reproductive development impairment [8]. Increases in the frequency of serious, chronic kidney disease were observed among male agricultural workers in some regions with heavy glyphosate use and "hard" water. And that the possible adverse effects of glyphosate exposure on kidney and liver warrant a focused, international research effort [9, 10]. Glyphosate can alter the functioning of hormonal systems and gene expression patterns at various dosage levels. Such effects will sometimes occur at low and likely environmentally-relevant exposures. Contemporary endocrine science has demonstrated that dose-response relationships will sometimes deviate from a linear increase in the frequency and severity of impacts expected as dose levels rise [11]. The timing, nature, and severity of endocrine system impacts will vary depending on the levels and timing of glyphosate exposures, this is pertinent as agrochemical users in Nigeria are indoctrinated in terms of dosage, rate, and timing of application.

Permethrin + Pyriproxyfen is used to kill a large range of pests; fleas, ticks, cockroaches, flies, and mosquitoes. The environmental protection agency (EPA) reviewed the pesticides register showed that permethrin stays a long time in the soil, very low amount stays in the water. Permethrin has some health risks; headaches, dizziness, nausea, shortness of breath, skin irritation, and redness of eyes when used at higher levels [12]. However, it is highly hazardous, with probable carcinogen in humans [13], and highly toxic to honey bees [14], aquatic, pollinator, and wildlife risk [5].

Cypermethrin is a pyrethroid insecticide, first synthesized in 1974, widely used to kill insects as it works quickly by affecting the nervous system, toxicity level in animals varies,for example, in rats includes tremors, seizures, and salivation, in cockroaches when exposed to little amount as 0.02 micrograms per gram causes brain paralysis, restlessness, and prostration. Cypermethrin is approved for use to manage agricultural insect pests. It is however listed as a highly hazardous pesticide in 2011 and 2019. It is classified as highly restricted use with mandatory risk-specific mitigation measures [3, 5]. It has highly aquatic toxicity, toxic on honey bees, and also with aquatic and pollinator risk [15]. Effect of cypermethrin in humans when exposed sometimes causes itching and tingling sensations. The half-life of cypermethrin in the environment takes about 30 days, soil microbes easily break it down because of the low potential to move in the soil but poses little to no risk when used responsibly [2].

The toxicity status of Chlorpyrifos is similar to cypermethrin except that it is not indicated in the EU database but UTZ classified it as highly restricted, may only be used under the specific condition with risk-specific mitigation measures, and is potential to be prohibited. Chlorpyrifos classified as highly hazardous in 2011 and 2019, poses inhalation risk to humans, high aquatic toxicity, highly toxic to bees [14], birds with aquatic pollinator, and wildlife risk [3].

Dichlorvos an organophosphate insecticide, also used as a public health vector control for animals, is registered worldwide for varieties of uses, majorly used as a post-harvest fumigant for control of various pests in food, the acceptable daily intake (ADI) for Dichlorvos was established as 0.004mg/kg bw and the acute reference dose was 0.1mg/kg bw. It can be applied with aerosols, fogging, and sprays equipment. It also breaks down rapidly in humid air, water, and soil, it takes longer time on wood when exposed to humans through food can be acutely toxic with typical cholinergic signs that are highly hazardous, dichlorvos is not teratogenic in mice and rats' half-lives of recovery is about 15days in human and 2 hours in rats [16].

Dichlorvos is not approved for use but found in open markets, it is restricted in use and meant to be phased out by the year 2024 (BCI). It is highly prohibited, may only be used under specific, defined conditions. The active ingredient is classified as highly hazardous to humans [17], it is fatal if inhaled according to the EU and globally harmonized system (EU, GHS). It is a possible and probable carcinogen [2, 7], with high aquatic toxicity and highly toxic to honey bees and birds [15, 18–20].

Dimethoate comes in different forms; dustable powder (DP), wettable powder (WP) soluble concentrate, its toxicity was evaluated in 1992 by (WHO), it is used to control a wide range of insects and pests, in cereals, citrus, coffee, cotton, fruits, grapes, potatoes, beetroot, tea, and vegetables. It can also be used to control flies because of its systemic nature and acaricide the solubility of dimethoate in water at 90% purity has 39.8 at 25°C after 4 hours, equilibrium. In rats, the toxicity of dimethoate is mostly acute, such as oral irritation, dermal sensitization, eye irritation in humans, WHO hazard classification of dimethoate is "class moderately hazardous," UN classification is "Toxic class 6.1," US EPA Classification is; (Formulation) 11, EC Classification; Risk Xn (R21/22) Reviews by WHOEHC (1986) concluded that when used in proper level and accordingly exposure of human through the air, food, or water can be negligible.

Nicosulfuron is used as post-emergence in forage maize, found to have low dermal and inhalation toxicity, can be slightly irritating in rats, and has not been evaluated by the FAO, JMPR, and WHO/IPCS, although it is currently under review, it is registered in the U.S.A, the WHO Classification of Nicosulfuron is U; unlikely to cause an acute hazard in normal use. This active ingredient does not meet the criteria established in the UN recommendations on the transport of dangerous goods and therefore is not considered hazardous for transportation purposes. It is also not co-formulated with other active ingredients; toxicity in rats includes acute dermal irritation and eye irritation [21].

Profenofos + Cypermerthrin is a co-formulated organo-phosphorous insecticide, studies have shown its toxicity levels on animals, plants, and even the environment's fate when it comes in contact. Profenofos was evaluated by JMPR in 1990, 1992, 1994, and 1995, toxicological, reviews were also conducted in 2007 when an ADI OF 0 to 0.03mg/kg bw and ARfD of 1mg/kg bw were established, profenofos is a clear liquid with weak odor, its solubility in water at 22°C is 2.8mg/l at a pH of 6.9, profenofos is slowly absorbed in metabolized, it was major residue when crops are harvested several weeks after the last applications, its residues are not expected to occur in succeeding crops. Reviewed by JMPR health risk shows that profenofos is unlikely to present a public health concern.

S-metolachlor is used for varieties of crops for control of grasses, for example, pigweeds (*Amaranthus* spp.); they are different commercial brands, herbicides that contain metolachlor as active ingredients although formulation and chemical composition may differ, some products metolachlor safeners are added some, no safeners added. A safener is added to metolachlor to reduce injury to crops, such as corn, but injuries to other crops solely depend on the amount of safener used or environmental concerns. Metolachlor has four different isomers but can be grouped into two, which are S-metolachlor isomers and R-metolachlor isomers, both are made from the same materials but S-metolachlor isomer is more active in herbicidal effects compared to R-metolachlor.

Lambda-cyhalothrin is a synthetic pyrethroid insecticide used in agricultural and public health to control a wide range of insects and pests at developmental stages, it is a nonsystemic chemical, does not stay long in the soil so has an only limited function when used as soil insecticide. Lambda-cyhalothrin can be applied by spree spraying and residual spraying. Additionally, the provided data on acute toxicity, skin irritation, and sensitization. The mutagenic study reviewed that Lambda-cyhalothrin is nonmutagenic, JMPR has defined an acceptable daily intake (ADI) of 0 0.02mg/kg bw, water solubility is 0.005 mg/l. The IPCS hazard classification of Lambda-cyhalothrin is moderately hazardous Class II (WHO). Lambda-cyhalothrin is approved for use in weed management but listed as highly hazardous in 2011 and 2019 [5]. It is to be phased out by the year 2024 and with highly restricted use, only be used under specific conditions, and according to the globally harmonized system (GHS), it poses a fatal risk to humans if inhaled. This active ingredient also poses a long-term health effect as an endocrine disruptor and as having reproductive toxicity [22].

The 2,2-dichlorovinyl dimethyl phosphate is another insecticide that is not listed in the active ingredients database of the EU. It is however listed as a highly hazardous substance in PAN as an endocrine disruptor, has highly aquatic toxicity, is highly toxic to honey bees, aquatic, and pollinator risk [5].

Carbendazim is a very common fungicide but was recently listed as highly hazardous in 2019 [5] and not on the approved list of EU pesticides. It is restricted, prohibited with exceptions for certain pests, in certain crops and regions, and may only be used under specific, defined conditions as recommended by Fairtrade. This active ingredient has a mutagenic effect on humans and it is a reproductive toxin according to EU and GHS [13, 23]. Carbendazim is a widely used systemic fungicide that is mainly used for protective and curative functions. It is used to control a large number of fungal diseases, such as mold, mildew, rot, and blight, in some crops, such as ginger, nuts, legumes, and even fruits. Additionally, carbendazim has been nominated for chemical program review under Australia Pesticide and Veterinary Medicines Authority (APVMA) because of its effect known to cause impaired human fertility and cause birth defects, the review made a conclusion it causes the above effects, the half-life of carbendazim is as long as 6 months, recommended warning for registered carbendazim products that it must contain the following stated warning "Contains carbendazim which causes birth death and irreversible male infertility, in laboratory animals, avoid contact with carbendazim" recommended usage level in drinking water is 0.09 mg/ls [24]. For safety operators mixing and loading carbendazim must wear gloves to avoid skin irritation, respirator face shield should be worn to prevent ingestion. Even with the use of these safety measures the risk cannot be mitigated, the use of carbendazim is no longer supported for occupational health and safety grounds [2].

Another active ingredient similar to carbendazim is mancozeb, also a fungicide with the recent addition to the highly hazardous list; also has restrictions of use, prohibited, risk-specific mitigation measures are mandatory and may only be used under specific, defined conditions according to FSC, RA, and Fairtrade standards. Mancozeb is a probable carcinogen to humans [13, 23], an endocrine disruptor, and has wildlife risk [3]. Mancozeb is used for a wide range of fungal diseases as protective fungicides for horticultural and agricultural purposes. Mancozeb is a member of the ethylenebisdithiocarbamate (EBDC) group of fungicides which maneb and metiram are some of the related active ingredients, used on crops, such as potatoes, apples, grapes, onions, tomatoes, and melons. Its effects on human health can be toxic because it is majorly harmful to thyroid organ, reviewed to cause thyroid toxicity, thyroid lesions, and thyroid tumors, the residual composition of mancozeb is not to a level of concern to the EPA and other effects, such as cancer risk, effects on terrestrial and aquatic species, are feasible by using restrictions [25].

The 2,2dichlorovinyle dimethyl phosphate is also known as (dichlorvos); it is a colorless to amber liquid, an agricultural chemical used to control insects, diseases, and eliminate storage pests and crops. Application of dichlorvos is mainly expelled into the air for household pesticides and it is usually distributed into the water for pesticide control and sprayed on land when used for agricultural purposes. Furthermore, it is eliminated by hydrolysis and biodegradation, some toxic effects on animals and humans include acute effects such as weakness, severe anemia, anticholinergic symptoms other effects on gastrointestinal tracts and nervous system in rabbits, it causes severe skin irritation. The current regulation in Japan for dichlorvos is Deleterious substance, Class I designated chemical substance.

Imidacloprid is a new insecticide that is related to nicotine chemically, just like nicotine, imidacloprid acts on the nervous system, it is used in large quantities in crops, pests, and turf grasses, when imidacloprid is exposed to animals or humans some of the effects includes, Apathy, emaciation, convulsion, labored breathing, when exposed for a long time it causes loss of weight and thyroid lesions in human. It can be acutely toxic in some animals, bird species, and plants by causing decreasing growth levels.

Hexaconazole is a systemic triazole fungicide that is used in the control of a wide range of diseases of crops example of some diseases are black and yellow Sigatoka diseases of banana, used on banana foliar to control diseases, The Health Effects Division Hazard Identification Assessment Reviews Committee (HIARC), evaluated the toxicological level of hexaconazole on human and animals is reviewed to have enhanced sensitivity to infants and children. In animals such as rats, the study revealed a decrease in body weight gains and decreased pub survival, although the aggregate exposure risk is limited to dietary exposure only, hexaconazole has low toxicity by oral, dermal, and inhalation mode of exposure, it can be slightly irritating to the eye and skin sensitization in animals.

Hexaconazole was found in the open market but not approved by the EU, classified as a highly hazardous substance, a possible carcinogen [26] very persistent/water, and highly toxic to bees [5].

Imidacloprid, a fungicide approved by the EU for the management of fungal diseases in crops, although approved, it is however prohibited with an exception for certain pests in certain crops and without exceptions by some other standards. The active ingredient may cause severe effects on humans and be highly toxic to honey bees and birds [5, 14].

A fungicide named dimethoate is not on the approved list of the EU, it is listed as highly hazardous in 2011 and 2019 [5]. Dimethoate is classified as a probable carcinogen and with reproductive toxicity according to globally harmonized system [13, 23]. This active ingredient is recommended as restricted with minima use and potentially to be prohibited according to FSC, RA, UTZ. It also has inhalation risk to humans, highly toxic to honey bees, birds, and aquatic, pollinator, and wildlife risk according to SAN [21].

## 9. Handling and disposal of agrochemicals

People are exposed to pesticides through varied means of handlings for domestic and agricultural purposes. Exposure can be through spray drift, residues in the environment, contaminated food, or drinking water and these can be directly or indirectly.

This exposure can also be through absorption through the skin, ingestion through food, or inhalation during the application or perceived from the environment. Exposure has an impact on the human body as related to the amount of pesticides exposed to (dose) and length of pesticides exposure (time). The health risks associated with pesticide use are a combination of toxicity and exposure. However, responsible pesticides use involves applying the right pesticide, in the right way, dosage, interval, and at the right time.

**Figure 1** shows a typical practice of some farmers on the use of pesticides on stored products in rural communities and poor urban areas. Pesticides were applied directly to the product to extend the shelf life in storage, especially against insect infestation. The pesticides were sprayed in overdose, at the wrong time as shown in **Figure 1** (around afternoon as depicted in the shadows) and the products were bagged immediately.

Apart from hazards of residue contamination in the food crops, the human and environmental hazards are also very loud. The humans were not in any way protected from spray drift on their skin and through inhalation or direct exposure. Likewise, was the volatility escape of the sprays into the environment, contaminating and polluting nearby produce and passersby. This practice showed wrongness in terms of quantity of agrochemicals applied, time of application, exposure of the crop, the farmers also unprotected and the environment been polluted.

The indiscriminate disposal of agrochemical contents into the soil, environment, and wrong handling are shown in **Figure 2**. Rural farmers use this method to prepare pesticides in containers, mixed with hands and occasionally tasted to "ascertain efficacy" of the pesticides.

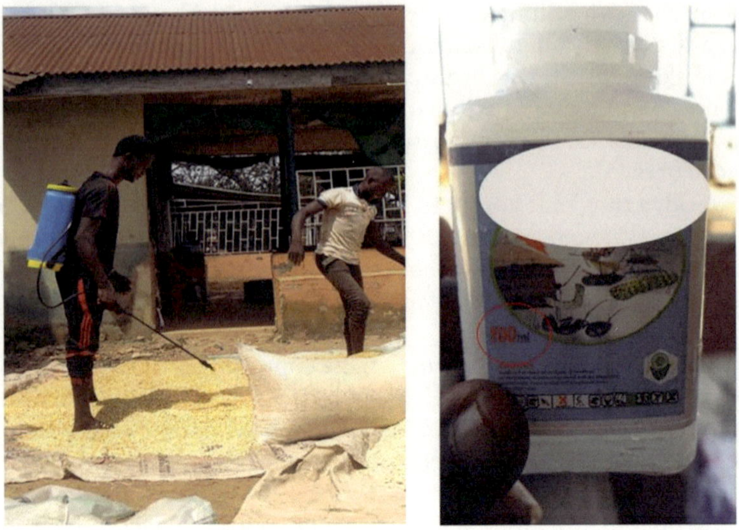

**Figure 1.**
*Over dose application of pesticides on stored product.*

**Figure 2.**
*Indiscriminate disposal and preparation of pesticide on farm.*

This practice proves the level of ignorance and literacy of potential risks agrochemicals pose to human health beyond a reasonable doubt. The pesticides residues contaminant in soils were usually washed into the streams during rains, the same water is used for domestic activities, bathing, and even drinking.

## 10. Conclusion

The study showed that many of the agrochemicals in open markets have some level of restriction of use or approval based on the recommendation of international standard organizations, with proved risks to humans, animals, and the environment. The general handling and indiscriminate use of these active ingredients in open markets and farmer's fields showed deficient knowledge and awareness of the potential danger they pose to crops, humans, and the environment.

## 11. Recommendation

Enlightenment programs on local broadcasting stations, such as radio, television, and marketplace campaign should be launched to create awareness of the risks and dangers associated with agrochemical use and misuse both for domestic and agricultural purposes. These avenues will reach the rural dwellers who are the most vulnerable to the potential risks. The relevant government/regulatory agencies should fulfill their mandate agrochemical related matters like control/enforcement, acceptable active ingredients, monitoring and safety measures as well as prosecution of offenders of national agrochemical laws.

**Author details**

Dele Omoyele Adeniyi
Cocoa Research Institute of Nigeria, Ibadan, Nigeria

*Address all correspondence to: modeleadeniyi@gmail.com

## References

[1] Gawarammana IB, Buckley NA. Medical management of paraquat ingestion. British Journal of Clinical Pharmacology. 2011;**72**(5):745-757

[2] United States Environmental Protection Agency. Prevention, Pesticides and Toxic Substances. 1997. EPA-738-F-96-018 (7508W)

[3] Lewis KA, Tzilivakis J, Warner D, Reen A. An international database for pesticide risk assessments and management. Human and Ecological Risk Assessment: An International Journal. Tokyo, Japa:n Makoto Irie, Ministry of Agriculture forestry and fisheries. 2018;**1375**(171). Website of the Stockholm Convention at http://www.pops.int; Website of the Rotterdam Convention at http://www.pic.int; Website of Montreal Protocol at http://ozone.unep.org DOI: 10.1080/10807039.2015.1133242. Profenofos

[4] Gotter A. Paraquat Poisoning. Medically reviewed by Judith Marcin. 2018. Accessed 24 October 2020. from: https://www.healthline.com/health/paraquat-poisoning

[5] Pesticide Action Network International. Impressum. PAN International List of Highly Hazardous Pesticides. © PAN International c/o PAN Germany, Nernstweg 32, 22765 Hamburg, Germany March; 2019. pp. 43

[6] Mensah PK, Palmer CG, Muller WJ. Lethal and sublethal effects of pesticides on aquatic organisms: The case of a freshwater shrimp exposure to Roundup. In: Larramendy ML, Soloneski S, editors. Pesticides: Toxic Aspects. Rijeka, Croatia: InTech Publications; 2015. pp. 163-185

[7] International Agency for Research on Cancer. Agents reviews by the IARC Monographs, Volumes 1-123 (by CAS Numbers), International Agency for Research on Cancer (IARC), Lion, France. 2018. Accessed October 24, 2020 from: http://monographs.iarc.fr/ENG/Classification/

[8] Vandenberg LN, Colborn T, Hayes TB, Heindel JJ, Jacobs DR, Lee DH. Hormones and endocrine-disrupting chemicals: Low-dose effects and nonmonotonic dose responses. Endocrinology Review. 2012;**33**(3):378-455

[9] Jayasumana C, Gunatilake S, Senanayake P. Glyphosate, hard water and nephrotoxic metals: Are they the culprits behind the epidemic of chronic kidney disease of unknown etiology in Sri Lanka? International Journal of Environmental Resources and Public Health. 2014;**11**(2):2125-2147

[10] Jayasumana C, Paranagama P, AgampodiS,WijewardaneC,GunatilakeS, Siribaddana S. Drinking well water and occupational exposure to Herbicides is associated with chronic kidney disease, in Padavi-Sripura. Sri Lanka Environmental Health. 2015;**14**(1):6-9

[11] Zoeller RT, Brown TR, Doan LL, Gore AC, Skakkebaek NE, Soto AM. Endocrine-disrupting chemicals and public health protection: A statement of principles from the Endocrine Society. Endocrinology. 2012;**153**(9):4097-4110

[12] Adam HP, John HA. Florida Department of agriculture and consumer services, commissioner Florida department of health, M.D, state surgeo general. 2013;**3**:20

[13] United State Environmental Protection Agency. Technical Overview of Ecological Risk Assessment Analysis

Phase: Ecological Effects Characterization, U.S. Environmental Protection Agency, Washington, DC. 2007. Accessed October 24, 2020 from :www.epa.gov/oppefed1/ecorisk_ders/toera_analysis_eco.htm(http://www.eu-footprint.org

[14] FOOTPRINT. The FOOTPRINT Pesticide Properties Data Base. Database collated by the University of Hertfordshire as part of the EU-funded FOOTPRINT project. 2007. (FP6-SSP-022704)'

[15] European Commission. Regulation (EC) No 1272/2008 of the European Parliament and of the Council of 16 December 2008 on classification, labeling and packaging of substances and mixtures, amending and repealing Directives 67/548/EEC and 1999/45/EC, and amending Regulation (EC) No 1907/2006.Official Journal of the European Union L 353/1. 2008

[16] Arpad A. Dichorvos (025) National food chain safety office, Budapest Hungary, Dichorvos in Drinking Water Background document for development of WHO Guidelines for drinking water quality, World Health Organization. 2007

[17] World Health Organization. The WHO recommended classification of pesticides by hazard and guidelines to classification 2009, International Program on Chemical Safety (IPCS) and World Health Organization (WHO), Geneva, USA. 2010

[18] European Commission. Towards the establishment of a priority list of substances for further evaluation of their role in endocrine disruption - preparation of a candidate list of substances as a basis for priority setting, European Commission, Delft. 2000

[19] European Commission. Commission Staff Working Document SEC (2004) 1372 on implementation of the Community Strategy for Endocrine Disrupters - a range of substances suspected of interfering with the hormone systems of humans and wildlife (COM (1999) 706), European Commission, Brussels. 2004

[20] European Commission. Commission staff working document on the implementation of the "Community Strategy for Endocrine Disrupters" - a range of substances suspected of interfering with the hormone systems of humans and wildlife (COM (1999) 706), (COM (2001) 262) and (SEC (2004) 1372). SEC (2007) 1635. European Commission (EC). Brussels, 2007

[21] Food and agriculture Organization of the United Nations, FAO/WHO Specification & evaluation for Agricultural Pesticides Reports; 2013:709

[22] European Commission. Commission Regulation 790/2009 amending, for the purposes of its adaptation to technical and scientific progress, Regulation (EC) No 1272/2008 of the European Parliament and of the Council on classification, labeling and packaging of substances and mixtures. Official Journal of the European Union L 235/1. 2009b

[23] United State Environmental Protection Agency. Annual Cancer Report. Chemicals Evaluated for Carcinogenic Potential, Science Information Management Branch, Health Effects Division, Office of Pesticide Programs U.S. Environmental Protection Agency (US EPA), Washington DC, USA. 2018

[24] Australian Pesticide and Veterinary Medicine Authority. Australian chemical review program, human health risk assessment of Carbendazim prepared by office of chemical safety and environmental health office of health protection of the department of health

and agency. 2009. Canberra, June 2008 revised

[25] United States Environmental Protection Agency. Prevention, pesticides and toxic substance (7508c). 2005: EPA 738.F.05.XX, Mancozeb fact September, 2005

[26] United State Environmental Protection Agency. Chemicals Evaluated for Carcinogenic Potential, April 26, 2006, Science Information Management Branch, Health Effects Division, Office of Pesticide Programs U.S. Environmental Protection Agency (US EPA); 2007b

Chapter 11

# Pesticides: Chemistry, Manufacturing, Regulation, Usage and Impacts on Population in Kenya

*Joseph O. Lalah, Peter O. Otieno, Zedekiah Odira and Joanne A. Ogunah*

## Abstract

Pesticide use in Kenya plays a critical role in socio-economic development because its economy depends heavily on agriculture, which contributes to 30% of the GDP and accounts for 60% of export earnings. For agriculture and public health vector control, the country relies on pesticides, most of which (95%) are formulated products imported from China, India and Germany as the top exporters. In this chapter, we present the chemistry, manufacturing, importation and regulatory processes regarding pesticides in Kenya as well as their usage and impacts. All the various categories, organochlorine, organophosphate, carbamate, pyrethroid, neonicotinod insectides, as well as fungicides, herbicides and biopesticides, which are used in the country, are considered. A total of 1,447 and 157, which include formulations and active ingredients, respectively, for use in agriculture and public health sectors, with sufficient information on their usages and toxicities, are listed on the Pest Control Products Board (PCPB) database that is available to the public. A significant number of studies have been conducted in major agricultural regions, which have characterized pesticides, their toxicities, the types of crops and pests, the usage and human and environmental health risk indices, since the 2000, but the reports have not made any impacts on pesticide regulation, as some of the very toxic active ingredients, belonging to the WHO Class I and II, are still reported by farmers. However, a recent call from NGO's made an impact in government and parliament, and a bill was introduced in 2020 with the aim of banning some of the toxic ones that have already been withdrawn from the EU market.

**Keywords:** pesticides, regulations, usage, toxicity, human, environmental, impacts, Kenya

## 1. Introduction

### 1.1 Why are pesticides important?

Human population development has been dependent on a steady increase in abundance of food supply over the years. The world population explosion began to be felt around 1600 AD; mainly as a result of two factors: (i) ability of man to control diseases and (ii) developments made in modern agriculture to increase food supplies [1]. Prior to 1800, there was little application of scientific information in agricultural production. Mass starvations occurred, whenever there were conflicts such as political conflicts, wars and climate change, which affected agricultural production and/or food flow. By about 1983, about 5 billion people existed, compared to the current global human population of approximately 8 billion. This population explosion to 8 billion over a period of just 38 years was possible because of developments in modern agriculture [1, 2] and ability of humankind to fight various diseases [3]. The fastest growth was realized in the 20th and 21st centuries when the population increased exponentially from about 2 billion in 1930 to about 7 billion in 2000 [2].

Kenya's *population*, which is equivalent to 0.69% of the total world population, was estimated at 53,771,296 people in 2020 [2], and this is expected to grow by around 1 million per year—3000 people every day—over the next 40 years, reaching approximately 85 million by 2050 [2]. The country will, therefore, rely on agriculture to provide food for the growing population. Agriculture contributes approximately 27% of Kenya's GDP and a large part of its rural population (approximately 80%) depends on subsistence farming as a source of food, employment and income [4]. The importation and use of pesticides are, therefore, foreseen to increase [3, 5].

Pesticides were originally introduced to control insects but have also nowadays been used to eradicate problems caused by nematodes, mites, rodents, birds, mollusks, parasitic fungi and weeds [1, 3]. Approximately 33% and 30% of food crops in the world are lost annually to pests and insects alone, respectively [1]. The losses occur in the field as well as during postharvest e.g. during storage or transportation. Tropical countries in Sub-Saharan Africa, which have a myriad of insects and disease pathogens, will have to continue relying on pesticides despite their negative impacts on the environment and human health.

Significant increases in different crop yields can be realized by using insecticides to control certain pests [1, 3]. In corn, 24.4%, 38.4% and 10.7% increases in yields have been achieved by controlling corn borers, leaf hoppers and corn root worms, respectively, using insecticides; whereas in wheat, 79%, 47%, and 29.5% increases have been realized by controlling brown wheat mites, cutworms and white grubs, respectively. In Irish potatoes, 45.6% and 42.8% increases in yields have been realized by controlling Colorado potato beetles and potato leafhoppers, respectively [3]. Bollworm and thrips can destroy cotton almost completely, reducing the yields to just 21.3% and 59.7%, respectively [3], if not controlled by insecticides. The FAO has estimated that 50% of cotton production in developing countries would be destroyed if there is no use of insecticides [3]. Pesticides not only reduce losses caused by pests and weeds but also increase profits for farmers by reducing the need for labor, specifically by using herbicides.

Many human diseases such as yellow fever and malaria, which are caused by mosquitoes, were eradicated or controlled in the past in industrialized countries by using pesticides [3]. The use of insecticides such as DDT contributed to the reduction of global annual malaria mortality rates from 6 million in 1939, to 2.5 million in 1965 and 1 million in 1991 [6]. Overcoming malaria is still a very big challenge for developing countries, especially in the Sub-Saharan African countries, partly because of failure to use pesticides effectively to control mosquito larvae, as recommended by the WHO [7]. Other diseases and their respective causes (as given here in parenthesis) have been controlled by use of insecticides including sleeping sickness (tsetse flies), anthrax (horseflies), bubonic plague (rat flea), dysentery (houseflies), filariasis, encephalitides, dengue fever, Chagas disease and West Nile virus (all these five caused by mosquitoes), hemorrhage and Q fevers (ticks and mites), bilharziasis (snails) and bronchial asthma (cockroaches) [3, 6, 8]. However, the agricultural sector consumes most of the conventional pesticides, e.g. approximately 77% in the USA [3, 5]. The situation is quite similar in Kenya, where most the conventional pesticides in form of insecticides, fungicides and herbicides are needed in the agricultural sector. Currently, some of the major classes of pesticides that have a significant stake in the global pesticide industry include organophosphate, carbamate, pyrethroid and neonicotinoid insecticides, fungicides and herbicides, respectively; and the organophosphates, carbamates, synthetic pyrethroids and neonicotinoids together account for 70% of the global insecticide sales [3, 5].

The cost of developing a pesticide active ingredient/compound is very expensive, ranging between US dollars 50 million and 100 million per active compound. These costs cover various aspects, including screening, synthesis, trials and regulation & registration; and the time period can take between 5 and 9 years before a product goes into commercial sale [3, 9, 10]. The developing countries such as Kenya, therefore, control a very small share of the pesticide industry, with Kenya importing most pesticides, which are already manufactured (95%) and only manufacturing a very small percentage (5%) of the products it needs [11].

## 2. Pesticide chemistry and biochemistry

Pesticides are classified in various ways, i.e. according to target pest, or according to their chemistry, chemical structures and particular functional groups on their molecules, respectively. The classification according to pests, including terminologies such as algaecides (developed to control algae), acaricides (mites), avicides (birds/avian), bactericides (bacteria), fungicides (fungi), herbicides (weeds/plants), larvicides (larvae), molluscicides (mollusks e.g. snail, slugs), nematicides (nematodes), termicides (termites), ovicides (eggs), pediculicides (lice), predicides (predators e.g. coyotes and wolves), rodenticides (rodents), slimicides (slime) and silvicides (trees and brushes or entire forest), are used and usually indicated on the labels of the products [1]. However, in the industry as well as among scientists and researchers, pesticides are grouped broadly, according to their chemistry, chemical structures and mode of action, into four main categories, i.e. insecticides, herbicides, fungicides and biological control compounds/products such as microbial pesticides, as discussed in the following section.

### 2.1 Insecticides

Insecticides are used to destroy insects and can be classified according to their chemical structure as well as their mode of action as (i) Stomach poisons—which are lethal only to insects, which ingest them and were tested on target organisms through oral exposure; (ii) Contact insecticides—which kill insects following external bodily contact and do not have to be ingested to impart expected toxic effects and (iii) Fumigants—which act on the insect through its respiratory system, by emitting poisonous vapors, which can be inhaled and enter into the target organism through the respiratory system [12]. An insecticide can act by one or a combination of two or three of these modes. These classifications are taken as the tested modes of toxicity (based on trials) at the point of registration of the product and are normally given on the labels on the containers. During development, all insecticides are subjected to standard toxicity tests as described in the EU or USEPA standard methods [3, 13] and are expressed as $LD_{50}$ or $EC_{50}$ values. The $LD_{50}$ is defined as the lethal dose of a compound that kills 50% of the target organism on exposure in a standard toxicity test procedure, in milligrams per kilogram weight of the test organism (mg/kg). The $EC_{50}$ is defined as the effective concentration of the compound in water that kills 50% of the target organism in a standard toxicity test and is expressed in mg/L, and is normally conducted for aquatic organisms. The toxicity tests are done for: insects—to show effectiveness (as insecticides), rats—to show potential hazards to mammals especially humans, birds/fishes/bees etc.—to show potential hazards to the environment or non-target organisms. Pesticides are, therefore, ranked as hazards according to the WHO, where Class I, II, III and IV pesticides, respectively, where Class I are the most toxic with the least $LD_{50}$ values [3].

Insecticides are subdivided into organochlorines, organophosphorus (or organophosphates), carbamates, pyrethroids, neonicotinoids, insect growth regulators (IGRs) and natural products (which include microbial insecticides), respectively. Brief descriptions of these seven categories, which are all popularly used in Kenya, are presented in the following sections.

#### *2.1.1 Organochlorine insecticides (OCs)*

The organochlorine insecticides are divided into three major classes, including the *DDT and its analogues*, the *benzene hexachloride (BHC) isomers* and the *cyclodiene compounds*, respectively. The DDT and its analogues include *DDT*, which is commonly known as *p,p′-DDT* (IUPAC nomenclature: 1,1,1-trichloro-2,2-bis (p-chlorophenyl) ethane). It was the first synthesized chlorinated insecticide, manufactured in 1873 but was recognized as an insecticide at the beginning of the world war in 1939 [1, 5]. After DDT, more and more organochlorine pesticides were discovered and synthesized in Europe and USA. When DDT is synthesized, the technical mixture contains a lot of impurities, with only seventy per cent (70%) of the mixture being the active ingredient p,p′-DDT, and others—21% is the ortho-isomer (o,p′-DDT), 1% the o,o′-DDT; 4% the p,p′-DDD and 0.04% the o,p′-DDD. These isomer impurities are formed during synthesis and account for 30% of weight of the technical mixture (Note: the 'o' means 'ortho' and 'p' means 'para' position on the benzene ring). These impurities potentiate the toxicity of DDT and also add to the environmental burden of DDT because they are also persistent. Other metabolites of DDT also form in the organisms or the environment, e.g. DDE, which is more potent and persistent than DDT (**Figure 1a**). The isomeric impurities have little value as toxicants, since they are just <10% as toxic as the p,p′-DDT. DDT is toxic to non-targets and has an $LD_{50}$ (oral, rats) of 250 mg/kg [3].

The other DDT analogues that were manufactured after DDT have various functional groups on the DDT molecule changed and include *methoxychlor*, *dicofol* and *chlorobenzilate*. As shown in **Figure 1a**, the various changes on the DDT molecule by researchers resulted in different pesticides, which were designed to be less toxic than DDT, with lower mammalian toxicities, e.g. methoxychlor has an $LD_{50}$ (oral, rat) of 600 mg/kg, while dicofol has $LD_{50}$ (oral, rats) of 595 mg/kg, chlorobenzilate

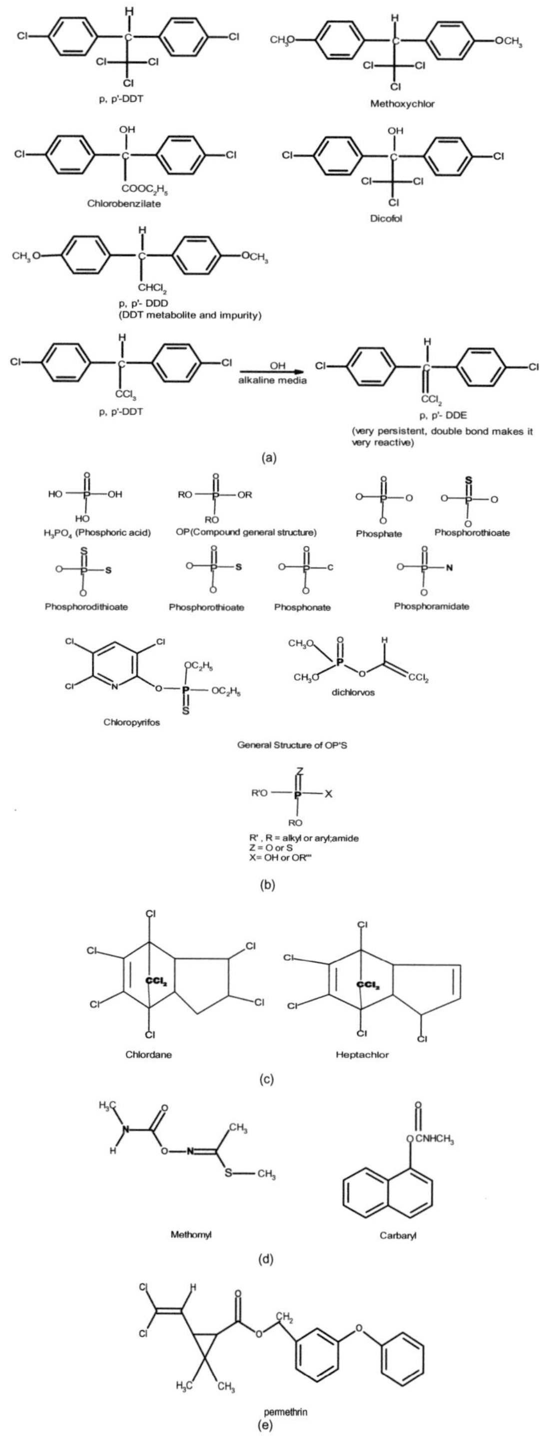


**Figure 1.**
*(a) DDT and its analogues. (b) The general chemical structure of OPs, the starting phosphates, and two OP insecticides. (c) Chemical structures of two cyclodiene insecticides. (d) Chemical structures of two carbamate insecticides. (e) Chemical structure of permethrin, a pyrethroid insecticide.*

(3888 mg/kg) and p,p′-DDD (3400 mg/kg) [3]. The DDT analogues have similar chemical and physical properties but because of the slight differences in chemical structures have differences in toxicity and specificity. Dicofol and chlorobenzilate, for example, have lower insecticidal activity, but better acaricidal activities. The DDT group are considered as persistent organic pollutants (POPs) and are environmentally persistent since they are non-polar, highly lipophilic, stable to photolysis, and have very low water solubilities and low vapor pressures [14, 15]. They cause endocrine disruptive effects because of their ability to mimic sex hormones [14, 15]. Although DDT has been banned globally, it is allowed restrictively for use in Kenya, for malaria vector control, only. *Methoxychlor* has been banned in Kenya, but *dicofol* and *chlorobenzilate* though not banned are not used in Kenya.

The second subclass of organochlorines, the *Benzene hexachloride isomers,* was first discovered in 1942 [3]. They are chlorinated saturated six-carbon cyclic alkane molecules, which can adopt a chair conformation [6]. On each C atom, there is a chlorine atom, located in axial or equatorial position, respectively, on the molecule, which gives the various five isomers (α, β, γ, δ, ε). Only the γ (gamma) isomer exhibits pronounced insecticidal activity. It is the main isomer and is known as γ-hexachlorocyclohexane (γ-HCH) or *lindane*. Lindane is odorless and is widely used even in agriculture. Other BHC isomers have unpleasant odor, and this gives an off flavor in root and tuber crops, limiting their use in agriculture. Although banned in most countries, lindane is still allowed restrictively for non-food-related use in Kenya, like seed dressing [3, 6]. Lindane is more toxic than DDT group and has an $LD_{50}$ (oral, rats) of 125 mg/kg. It has a higher vapor pressure ($9 \times 10^{-6}$ mmHg) than DDT and, therefore, is slightly more water-soluble (10 ppm in pure water) [16].

The third subclass, the *cyclodiene compounds* or *cyclodiene insecticides* were discovered in the USA after the War II (around 1948). They are cyclic hydrocarbons and include *aldrin*, *dieldrin*, *chlordane* (two isomers α and β-chlordane), *heptachlor*, *endrin*, *endosulfan*, *chlordecone*, *mirex* and *toxaphene* (**Figure 1b**). Although others are synthesized, toxaphene is strictly a chlorinated terpene, produced by passing chlorine into camphene, a natural product [3, 6]. Like other OCs, all these cyclodienes are heavily chlorinated compounds. Some of the cyclodienes have higher mammalian toxicity than the DDT group but the $LD_{50}$s (oral, rat) vary widely, ranging from aldrin 38–67 mg/kg, dieldrin 37–87 mg/kg, chlordane 367–515 mg/kg, endrin 7–15 mg/kg, heptachlor 147–220 mg/kg, endosulfan 18–43 mg/kg, chlordecone 114–140 mg/kg, mirex 306 mg/kg and toxaphene 69 mg/kg [3, 6]. In this group, *aldrin, heptachl*or and other *cyclodienes* have long residual lives in soil, expressed in terms of 'half-life', and; therefore, cyclodienes were important agents for controlling termites and soil insects. They are very stable, lipophilic and have low vapor pressures (range of $10^{-5}$ to $10^{-7}$ mmHg). They have been banned or severely restricted because of environmental persistence and non-target toxicity in most countries, and even in Kenya some of them such as *aldrin, dieldrin, chlordane*, *heptachlor* and *endrin* have been banned [11]. The principal site of action of organochlorines is the nervous system, where they bind to the sodium channel and cause delayed Na-inactivation, resulting in a prolonged delay in Na-inactivation and subsequent interference with nerve impulse functions.

### *2.1.2 Organophosphorus insecticides (OPs)*

After the OCs, the trend was to (i) avoid persistence (ii) build biodegradability and (iii) have a narrow spectrum of activity (more specificity). Therefore, the OCs have

| OP subclass | Examples in the subclass |
|---|---|
| Phosphates | Dichlorvos, naled, dicrotophos, mevinphos, chlorfenvinphos, crotoxyphos. |
| Phosphorothioates | Ethyl parathion, methylparathion, chlorpyrifos, diazinon, temephos |
| Phosphorodithioates | Malathion, dimethoate, phorate, azinphos-methyl, methidathion, phosmet, azinphos-ethyl |
| Phosphorothiolates | Oxydemeton-methyl, pyraclofos |
| Phosphoramides | Acephate and methamidophos |

**Table 1.**
*Sub-classes of OPs and specific examples of active ingredients.*

now been replaced by OPs, carbamates, neonicotinoids and pyrethroids, which have these desirable properties. The OPs are very popular in Kenya and, as insecticides, dominate the Kenyan synthetic pesticide market. In general, they are more toxic to insects and mammals than OCs but are readily biodegradable. They were discovered as by-products of chemical warfare research involving the development of nerve gases such as *sarin, soman and tabun*, in Germany during War II [1, 3, 17], and esters of *phosphoric acid*, *thiophosphoric acid* (*phosphorothioate*), *phosphorothiolate*, *phosphorodithioate* or *phosphoramidate* are given in **Table1** and their structures in **Figure 1c**.

These starting organophosphoric acidic compounds for synthesis of OPs indicate the six subclasses of OP insecticides; e.g. when H atoms of phosphoric acid are replaced with organic radicals such as methyl, ethyl or phenyl, the compounds obtained are organophosphates. On the other hand, oxygen can be replaced with S, C, or N to yield different derivatives. *Phosphates* such as *dichlorvos* are few but many other subclasses also become *organophosphates* during metabolism by various organisms, for example by changing an S atom by an O atom through oxidation. In the subclass *phosphorothioates* (e.g. *chlorpyrifos* and *diazinon),* an S atom is double bonded to phosphorus, while in the subclass *phosphorodithioates* (e.g. *malathion* and *dimethoate)*, the molecule contains two sulfur atoms in the phosphoric acid part. In *phosphorothiolates,* there is a single bond between S and P atoms, and in p*hosphoramides* (e.g. *acephate),* there is a P atom bound to N atom (**Figure 1c**) [3, 6].

Organophosphates have different physical and chemical properties from the organochlorines. Overall, they have moderate-to-considerable water solubility, with some such as *oxydemeton-methyl*, *trichlorfon* and *phosphoramidates*, being very soluble. They also have moderate-to-considerable vapor pressures (generally ranging between $10^{-3}$ and $10^{-5}$ mmHg) and therefore are less volatile than most OCs, although some such as *naled, parathion* and *dichlorvos* are very volatile [6, 17]. The OPs are degradable in the environment, e.g. in water, soil and other compartments and are easily metabolized in the organisms, with the most common chemical reactions being hydrolysis, catalyzed by water and esterases. Insects, mammals and other organisms have esterases, which can metabolize OPs e.g. *malathion carboxylesterase,* which has been shown to *decarboxylate malathion* in rat liver [18, 19]. Another common reaction of OP compounds is oxidation of the P=S moiety on an OP molecule to P=O, which is mediated by the cytochrome P450 monoxygenases. This oxidation is referred to as oxidative desulfurization, e.g. conversion of *malathion* to *malaoxon,* which occurs in insects and mammals [3, 18]. The combination of physical (e.g. *water solubility*) and chemical (*hydrolysis and oxidation*) properties make the entire OP class of insecticides biodegradable and more easily excreted.

### 2.1.3 Carbamate insecticides (CBs)

Carbamates are esters of carbamic acid, $HOOCONH_2$. The 3 H atoms in the molecule can be replaced by aliphatic or aromatic radicals to become carbamate insecticides. However, the second H on the nitrogen (N) is not replaced in making CB insecticides because the monoalkyl structure (NRH) is more toxic than the N-disubstituted compound (NRR″) [20]. Carbamic acid is similar in chemical structure to the pharmaceutical agent, physostigmine (eserine), whose synthesis started the curiosity on carbamates. Physostigmine is a poison, an acetylcholinesterase (AChE) inhibitor. A typical carbamate structure is represented by *carbaryl* (1-naphthylmethyl carbamate), which was the first carbamate insecticide to be synthesized. Substituted phenyl-N-methyl carbamate insecticides can be synthesized by addition of methyl isocyanate ($CH_3NCO$) to various phenols, some of them with substituted alkyl or phenyl functional groups (R), as shown:

$$RC_6H_5OH + CH_3NCO \circledR RC_6H_5OOCNCH_3 \left(R = \text{alkyl or phenyl}\right) \quad (1)$$

During this synthetic process, all reagents and solvents are kept free from water because water reacts readily with methylisocyanate. The reaction is usually vigorous and thermic generating a lot of heat and can cause the release of methyl isocyanate, which is extremely toxic, from the reaction vessel. This is what caused the Bhopal accident in 1984, in which methyl isocyanate gas leaked from a Union Carbide factory in Bhopal, India, killing approximately 3800 people [1]. Making various changes to the carbamyl functional group ($-O-(C{=}O-NH_2$) by varying R (alkyl or aryl) groups researchers resulted in many different CB insecticides (**Figure 1d**). The chemical structures of two CBs, *methomyl* and *carbaryl*, are shown in **Figure 1d**. Carbamates, like OPs, act by binding to and inhibiting *acetylcholinesterase* (AChE), resulting in a buildup of acetylcholine at the synapse, which causes excessive neuroexcitation, paralysis and death. The second mode of action is by binding and interference with neuropathy target esterases (NTE) located in the nervous system in insects.

Carbamates are popular in Kenya, for example, *carbofuran*, *aldicarb*, *propoxur*, *carbaryl*, *methomyl*, *oxamyl*, *carbosulfan* and *pirimicarb* are often used in the agricultural sector [11]. Some of the carbamates are very toxic to non-target, with very low $LD_{50}$ (oral, rats) values, e.g. *aldicarb* ($LD_{50}$: 1 mg/kg), *methomyl* (17–26 mg/kg), *oxamyl* (54 mg/kg) and *carbofuran* (5–13 mg/kg), while others such as *carbaryl* (500–700 mg/kg) and propoxur (95–104 mg/kg), are less toxic [3]. Carbamates are also very toxic to birds, i.e. *carbofuran* ($LD_{50}$ 25–39 mg/kg in birds), *carbosulfan* (10 mg/kg), *propoxur* (4–120 mg.kg), *aldicarb* (1.78–5.34 mg/kg) and *methomyl* (10–42 mg/kg), except *carbaryl* ($LD_{50} > 2000$ mg/kg). *Carbofuran* and *carbaryl* were popularly used in field crops including rice and maize farming, from 1980's to 2000 in Kenya. Due to their high toxicity, *aldicarb, carbofuran* and *carbosulfan* have recently been misused by pastoralists and farmers against wildlife, especially predators, which has led to carbofuran withdrawal pending banning in Kenya [21, 22]. Granular forms of carbofuran and aldicarb, which are fairly soluble in water, can be picked by small organisms such as worms and grasshoppers in farm fields and through food chain transfer, larger scavenger birds and other insect-eating species get poisoned [21, 22]. Liquid formulations of carbofuran are considered safer and are still allowed in other countries despite the ban on granular formulations [21].

## 2.2 Herbicides

### *2.2.1 Herbicide chemistry and biochemistry*

By definition, a herbicide is a compound that is capable of either killing or injuring plants (or weeds) and can control their growth. Herbicides are used to control weeds in farms as well as lawns, roads and other facilities, and their use has led to great reductions in agricultural production costs. Herbicides can be: (i) *selective,* killing only a particular group of plants such as the leafed plants or grasses or (ii) *non-selective*, making the ground barren of all plant life. They can be formulated either in (i) *granular form*, which is worked into the soil prior to planting the crop in a *preemergence* application or (ii) *liquid spray form,* which may be applied best at various stages after planting, *postemergence or preemergence*, the choice between (i) and (ii) depending on the particular chemical, weed, soil type and crop cultivated [3]. However, the use of herbicides is very intricate and several factors must be considered, e.g. it can destroy a lawn or the plant crops, which are meant to be protected against weeds. Therefore, there is a need to consider wind direction and proximity to wanted plants, when applying herbicides. Different species of plants in the same class may respond differently, some requiring one application and others up to 3 applications before being controlled. Herbicides can become effective either by (i) *direct contact* with plants or (ii) by movement through the entire plant following absorption (called *systemic action*) [23].

Generally, herbicides may, therefore, be classified into a number of groups, either (i) based on the *chemical structure* or (ii) based on when and how it is applied, e.g. *preplanting*—applied to soil before crop is seeded, *preemergenc*e—applied to soil before usual time of appearance of unwanted weeds/vegetation or *postemergence*—applied to soil or foliage after the germination of the crop and/or weeds; or (iii) based on mode of exposure, e.g. as *contact herbicide*—which acts by impinging on plant foliage; or *translocated/systemic herbicide* – which are absorbed via the soil or through foliage into the plant xylem and Phloem; or (v) based on *mode of toxicity* in plants e.g. as *selective herbicide*—toxic to some species only or *non-selective,* which kills all plants (**Table 2**). There are two modes of toxicity of herbicides, the first one applies to *non-selective* herbicides, which interfere with photosynthesis and thereby starve the plant to death, with loss of its green color and withering due to lack of energy to carry out the life processes. The second one applies to *selective* herbicides, which act like hormones or biochemical catalysts that control a particular chemical change in a particular type of plant organism at a particular stage/state of its growth. Most selective herbicides today are *growth hormones*, which cause abnormal growth in a plant and swelling of cells, resulting in the leaf becoming so thick that nutrients and water cannot be absorbed [23].

For example, benzoic acids act as growth hormone herbicides, and move both from leaves to the terminal meristems of leaf, shoot and root, and also move in the transpiration stream, and this permits them to also be soil-applied [24]. The majority of herbicides act by inhibiting photosynthesis I and II (**Table 2**). Various chemicals such as calcium cyanamide (CaHCN), borates, arsenates, copper sulfates, sulfuric acid, and chlorates, were used as weed killers, and some formulations such as aqueous solutions of sodium chlorate $NaClO_3$ (40%) and sodium metaborate $NaBO_2$ (50%), respectively, are still used as *non-selective* herbicides [3]. The discovery of selective herbicides started in 1935, starting with nitrophenol [3], and later, more work was directed towards auxins or hormones, as selective herbicides.

| Chemical class | Mode of toxicity | Examples |
|---|---|---|
| Phenoxy acetic acids | Production of high levels of RNA causing abnormal fast growth and death. | 2,4-D; 2,4,5-T; MCPA (4-chloro-o-toloxy acetic acid). |
| Triazines | Inhibition of photosynthesis I and II. | Chloro-s triazines e.g. simazine, atrazine, cyanazine; thiomethyl-s-triazines e.g. ametryn, prometryn; methoxy-s-triazines e.g. prometon. |
| Arylcarbamates | Inhibition of photosynthesis I and II. | Propham, chlorpropham. |
| Ureas (substituted ureas). | Inhibition of photosynthesis I and II | Monuron, diuron. |
| Dinitrophenols | Inhibition of respiration by blockage of electron transfer processes (NADH → $NAD^+$ + $e^-$ or ATP → ADP + Pi). Desiccants | 2,4-dinitrophenol; DNOC; dinoseb, dinoterb. |
| Bipyridyl derivatives | Inhibition of growth of seedling | Paraquat; diquat. |
| Acetanilides | Inhibition of protein synthesis | Alachlor; propanil. |
| Dinitroanilines | Inhibition of protein synthesis and cell division | Dichlobenil; trifluralin; pendimethalin. |
| Amides (chloroacetamides). | Growth stimulants ('auxins', induces light absorption and causes rapid overgrowth). | Metalochlor; acetolachlor; propachlor; butachlor. |
| Aryl aliphatic acids (chloroaliphatic acids). | Inhibition of carotenoids synthesis (destroys chlorophyll) | Chloro-substituted benzoic acids: Dicamba; chloramben; naptalam. |
| Hydrazines | Inhibition of cell nucleus division. | Metribuzine; fluconazole; triadimefon; metamitron; metazachlor. |
| Alkyl N-aryl carbamates | Inhibition of respiration by blockage of electron transfer processes (NADH → $NAD^+$ + $e^-$ or ATP → ADP + Pi). | Chloropropham; propham. |
| Halophenols | Inhibition of protein synthesis. | 2,4,5-trichlorophenol; 2,3,4.6-tetrachlorophenol. |
| Aliphatic chlorocarboxylic acids | Plant growth regulator (inhibit protein synthesis) | Trichloroacetic acid (TCA), dalapon. |
| Glycine derivatives | | Glyphosate. |
| Non selective | Non selective; photosynthesis inhibition; desiccants etc | Inorganic agents: copper sulfate; sodium borate; organic: bentazon. |

**Table 2.**
*Herbicide classes and the corresponding modes of toxicity.*

In Kenya, *2,4-D* (2,4-dichlorophenoxyacetic acid), a selective phenoxyacetic herbicide, is still one of the most widely used, while *2,4,5-T* (2,4,5-trichlorophenoxy acetic acid), was highly effective but it is no longer used because it was banned in most countries due to non-target toxicity caused by dioxins, which are inherent in the technical mixture [3, 25]. The active ingredients of the various classes of herbicides presented in **Table 2** are very popular in Kenya, including *2,4-D, atrazine, glyphosate, diuron, metribuzine, hexazinone, paraquat, alachlor, metolachlor and fluconazole*, which are commonly used in cereal (maize and wheat), coffee, tea, sugarcane and horticulture. The enhanced efficacy and popularity of *atrazine* is because corn and certain types of crops are unaffected by it, rendering it harmless, yet killing weeds [3]. Herbicides are intensively used in certain crops in Kenya e.g. large-scale farming of maize and sugarcane, where large farm acreages are involved e.g. in Trans Nzoia maize farms and Nzoia Nucleus Estate sugarcane farms. In 2010, approximately 10,500 kg of various types of herbicides were used in a total acreage of 18,000 Ha of sugarcane farms in Nzoia [26]. Generally, in Kenya, even though herbicide use is increasing, insecticides are still being imported in higher amounts, which is different from the USA where 59% of all pesticides used are herbicides [3].

### 2.3 Pyrethroids

Synthetic pyrethroids entered the market in 1980s and by 1982, 30% of worldwide insecticides (in terms of sales) in the market were pyrethroids [3, 25, 27, 28]. They arose from a much older class of botanical insecticides, the pyrethrum. Pyrethrum is a mixture of five (5) insecticidal esters, *pyrethrin I, pyrethrin II, cinerin I, cinerin II and jasmins*, which are all extracted from dried pyrethrum flowers [3]. The *chrysanthemum* variety of pyrethrum grown in Kenya yields the highest proportions of active ingredients. In 1965, the world production of pyrethrum was 20,000 tons, with Kenya accounting for 10,000 tons. However, pyrethrum production dwindled around the 1990s due mainly to competition with synthetic pyrethroids. It is however currently being revived again [3, 25, 27, 28]. The increase in usage of pyrethrum extracts amidst plenty of other various types of insecticides (e.g. OPs and CBs) lies in the fact that it has rapid knockdown effect or paralytic action on flying insects. In addition, pyrethrum extracts have lower mammalian toxicity due to their more efficient enzymatic biodegradability, and good selectivity due to low toxicity in some insects. Due to high demand, chemists synthesized analogues of pyrethrum extracts, called *synthetic pyrethroids*, with better stability in air, more persistent residual effect, better selectivity to target insects, lower mammalian toxicity and cheaper costs. The term '*pyrethroids*', therefore, includes both the *pyrethrum flower extracts* and the *synthetic* analogues. The active ingredients in the *synthetic* analogues are called *Pyrethrins*. Pyrethrin consists of esters, namely *Pyrethrin I and II and Cinerins I and II*, each of which are comprised of a combination of two different alcohols, *pyrethrolone* and *cinerolone*, respectively, and two different carboxylic acids - *chrysanthemic* and *pyrethric acids*, as follows: (a) Pyrethrin I (an ester of chrysanthemic acid + pyrethrolone); (b) Pyrethrin II (an ester of pyrethric acid + pyrethrolone); (c) Cinerin I (and ester of chrysanthemic acid + cinerolone and (d) Cinerin II (an ester of pyrethric acid + cinerolone).

Pyrethrin I is the most active ingredient of the pyrethrins for lethality. Pyrethrin II possesses remarkable knockdown properties for a wide range of household, veterinary and postharvest storage pests. The esters formed from the alcohols and respective carboxylic acids are the different active ingredients used in pyrethroid insecticide formulations, whose composition includes synergists and other adjuvants [3, 25, 27, 28]. The various changes in functional groups of pyrethrins and alcohols have resulted in different chemical structures of synthetic pyrethroids (**Figure 1e**). Based on their chemical structures, there are two types of pyrethroids, Type I pyrethroids (*e.g. permethrin, resmethrin, tetramethrin, allemethrin, bifenthrin and metofluthrin*) and Type II pyrethroids (*e.g. cypermethrin, fenvalerate, esfenvalerate, deltamethrin, fenprothrin, lambda-cyhalothrin, tefluthrin, cyfluthrin, acrinathrin and imiprothrin*). The main structural difference between Type I and Type II pyrethroids lies in the fact that Type II synthetic pyrethroids contain a cyano (C=N) group, whereas Type I do not. Type I general structure can be abbreviated as $R_1(C3)C{=}O(OR_2)$ and that of Type II as $R_1(C3)C{=}O(C(CN)R_2)$, where $R_1$, $R_2$ are alkyl or phenyl groups, C3 is a rigid cyclic propane and CN is the cyano group. Therefore, distinct chemical structures of synthetic pyrethroids convey selectivity towards certain insect species and mammals [3].

Synthetic pyrethroids have unique properties because of structural differences, which are seen in form of *stereoisomerism, i.e. geometric (cis-trans) and enantiomerism (or optical isomerism), e.g.* a technical mixture of *permethrin* contains 40% cis and 60% *trans* isomers, with the *cis* isomer being five times more toxic against tobacco budworms [3, 27, 28]; and the active isomer in the *deltamethrin* is the *dextro (+)-cis*-deltamethrin [3]. They have low water solubility, low vapor pressures ($10^{-6}$ to $10^{-7}$ mmHg) and high efficacy, being very effective against most agricultural pests at low rates, especially the Type II compounds, which are more effective than organophosphorus or carbamate insecticide [3]. Apart from their application in agriculture, synthetic pyrethroids are frequent components of household sprays, flea preparations for pets, and plant sprays for green houses, among others. Currently pyrethroids are used widely in Kenya in the domestic, public health vector control, as well as agricultural sectors, where both Type II and I are widely used. Most Type I pyrethroids belong to Category WHO Class III pesticides (oral $LD_{50}$ (rats) of 500–5000 mg/kg range), Type II pyrethroids mostly are more toxic and belong to Category WHO Class II pesticides (oral $LD_{50}$ (rats) of 50–500 mg/kg range) and just a few belong to WHO Class I, according to the WHO rating which is based on $LD_{50}$'s oral rats [3]. Pyrethrum (the extract) is a safe insecticide (oral $LD_{50}$ 1500 mg/kg in rat) and very fast-acting on insects, causing immediate paralysis. Both the natural pyrethrins and synthetic pyrethroids were more active as *contact* than *stomach* poisons, although more recently some of the synthetic pyrethroids tend to show particular potency when ingested and less susceptibility to biotransformation by insects and mammals [3, 27, 28].

### 2.4 Other botanical insecticides

There are six botanical insecticides currently available in the market. These are *pyrethrum, nicotine, rotenone (rotenoids), azadirachtin, sabadilla and ryania;* which are naturally occurring agents of plant origin that have been used to control insect pests. Despite many formulations of synthetic insecticides being present in the market, the botanical insecticides are still found in the market and are now becoming popular in Kenya, especially in the horticulture sector, because they are perceived to have eco-toxicological advantages compared to traditional synthetic insecticides. The advantages include less negative impacts on ecology, low human toxicity and less environmental persistence [3, 29–31]. Botanical insecticides are composed of secondary metabolites such as alkaloids, amides, chalcones, flavones, phenols, lignans, neolignans or kawapirones. They act as repellents with unpleasant odors or irritants, growth regulators and some have deterrence on oviposition and feeding, as well as biocidal activity [29, 30].

*Nicotine* was first used as botanical insecticide in 1763. It is highly toxic to both target and non-target species, with moderate to high toxicity in vertebrates (oral $LD_{50}$ in rats: 55 mg/kg) and is toxic to insects such as bugs, beetles and cockroaches ($LD_{50}$ ranging from 190 to 650 mg/kg) [31, 32]. It is an alkaloid extracted from leaves of tobacco plant by Soxhlet (with solvents such as toluene) or with alkali using steam distillation and is used in home gardens and greenhouses for controlling sucking insects such as leafhoppers, aphids, scales, thrips and white flies, and therefore is also used in horticulture e.g. in Naivasha, Kenya [11]. Its demerits include high mammalian toxicity, ready absorption by skin and, therefore, increased exposure.

Nicotine sulfate and other salts in the form of crystals such as nicotine benzoate, oxalate, salicylate and tartrate, as well as fixed nicotines (water-insoluble salts such as nicotine tannate and nicotine bentonite) are stable and have been used as insecticides, baits and for control of ectoparasites in livestock, respectively [32]. Nicotine is also used most commonly as a fumigant and as a contact spray in greenhouses [32]. Preparations of tobacco teas from tobacco products sold for smoking and chewing as homemade preparations for use against pests can also be used. Nicotine poisons insects and mammals by a similar mode of action, i.e. inhibition of acetylcholine esterase by mimicking acetylcholine which binds to postsynaptic receptors [3, 33], and since its breakdown is not catalyzed by acetylcholinesterase, it causes repeated stimulation of the receptor.

*Rotenone is* present in the roots of *Derris spp* plant and similar Leguminosae family of plants found in Malaysia, the East Indies and other East Asian countries [31]. It is an alkaloid extracted by solvent extraction (e.g. Soxhlet), purified and crystallized; often added in combination with other insecticides. It is a selective insecticide with acaricidal properties used against garden insects, lice and ticks on animals; such as headlice (by topical application). It is very toxic to fish and can control unwanted fish species in lakes, streams and reservoirs, which are used for power generation [31]. The $LD_{50}$ is 132 mg/kg oral in rats, obtained by administering crystalline rotenone. Rotenone is toxic not only to insects and fish but also to humans and animals, with oral $LD_{50}$ in rats being approximately 60–135 mg per kg of body weight. Liquid preparations of derris or Derris dust can also be used. It acts by blocking electron transport in mitochondria, inhibiting oxidation linked to NADH, by binding to NADH dehydrogenase thereby interfering with electron transfer, and is referred to as mitochondrial complex I inhibitor [31].

*Azadirachtin* is a secondary metabolite belonging to the limonoid group present in neem seeds. This compound is found in the seeds (0.2–0.8 percent by weight) of the neem tree, *Azadirachta indica*. *Azadirachtin* is the main compound of the neem oil with insecticidal activity and can be found in its fruits and leaves. *Azadirachtin* is the active ingredient in many pesticidal products or formulations in the market, including TreeAzin and Terramera Cirkil [31, 34, 35], and it has been used as a biopesticide in Kenya. Azadirachtin has various modes of activity, including being a broad-spectrum insecticide, and acting as a feeding deterrent, insect growth disruptor (IGD) and sterilant, respectively, and is used to control various agricultural pest species, including Coleoptera, Hymenoptera, Diptera, Orthoptera and Isoptera [29, 30].

*Sabadilla* use as a pesticide dates back to 1819 when a basic substance from sabadilla seed was isolated [33]. *Sabadilla* is a plant that grows in countries such as Central America and Mexico. It is toxic and is used in farming as an insecticide since it contains alkaloid compounds including *veratran, cevadine, veratridine, sabadine and sabadiline,* which have insecticidal activity. The veratrine alkaloids comprise approximately 0.3% of the weight of aged sabadilla seeds; of these alkaloids, cevadine and veratridine are the most active insecticidally and have been tested successfully in citrus thrips [31]. Sabadilla alkaloids from the dried ripe sabadilla seeds of a member of the lily family, *Schoenocaulon officinal*e, are often used and considered as generally safe and non-persistent insecticides. Veratrine, which is the term now used to describe the alkaloid mixture from sabadilla, has long been known for its toxicity to certain species of insects. The powdered seed itself or kerosene extract of it has been tested and used as an insect repellant. Sabadilla alkaloids have also been formulated as a wettable powder and then mixed with water and applied by either aerial or ground equipment on citrus, avocados and mangos. In making a commercial formulation, the active ingredients of sabadilla are synergized by piperonyl butoxide (PBO) and N-octyl bicycloheptene dicarboximide (MGK 264) [3, 33]. The mode of action of sabadilla is similar to that of the pyrethrins, as it affects the voltage-dependent sodium channels of nerve axon [33], i.e. affect nerve cell membrane function by binding to the sodium channel causing loss of nerve function, paralysis and death.

*Ryania* insecticide preparations are derived from the *woody stem tissue* of the shrub *Ryania speciosa* (family *Flacourtiaceae*), a plant that is native to South America and has been used in the USA since 1940s. A mixture of components is present in extracts or powders of this plant material, and eleven compounds with insecticidal activity have been identified [31], the most abundant active constituents of these alkaloids (*ryanoids)* being *ryanodine* and *dehydroryanodine.* Most commercial formulations are crude dust (50% ryania powder), though the constituent alkaloids can be extracted in water, alcohol, acetone, ether or chloroform to produce liquid or wettable powder formulations. Ryania extracts or powders have very low mammalian toxicity ($LD_{50}$ rats ranging from 750 to 4000 mg/kg), but the active ingredients are much more toxic to mammals [33]. Ryania's toxicity to insects can result from contact or ingestion; it is synergized by PPO and used most often for control of caterpillar pests of fruits and foliage, the codling moth and thrips in fruit trees (apples, pears, citrus), as well as European corn borer in corn, by organic farmers [33]. Like *rotenone,* ryania persists longer in the field after application than most other plant-derived insecticides, with residues giving some degree of residual control for up to 3–5 days after application on plant surfaces. The mode of action *of ryania is by Ryanodine effects on* the calcium cation ($Ca^{2+}$) release channel in muscle, resulting in poisoning of insects and mammals by a sustained contraction of skeletal muscle without depolarization of the muscle membrane, causing cardiac arrest and then eventual paralysis [33]. The binding of raynodine changes the structure of the $Ca^{2+}$ channel and prevents its complete closure. This binding affects the cardiac and skeletal muscles.

### 2.5 Neonicotinoids

Neonicotinoids (meaning "new nicotine-like insecticides"), also known as chloronicotinyls, are synthetic analogues of nicotine, but unlike nicotine, they are relatively non-toxic to mammals. Neonicotinoids are a new class of insecticides with widespread use in veterinary and crop production, and include *imidacloprid, acetamiprid, dinotefuran, thiamethoxam, clothianidin, amitraz and chlormideform*. *Imidacloprid* ($LD_{50}$ 450 mg/kg (oral rats)), *acetamiprid* (417 mg/kg), *thiamethoxan* ($LD_{50}$ > 5000 mg/kg) and *thiacloprid* ($LD_{50}$ 836 mg/kg), are all systemic insecticides, which have been used widely in agriculture against sucking insects such as aphids, leafhoppers, planthoppers, thrips, white flies [3], as well as soil insects, termites and biting insects, in Kenya.

Neonicotinoids first entered the market in the early 1990s and appeared to address the concerns associated with some earlier pesticide compounds, because they are effective, possess a high degree of selectivity to insects and have low mammalian toxicity, making them safer for human use than the organochlorines, organophosphates and carbamates [3]. Therefore, they soon became some of the most widely used insecticides in the world by 2014. Neonicotinoids are used to manage many honeydew-excreting pests, which are primary pests in most agricultural systems, including field crops, vegetables, fruit and nut production, tree plantations and urban forests, and therefore they have a strong potential to reach non-target pest species, which are essential in agriculture, such as bees [3]. They are most often applied as a seed coating and are absorbed into plant tissues, localizing the protectant and reducing contamination to the environment. The insecticide's ability to translocate into plant tissues could keep environmental concentrations low and minimize exposure to sensitive non targets such as quail and other wildlife, but experimental data suggest that environmental concentrations are usually higher than anticipated [36, 37]. It was estimated that approximately 5% of the pesticide a.i. applied as a seed coating would be absorbed by the plant while the rest (95%) would be blown away during sowing, which has led to their deposition in the surrounding soil and water, leading to soil residue concentrations up to 1000 ppb, in some cases [36, 37]. Compared to OPs and carbamates, neonicotinoids differ in that they are more strongly attracted to acetylcholine esterase receptors in the invertebrate's nervous system than the vertebrate ones, making them more specific. As a mode of toxicity, neonicotinoids are neurotoxins, which target insect nicotinic acetylcholine receptors (nAChRs). By 2018, neonicotinoids made up ~30% of insecticide sales worldwide [36, 37]. However, due to their adverse impact on pollinators such as honey bees and bumble bees, as well as aquatic invertebrates, some neonicotinoids are being banned by the EU, and other countries may also follow suit in future [36]. Neonicotinoids have become popular in Kenya and are already widely used in the horticultural sectors [4]; as many are registered by the PCPB.

### 2.6 Insect growth regulators (IGRs)

IGRs are chemical substances that disrupt insect growth and development, resulting eventually in death. They are pesticides that affect insects' ability to grow and mature normally, rather than killing them outright as 'conventional' insecticides do. Currently, there are 5 IGRs, namely *juvenoids, benzoylphenylureas, diacylhydrazines, triazines and thiadiazines*, respectively. The IGRs have low mammalian toxicity, and there are many of their formulations in the market, including the Kenyan market [3, 11]. They are very useful in controlling disease vectors such as mosquitoes, specifically mosquito larvae [3, 31]. Many IGR products can also be mixed with other insecticides that kill adult insects. Several features of IGR make them attractive as alternatives to broad-spectrum insecticides; i.e. they are more selective, less harmful to the environment and more compatible with integrated pest management. Because IGRs act on systems unique to insects, they are less likely to affect other organisms. Some of the modes of action include acting as antijuvenile hormone agents by blocking juvenile hormone production, mimicking hormones and therefore interfering with stages of growth or life cycle, from eggs to larvae, to pupae, and to adults; and inhibiting chitin synthesis by preventing development of exoskeleton, respectively [31]. Examples of known juvenoid active ingredients are *methoprene, hydropene, fenoxycarb, pyriproxyfen* and *diflubenzuron (a benzoylphenylurea)*. *Benzoylphenylureas* (e.g. *diflubenzuron*) and *diacylhydrazines* are known to prevent chitin synthesis by inhibiting chitin synthetase. Thiadiazines (e.g. *buprofezin*), diacylhydrazines (e.g. *halofenozide*, *methoxyfenozide* and *tebufenozide*) and benzoylureas (e.g. *novaluron*) disrupt or mimick insect growth hormones, inhibit chitin synthesis, prevent molting and metamorphosis, respectively [6].

*Methoprene* ($LD_{50}$: 34,600 mg/kg oral rats) is a larvicide juvenoid, which mimicks juvenile insect hormones, since it is similar in chemical structure to them. It has been used to control mosquitoes (in flood waters, effective at 2–4 instars stage), cigarette beetles and fleas. It is not toxic to the pupal or adult stages, with treated larvae able to pupate but adults do not hatch from the pupal stage [38]. The optically active juvenile hormone analogue, S-(+)-methoprene is synthesized by a chemical procedure [3, 39]. *Hydropene* is also a juvenoid, which is registered for use against cockroaches and mosquito larvae, with an $LD_{50}$ > 34,000 mg/kg (oral, rats). It disrupts normal development and emergence of insects by mimicking juvenile hormones [3, 6]. It may also cause adult sterility, physical abnormalities, desiccation, and premature death [6]. Its products are used in a variety of sectors, with commercial formulations including aerosols, liquids and impregnated materials (i.e. bait stations) [6]. *Fenoxycarb* is a carbamate insect growth regulator, with low toxicity to bees, birds and humans, but is toxic to fish [6]. The oral $LD_{50}$ for rats is greater than 16,800 mg/kg and is used in fire ant flea baits, and for control of mosquitoes and cockroaches, as well as butterflies, moths, beetles, and scale and sucking insects on olives, vines, cotton and fruit, where it is often formulated as a grit or corncob bait. Fenoxycarb blocks metamorphosis into adults and larval molting. *Pyriproxyfen* affects a target if touched or eaten, but it is rarely toxic to adult insects. It disturbs egg-laying, egg-hatch and keeps young insects from growing into adult form, and has been used against fleas, cockroaches, ticks, ants and mosquitoes [6]. *Diflubenzuron* is a synthesized active compound, an acaricide/insecticide and IGR used to control many leaf-eating insect larvae in agricultural, forest and ornamental plants, as well as mosquito larvae in standing water, using various formulations such as emulsifiable and solution concentrates, flowable concentrates, wettable powders and pellets. Some of its benefits include being relatively non-toxic to avian species, small mammals, freshwater fish, marine/estuarine fish and bees on an acute oral dietary basis [3, 6].

## 2.7 Microbial pesticides

### *2.7.1 (a) Entomopathogenic bacteria*

Entomopathogenic fungi are often relied on as important components of integrated pest management in tropical agriculture, either as biopesticides or as naturally occurring soil microbes conserved in the environment. As pest control products, they are becoming very significant, especially in mosquito larval control. The entomopathogenic fungi, *Metarhizium anisopliae* and *Beauveria bassiana,* have demonstrated effectiveness against *anopheline* larvae in the laboratory, but effective formulations from such fungi, which are not sensitive to UV radiation, high temperatures and water not are needed [40, 41]. They are being manufactured and used in Kenya [11]. Metarhizium *robertsii*, formerly known as M. *anisopliae*, and even earlier, as Entomophthora *anisopliae* (*basionym*) is a fungus that grows naturally in soils throughout the world and causes disease in various insects by acting as a parasitoid [42]. A parasitoid is an insect whose larvae live as parasites in a host that eventually kill its host (typically another insect). Many isolates of parasitoids, which can be injected or exposed to insect hosts, have long been recognized as new Metarhizium species, such as M. *robertsii*, M. *majus* and M. *acridum*, respectively. Metarhizium *taii* was placed in M. var. *anisopliae*, but has now been described as M. *guizhouense*. The commercially important isolates, M.a.43 (or F52, Met52, etc.), which infect Coleoptera and other insect orders have now been assigned to Metarhizium *brunneum*. This technique, which involves using various fungi has been practiced in other countries and is being used in Kenya, as various products have been registered by the PCPB and manufactured by some local companies [11].

### *2.7.2 Bacillus thuringiensis*

*B. thuringiensis* Berliner (known as Bt) is an insecticidal bacterium discovered in early 20th century. There are several dozen recognized sub-species of *B. thuringiensis*.

The sub-species commonly used as insecticides include *B. thuringiensis* (Bt) sub-species *kurstaki* (Btk), sub-species *israelensis* (Bti) and sub-species *aizawa,* respectively. During sporulation, many Bt strains produce crystal proteins, called *delta endotoxins*, that have insecticidal action. Commercial Bts are powders containing a mixture of dried spores and crystalline δ-endotoxin, though some contain only the toxin component. Both spores and the toxin crystals are produced within the bacterial vegetative cell of the Bt [3, 40, 41]. Currently, there are 6 strains of Bt, which possess specific activity against different insects species e.g. for control of insects such as lepidopterous on crops, e.g. corn, fruits, tobacco and vegetables, as well as mosquito larvae, but Bt has very low toxicity in mammals ($LD_{50}$ in mammals is >5000 mg/kg). They are used as biopesticides, in form of sprayable products and currently take about 2% of the global pesticide market.

Bt still finds low use in many countries such as Kenya because of high costs, lower efficiency, poor control of sucking and borers insects (e.g. in maize where there is large need), limited persistence and narrow spectrum; but a significant amount of them are being imported and registered by the PCPB [11], indicating their use in agriculture and vector control in Kenya. The advantages of Bt include their environmentally friendly nature compared with other synthetic pesticides as well as their ability to be adopted in new biotechnology. Bt toxins genes have been inserted into chromosomes in some plants and therefore such plants are resistant to attack by insects as they grow. Such crops are called transgenic crops and are available in the market e.g. Bt-corn (a genetically modified crops/organisms (GMO)). Bt corn produces Cry 1Ab toxin, which is used to control European corn borers; such plants which have been genetically engineered to contain δ-endotoxin are called *plant pesticides*. The major concern with Bt (and GMOs in general) is the potential impacts on non-target insects (e.g. beneficial insects such as bees), and interference with natural processes, such as change in biodiversity, which are still not yet fully known.

The δ-endotoxin is a cytolytic pore-forming toxin with insecticidal action, e.g. Cry 1AB toxin, which is a crystal protein with helical structure. When insects ingest it, it gets activated by proteolytic cleavage and once activated it binds to the mid-gut epithelium cells of targeted pests resulting in their rupture and causing cell death. Other organisms (including humans, other animals and non-targeted insects) that lack the appropriate receptors in their gut cannot be affected by the *cry* protein, and therefore are not affected by Bt [41]. Various types of δ-endotoxin can be found in various hosts e.g. cry 1 protein, cry2 protein, cyt protein, vip 1 protein [41]. To be effective, Bt must be eaten by insects during their feeding stage of development, when they are larvae. It is ineffective against adult insects. More than 150 insects, mostly lepidopterous larvae, are known to be susceptible in some way to Bt. Different strains of Bt have specific toxicities to particular types of insects: Bt *aizawai* (Bta) is used against wax moth larvae in honeycombs; Bt *israelensis* (Bti) is effective against mosquitoes, black flies and some midges; Bt *kurstaki* (Btk) controls various types of lepidopterous insects, including the gypsy moth and cabbage looper; and a newer strain, Bt *San Diego*, is effective against certain beetle species and the boll weevil [41]. Due to its short biological half-life and its specificity, Bt is less likely than chemical pesticides to cause field resistance in target insects. It is moderately persistent in soil, with a half-life of about 4 months in suitable moderate conditions [6, 41]. Bt spores can be released into the soil from decomposing dead insects but can get rapidly inactivated in soils that have a pH below 5.1 [41].

### *2.7.3 Abamectin*

A microbial pesticide, is a bacterium containing a mixture of endotoxins, *avermectin B1a* (>80% by wt) and *avermectin B1b* (<20%) as active ingredients [3]. The toxins are macrocyclic lactones derived from the *Actinomycete i.*e. *Streptomyces avermitilis* (a soil microorganism). The lactones are natural fermentation products of this bacterium. Abamectin ($LD_{50}$ 300 mg/kg oral rat) is used against insects and mites on vegetable, fruit, ornamentals and fire ants (at home) and is now being used in horticulture in Kenya. The two components, **B1a** and **B1b** have very similar biological and toxicological properties and act as insecticides by affecting the nervous system of and paralyzing insects, and on exposure to high concentrations in humans, symptoms similar to OP poisoning are shown [3, 33]. It is highly toxic to insects and fish, extremely toxic to aquatic invertebrates, but non-toxic to birds, with $LD_{50}$ in bobwhite quail being >2000 mg/kg. Abamectin is rapidly degraded in soil, and at the soil surface, if subjected to photodegradation, with half-lives ranging from 8 hours to 1 day [3].

### *2.7.4 Spinosad*

Spinosad is a bacterial fermentation product, a natural substance made by a soil bacterium *Saccharopolyspora spinosa* that is toxic to insects. It is a mixture of two chemicals or metabolites called spinosyn A and spinosyn D. Spinosyn A & D have the most insecticidal activity and are used to control a wide variety of pests, including thrips, armyworms, codling moths, cutworms, leafminers, spider mites, mosquitoes, ants, fruit flies and others. Spinosad has been registered for use in pesticide formulations by the US Environmental Protection Agency (EPA) since 1997 [3] and is already being used in fruit and vegetable farming in Kenya [43]. Currently, they are found in over 80 registered pesticide products, many of them being used on agricultural crops and ornamental plants, where they are important in IPM to avoid food residue problems. Other spinosad products are used in and around buildings, in aquatic settings, and as seed treatments. The products are commonly used as sprays, dust, granules, and pellets. They are neuroactive and have same mode of action such as neonicotinoids but affect different binding sites.

### *2.7.5 Wolbachia*

Wolbachia are obligate endosymbiotic bacteria that infect many insects, living in all orders of insects and other invertebrates, including some species of mosquitoes [44]. Although it is believed that *Wolbachia* does not naturally infect *Anopheles* mosquitoes, which are the species that spread malaria to humans, their prevalence, though sparsely in *Anopheles arabiensis* and *Anopheles funestus*, which are the two main malaria vectors, were reported recently in Tanzania [45]. Factors influencing *Wolbachia* transferring into new species are still being investigated, but the biocontrol technology has already been tried in Brazil [46]. It has not yet been tried in Kenya. Artificial infection with different *Wolbachia* strains can significantly reduce levels of the human malaria parasite, *Plasmodium falciparum*, in the mosquito, *Anopheles gambiae*. In addition, it was found to reduce levels of *Plasmodium falciparum* that could be transferred to humans and, therefore, suppressed malaria infections [47]. The procedure involves infecting or exposing *A. gambiae* mosquitoes, or any disease vector insect, with different *Wolbachia* strains (e.g. wMelPop, wAu, wInn, wMeICS and wAlbB). After infection, *Wolbachia* strains disseminate widely inside the mosquitoes and infect diverse tissues and organs, affecting the host by manipulating its immune response, inhibiting its replication, reducing the parasite (*Plasmodium falciparum*) levels in the mosquito gut or killing the mosquitoes within a day (as was found in *A. gambiae* exposed to wMelPop strain) after the mosquitoes were blood-fed, including other transfers [47–49]. There is a vast diversity of *Wolbachia* strains available in natural populations of insects related to mosquitoes.

### 2.8 Fungicides

Fungicides are used against fungi (.e.g. mildews, rusts, smuts, mushrooms), parasitic plants and many allied forms capable of destroying wood, timber, leather, fabrics, glass, industrial products (e.g. paint and adhesives) and higher plants [50, 51]. Fungal attacks can cause problems of very significant importance not only to materials, the environment and aquatic organisms but also to humans. A good example of devastating fungal effects is normally seen in *Aspergillus ssp* fungi, which attack grains producing aflatoxins, and is a common problem in Kenya. Aflatoxins, which belong to the class of mycotoxins, cause acute lethal toxicity problems and, in the long term, carcinogenicity in humans. Fungicides for plant protection act by direct contact and often injure the host as well as the fungus. They can be described as *protective*, *curative* or *eradicative*; where *protective fungicides* are applied before appearance of infestation to prevent it, *curative fungicides are* applied when infestation has already begun to invade the plant, thus they penetrate the plant cuticle, and destroy young fungal mycelium growing in the epidermis of the plant to prevent further development, and *eradicative fungicides* kill and also prevent sporulation, *i.e.* control fungal development following appearance of symptoms usually after sporulation, killing both new spores and the mycelium by penetrating the cuticle of the plant to the subdermal level [50, 51]. These modes of activity are established during product development and are often indicated on the labels. In agriculture, fungicides are used as foliar, soil or seed dressing, respectively.

#### *2.8.1 Inorganic fungicides*

Inorganic fungicides include elemental sulfur and alkyl/aryl compounds of heavy metals e.g. copper (Cu), mercury (Hg), tin (Sn) (e.g. organo mercury $Hg(CH_3)_2$ and organotin tin $Sn(CH_3)_4$). Heavy metal fungicides are not popular anymore due to their environmental persistence, ability to biomagnify in food chain and toxicity, and have been banned in the EU, where organotin compounds additives are no longer allowed in paints used in ships, where they were widely used [52, 53]. Methyl mercury fungicides were used in storage of cereal grain storage but were banned following two accidents of severe poisoning reported in Iraq and Minamata in Japan, respectively [53]. These heavy metal fungicides are not used in Kenya. Three inorganic fungicides, *Bordeaux mixture, lime sulfur and copper oxychloride*, respectively, are registered by the PCPB and are used to control molds and mildews in fruit and vegetable farms in Kenya [54].

#### *2.8.2 Organic fungicides*

Organic fungicides are commercially produced by chemical synthesis and are commonly used for control of vegetable blights, especially in potatoes and tomatoes, as wood preservatives. Examples of organic fungicides include *dithiocarbamates*, *chlorinated phenols* (e.g. *pentachlorophenol* 5%), *formalin* (40% formaldehyde) and *coal-tar creosote* (which is used to preserve fencing posts and wooden rail truck ties).

Fungicides are also used widely in large quantities in agriculture and domestic sectors in Kenya due to frequent damp weather conditions which encourage microbial growth. Maize, fruits and vegetable farmers in Kenya, use a number of fungicides, with main active ingredients including *carbendazim, tebuconazole, metalaxyl, mancozeb, azoxystrobin, difenoconazole, fludioxonil, epoxiconazole, trifloxystrobin and mefenoxam* [50, 51, 55, 56].

## 2.9 Other pesticides available in Kenya for specific uses

### *2.9.1 Petroleum products*

The use of emulsions of certain petroleum oils with water for use as fruit tree sprays against insects such as scale insects, red spider mites, aphids and mosquito larvae has been known [57], and kerosene products are still being imported and are registered for by the PCPB [11].

### *2.9.2 Rodenticides*

Rodenticides are used to control certain pest animals e.g. mice, rats, groundhogs, bats, squirrels and field rodents, which can cause extensive damage to crops property or spread disease [58, 59]. In food storage, cereal farming, food handling and distribution and rodenticides are important e.g. *thallium sulfate, zinc phosphide ($Zn_2PH_3$), strychnine, and red squill, fluoroacetate* ($CH_2FCOONa$), *fluoracetamide* ($CH_2FCONH_2$) and *ANTU* (Alpha-naphthylthiourea) are used in Kenya. Other rodenticides include *fluoroacetate* ($CH_2FCOONa$), *warfarin, fluoracetamide* ($CH_2FCONH_2$) and *ANTU* (Alpha-naphthylthiourea). Organic rodenticides, such as *difenacoum, brodifacoum, difethialone, flocoumafen and bromadiolone*, are toxic to mammals and extremely toxic to birds (e.g. *brodifacoum* $LD_{50}$ values of 0.31, 0.72, and 19 mg $kg^{-1}$ in ducks, gull and quails, respectively) are not registered in the PCPB database.

### *2.9.3 Fumigants*

Fumigants act on insects through respiratory system by emitting vapors, but also kill nematodes, weed seeds, fungi, in soil, silos for stored grains, and fruits and vegetables. Often treatment is carried out in enclosures since they are volatile. Fumigants, such as carbon tetrachloride, ethylene dichloride ($CH_2ClCH_2Cl$), ethylene dibromide ($CH_3CH_2CH_2Br$), methyl bromide and carbon disulfide, have been used as liquid fumigants in commodities e.g. grain storage but have been banned due to human toxicity and ozone depletion properties. They have been replaced with others such as $CO_2$, phosphine ($PH_3$; a liquid, storage of grains) and sulfuryl fluoride ($SO_2F_2$, termite control), which are not listed by the PCPB. However, *malathion* dust (2%) and *pirimiphos-methyl* (actellic) dust formulations, respectively, are registered and are used in bulk grain storage in silos [60, 61].

### *2.9.4 Avicides*

Avicides are used against certain birds when they become pests, such as quail birds on rice farms. The red-billed quelea (*Quelea quelea* Linnaeus) is the most important avian pest of small grain crops in Africa, causing damage up to the equivalent of US$ 88.6 million per annum [62]. It is controlled by *fenitrothion, fenthion* (Queletox) and *cyanophos,* which are both highly toxic to non-target and costly [62] and have been used in Kenya by aerial or ground spraying. An avicide can be used as a repellant e.g. Avitrol (4-aminopyridine) or reproductive control e.g. Ornitrol, a derivative of cholesterol, which produces temporary sterility in pigeons but has no effect on mammals. *Fenitrothion* and *fenthion* are listed in the database confirming their use in Kenya.

#### *2.9.5 Nematicides*

Nematicides are used against nematodes, which can infest plant root systems and damage roots and/or encourage other microorganisms e.g. fungus to attack plants. Fumigation with *1,3-dichloropropene* can control these, although conventional pesticides such as some OPs have both insecticidal and nematicidal properties.

#### *2.9.6 Molluscicides*

Molluscicides also known as snail baits, snail pellets or slug pellets, are pesticides against gastropods such as mollusks, which are usually used in agriculture or gardening. These organisms can damage crops or other valued plants by feeding on them or exposing disease pathogens, which they carry on their bodies to humans (e.g. in vegetables, or *bilharzia* in freshwater) [63]. Synthetic *niclosamide* is mostly used although others such as *metaldehyde* have also been used against mollusks [63].

### 2.10 Metabolism, detoxification and excretion of pesticides

Insecticides are toxic to target insects as well as humans. However, like other xenobiotics, there are mechanisms of degradation and metabolism in both species, which are mediated by various enzymes and are responsible for reducing their toxicity and excreting them. Apart from killing the target pests such as insects, pesticides are just like any other chemical, which the human is inevitably exposed to through air, food and water, and exposure to them can lead to acute toxicity, long term-diseases or excretion. The ability of pesticides and other xenobiotics to cause long-term diseases or endocrine disruption is statistical and dependent on many factors, but the human body has inherent mechanisms to detoxify them or reduce their toxicity. Studies on pesticide metabolism, detoxification and excretion by insects and mammals, with reference to OCs, OPs and carbamates, which have been most studied, have made us understand how organisms naturally deal with toxic pesticides [3].

#### *2.10.1 Metabolism, detoxification and excretion of pesticides: OCs, OPs, CBs*

In insects and mammals, hydrophobic compounds such as OCs undergo various metabolic reactions, which make them more water-soluble and ready for excretion through urine or other matter. These reactions include hydrolysis, oxidation and reduction, followed by conjugation to more polar metabolites or biomolecules, such as sugars, amino acids, glutathione, phosphates and sulfates, which make them even more hydrophilic and, therefore, excretable. The reactions are mediated by various enzymes. *Cytochrome P450 monoxygenases* (a group of enzymes) located in the microsomes in the mitochondria are responsible for oxidation reactions in mammals, birds, fish, mollusks and insects and can transform various functional groups or moieties of the pesticide molecules, through various chemical changes such as *epoxidation, demethylation, hydroxylation, oxidation and reduction*. All other pesticides including

OPs, carbamates and other xenobiotics also undergo similar biochemical changes that make them more water-soluble for excretion. From the onset, the chemical structures of various pesticides, as shown by examples in **Figure 1a-e**, determine the kind of biochemical reactions, which are expected to occur in the environment and organisms. The metabolic pathways for these biochemical reactions have been elucidated and can be found in Hodgson and Levi [64], Usmani et al. [65], Yu [3] and Jing et al. [66]. The OPs, carbamates, neonicotinoids, pyrethroids, herbicides and fungicides are not as bioaccumulative as the OCs in the organisms and in the environment because of the nature of their chemical structures [3]. They undergo more rapid metabolism and get excreted more [3]. These descriptions of metabolic pathways can be understood by making references to the specific chemical structures (**Figure 1a-e**).

#### *2.10.1.1 Epoxidation*

Epoxides are formed by oxidation of double bonds, which can occur on phenyl rings, or alkene part of the organic molecules, and are mediated by microsomal Cyt P450 monoxygenases. Epoxides can be environmentally persistent and highly reactive and can form adducts with cellular macromolecules e.g. proteins, RNA and DNA, often resulting in chemical carcinogenesis and, therefore, not directly advantageous for the organism. However, they are more water-soluble for excretion and can also be further hydrated by *epoxide hydrolases*, which catalyze addition of $H_2O$ molecules into the epoxide ring to yield trans-diols [3].

#### *2.10.1.2 Hydroxylation*

Hydroxylation can occur on H attached to an aliphatic or aromatic carbon atom (represented as: $-C-H \rightarrow C-OH$), and is mediated by Cyt P450's, usually resulting in a more polar and water-soluble product, and is considered a detoxification process in insects and mammals.

#### *2.10.1.3 N-dealkylation*

Microsomal N-dealkylation is a commonly observed metabolic reaction for xenobiotics containing secondary and tertiary amines, including OPs insecticides, releasing an aldehyde/ketone and alcohol. It is catalyzed by certain cytochrome P450 monooxygenases and peroxidases.

#### *2.10.1.4 O-dealkylation*

O-dealkylation of alkyl groups of the ester or ether structures on pesticide molecules occurs frequently in insects and mammals, e.g. in OP insecticides and other pesticides, mediated by cytochrome P450s, results in formation of an alcohol and acetaldehyde and detoxification.

#### *2.10.1.5 Desulfurization*

Desulfurization is also known as phosphorothioate oxidation, e.g. the OP insecticides with P=S get oxidatively desulfurized by Cyt P450 monoxygenases to give P=O analogues, resulting in activation because it gives a metabolite, which binds more strongly to AChE and, therefore, a potent inhibitor of the enzyme AChE. Examples include parathion and malathion, giving paraoxon and malaoxon, respectively, which are more toxic than the sulfur analogues.

#### *2.10.1.6 Sulfoxidation*

Sulfoxidation of many thioether ($R_1C{-}S{-}CR_2$)-containing insecticides, such as OPs, are oxidized by Cyt P450 monoxygenases to their corresponding sulfoxides (S=O). Usually, it is an oxidative activation leading to increased anti-AChE activity (i.e. it is an inhibitor), as demonstrated in phorate. Sulfoxides are compounds containing a sulfoxide functional group, with the structure RS(=O)R′ (R,R′ = alkyl/aryl; S is joined to O (i.e. −SO)). Oxidation of certain sulfur and nitrogen-containing insecticides is also performed by another group of microsomal enzymes known as flavin-containing-monoxygenases (FMOs), e.g. sulfoxidation of phorate by FMO was demonstrated in mammalian liver [3]. Like Cyt P450's FMOs also require NADPH and oxygen for their activity, but FMOs are only involved in catalysis of oxygenation reactions.

#### *2.10.1.7 Hydrolysis*

OPs and carbamates and others containing ester linkages are susceptible to hydrolysis. *Esterases* (e.g. carboxylesterases) are *hydrolases* that split ester compounds by addition of water to yield an acid and alcohol (i.e. $R'COOR + H_2O \rightarrow R'COOH + ROH$; R, R′ = alkyl, phenyl). Carboxyl-esterases have been classified into three categories (A, B and C) on the basis of differential patterns of inhibition by organophosphates, as discussed in detail in other texts [3]. **A-esterases** are typical aromatic esterases, which hydrolyze phenyl acetate and phenyl butyl acetate groups but not aliphatic esters. A-esterase levels of activity in plasma and liver of birds are much lower than those of mammals, the reason why birds are much more susceptible than mammals to OPs such as *pirimiphos-methyl and diazinon*. **B - esterases** are aliphatic and aromatic esterases e.g. *carboxyl esterases and lipases (in lipids)* as well as *acetylcholine esterases(AChE)*. B - esterases e.g. *AChE* are *sensitive to OP and carbamate compounds* and hydrolyze both aliphatic and aromatic esters but not choline esters. B esterases are used as non-destructive biomarkers for exposure to anticholinesterase insecticides. Two types of esterases, *carboxylesterases and phosphatases* (or phosphorotriester hydrolases), are involved in metabolism of insecticides, e.g. hydrolysis of *malathion* to yield α and β- monoacids and ethanol by *carboxylesterases*. Carboxylesterase-mediated metabolism is one of the major mechanisms involved in *insecticide resistance*, and *multiple carboxylesterase* genes have been identified which are involved in pyrethroid insecticide resistance in housefly, just like glutathione s-transferase [3]. **C -esterases** preferentially *hydrolyze acetyl esters* and are, therefore, also called *acetyl esterases*, and split *acetylcholine esters* at higher rates than both *aliphatic and aromatic esters*, the latter at lower rate than aliphatic or not at all, typical substrates being 4-nitrophenyl acetate, propyl chloroacetate and fluorescein diacetate. *Phosphatases*, use water to cleave a phosphoric acid monoester into a phosphate ion and alcohol, and detoxify many OP insecticides especially the *phosphate* group, in insects and mammals, e.g. *paraoxon* can be hydrolyzed to diethyl phosphoric acid and p-nitrophenol in houseflies [3].

#### *2.10.1.8 Reduction*

Insects contain *reductases* that catalyze reduction of xenobiotics. Reduction is less common than oxidation, and there are three types of reduction:

*Nitro reduction* ($RNO_2 \rightarrow RNH_2$), i.e. nitro group reduction to ammine group on the pesticide molecule.

*Azo reduction*—reductive cleavage of azo linkages on a pesticide molecule ($R-N=N-R_1$), resulting in formation of an ammine, e.g. for aromatic ammine ($Ar-N=N-Ar' \rightarrow ArNH_2 + Ar'NH_2$). The *Azo* group ($RN=NR$) *reduction* is similar to nitro *reduction* in many ways, i.e. it, too, is mediated both by *cytochrome P450* and by *NADPH-cytochrome P450 reductase.*

*Aldehyde or ketone reduction*—Reduction of aldehydes and ketones (hydrogenation) forms various metabolites, including primary alcohols (for aldehydes) and secondary alcohols (for ketones) mainly. Cytosolic a*ldehyde dehydrogenases,* as well as the *NADPH-dependent aldehyde reductases* widely distributed in insects and animals and their role in detoxification and insecticide resistance, have been discussed by Jing et al. [66]. Nitro group reduction, azo group reduction and aldehyde/ketonic group reduction, have all been found in insects; e.g. reduction of *parathion* to *amino parathion* and *trifluralin* reduction to *amino trifluralin*, in housefly cytosol, NADPH–cytochrome P450 reductase has mediated the resistance of *Aphis* (*Toxoptera*) *citricidus* (Kircaldy) to Abamectin by Jing et al. [66]. OPs and carbamates have various functional groups, which can be attacked, e.g. malathion can be attacked by two types of enzymes the *carboxylesterases* and the *Cyt P450 monoxygenases* for example demethylation (removal of methyl group from $CH_3O$-P moiety) by *Cyt P450 monoxygenases* to give other polar metabolites, and all carbamates have at least three sites that enzymes can attack; i.e. N-alkyl (methyl) group, the *ester linkage* and the *alcohol* or *phenol* group, respectively, the most important reaction in all carbamates being *hydrolysis,* which occurs in insects and mammals. Other important reactions in carbamate insecticide metabolism would be *hydroxylation* of both ring and N-methyl and *epoxidation* to give *diols* ultimately, followed by conjugation and excretion. The oxidation, reduction, hydrolysis, epoxidation, hydroxylation, dealkylations, desulfuration, and sulfoxidation, which are involved in changing pesticide molecules to become more polar for excretion, as discussed above, are primary reactions called **Phase I** reactions or Phase I metabolism. Products of Phase I metabolism, if not excreted, can then be subjected to **Phase II** reactions or Phase II metabolism. In Phase II reactions, the phase I products are further metabolized by getting them conjugated to various endogenous molecules, e.g. Phase II conjugation with glucose (sugars), amino acids (AAs), glutathione (GSH), phosphate and sulfate. The metabolism of insecticides involving Glutathione (GSH) binding or conjugation, which is mediated by glutathione-s-transferases (GSTs) is well known as a mechanism for detoxification of pesticides in various insects, demonstrated first in housefly [3].

Conjugations are Phase II reactions and are mediated by various enzymes, leading to products which are more polar, less toxic and more readily excreted, therefore, Phase II metabolism leads to detoxification. There are three types of Phase II metabolism, known as **Type I, Type II and Type III**, respectively, depending on the types of functional groups of the metabolites which are involved. The chemical functional groups required for Type I (of Phase II reactions) include $-OH$, $-NH_2$, $-COOH$, $-SH$ conjugation with *glucose*, *sulfates* and *phosphates*. Type II (of Phase II reactions) involves $-COOH$ groups binding with *amino acids*, i.e. amino acid conjugations; and Type III of Phase II conjugation involves halogens, alkene, -$NO_2$, epoxides, ethers functional groups and their conjugation with *Glutathione*, i.e. glutathione conjugation. Glucose conjugation is found in insects and plants but is rare in mammals. Mammals use glucuronic acid instead of glucose for excretion of xenobiotics. Glucose conjugation involves binding of Phase I metabolites to α, D-glucose, mediated by glucosyl transferase:

α, D-glucose + ROH → (*glucosyl transferase)* → RO. β-D-glucose (in insects/plants), where R = alkyl or phenyl group and ROH is a metabolite. Glucuronic acid conjugation, which occurs in mammals, involves Phase I metabolite conjugation to UDP-α-D-glucuronic acid (UDPGA); whereby D-glucose is first activated with *uridyl triphosphate* (UTP), mediated by *uridyl diphosphate glucose (UDPG) pyrophosphorylase*, to form *uridyl diphosphate-α-D-glucose (*UDPG), which is then oxidized to *uridyl diphosphate-α-D-glucuronic acid* (UDPGA). It is the UDPGA that binds to the Phase I metabolite, mediated by *UDP-Glucuronyl transferase*, for excretion, summarized as:

$$\begin{aligned} &\text{D-glucose-Pi} + \text{UTP} ® (\text{UDPG pyrophosphorylase}) \\ &\qquad ® \text{UDP-}\alpha\text{-D-glucose}(\text{UDPG}) + \text{PPi} \end{aligned} \tag{2}$$

$$\begin{aligned} &\text{UDPG}) + 2\text{NAD} + H_2O ® (\text{oxidation}) \\ &\qquad ® \text{UDP-}\alpha\text{-D-glucuronic acid}(\text{UDPGA}) + 2\text{NADH}_2 \end{aligned} \tag{3}$$

$$\begin{aligned} &\text{UDPGA}) + \text{ROH} ® (\text{UDP-Glucuronyl transferase}) \\ &\qquad ® \text{RO-}\beta\text{-glucuronic acid} + \text{UDP} + H_2O \end{aligned} \tag{4}$$

In Phase II metabolism, sulfate conjugation requires ATP and *sulfotransferase* and phosphate conjugation, which occurs in insects but is rare in mammals, requires *phosphotransferase*. In amino acid conjugation, *glycine* is most frequently used, and Glutathione conjugation is mediated by a group of enzymes, the *glutathione-s-transferases* (GSTs). GSTs are involved in conjugation of various metabolites, e.g. binding *to epoxide, unsaturated compounds, aldehydes, ketones, lactones, nitriles, nitro compounds, phosphorothioates and phosphates* [3]. Phase I metabolism is responsible for decreasing biological activity and toxicity of toxicants and Phase II metabolism is responsible for detoxification or excretion.

## 3. Pesticide importation, regulation and manufacturing in Kenya

### 3.1 Pesticide regulation and importation in Kenya

#### *3.1.1 The role of PCPB*

The Pest Control Products Board (PCPB) was established in 1984 under Cap 346 Laws of Kenya, to regulate the use of pest control products (PCPs) and safeguard human and environmental health from the undesirable risks associated with PCPs. Pesticide regulation includes policy making and changing (with involvement of the PCPB, government agencies, non-governmental organizations (NGOs) and Parliament), adherence to International Conventions that Kenya is a signatory to (such as UNEP), and prosecution, respectively. The importation, registration, distribution and sale, as well as law enforcement against misuse, are implemented by the PCPB. The role of PCPB, i.e. the Board, in pesticide regulation and its mandate are prescribed in PCPB Act Cap 346 Laws of Kenya of 1984 [67, 68] and include issuance of import/export permits, assessment of safety, efficacy and quality of PCPs, assessment of suitability of premises, advising the Cabinet Secretary/Minister, monitoring and adherence to standards in the entire pesticide industry/trade, supervision of disposal of obsolete PCPs, keeping records of importation and information on specific uses, creating awareness and investigation and prosecution of contravenors of the Act. Since its establishment, the Board has registered many pest control products for use in public health, livestock and agriculture, and provided important information for labeling, and this is all available to the public on the PCPB website. It is an offense under the Pest Control Products Act to import or sell in Kenya any PCP unless it has been registered by the Board.

In undertaking the regulation of PCPs, the PCPB undertakes evaluation and registration of imported pesticide products and those manufactured in the country for safety, efficacy and quality, before registration. In addition, it regulates trade of pest control products through inspection, licensing and product certification. Any other uses of the products outside those specified in the registration are not authorized unless the product is reviewed and given a label extension [67, 68]. The PCPB registration numbers of products are given and continue to be amended as prescribed in the Pest Control Products Act under the labeling, advertising and packaging Amendment Regulations L.N. 127/2006. To carry out its mandate, the PCPB, is thus, composed of three technical departments namely registration, compliance and enforcement and Analytical Departments, respectively, with clearly defined roles available on their website [67–69].

### *3.1.2 Laws and regulation of pesticides*

The regulation of pesticides is governed by the Pest Control Products Act Chapter 346 laws of Kenya, which was enacted in 1984 and became operational in 1985 [67]. There are other pieces of legislation in the Pest Control Products Act (Revised Edition 2012) and the Pest Control Products Act (Subsidiary Legislation), available freely on internet. The Pest Control Products Board (PCPB) was established under the Act to oversee its implementation. The Act regulates importation, exportation, manufacture, repackaging, warehousing and distribution. Some important Clauses in the Act include all aspects of manufacturing, storage, distribution, packaging, labeling, sale, importation and exportation, as stated therein, and each piece of legislation is given a number L.N. (L.N. meaning legal notice), e.g. L.N. 45/1984: licensing of premises regulations, L.N. 46 and 109/1984: registration regulations, L.N. 125/2006: the pest control products (importation and exportation) (Amendment) Regulations, etc. The Acts and these pieces of legislation can be retrieved freely from the internet or bought from the Government Printer in Nairobi. The Minister/Cabinet Secretary in charge of the Ministry of Agriculture in consultation with the Board is empowered to make subsidiary legislations (Regulations), which are then printed by the government printer as legal notices (L.N) in the Kenya Gazette.

### *3.1.3 Pesticide importation*

Kenya is among the largest consumers of pesticides in Africa besides South Africa, Nigeria and Ethiopia [70]. It is an agricultural economy, and therefore farmers use a significant amount of pesticides every year in different parts of the country in order to enhance agricultural productivity. Pesticide imports have increased steadily from about 9.52 thousand metric tons in 2009 to about 14.6 thousand metric tons in 2019. Currently, the PCPB has listed about 1447 formulations and active ingredients registered for use [11]. Most of the products have been insecticides (43%), followed by fungicides (22%) and herbicides (18%), but this changed in 2021, when the volume of total imports further rose to 20.5 thousand metric tons, with a significant increase in fungicides to 6.9 thousand metric tons (34%), herbicides to 7 thousand metric tons (3.14%) and a decline in volume of insecticides to 4.8 thousand metric tons (23.4%). The insecticides include those used in public health and in mosquito nets. The consumption of biopesticides is still very low, with just about 311 metric tons imported in 2020/2021 financial year. In the 2021/2022 financial year, approximately 267 active ingredients of pest control products were imported into the country, either as formulated products or technical grade material (a.i.) for formulation locally, respectively. In some instances, the active ingredients were of mixed form containing more than one active ingredients [11]. On average, 5% of the volume of pesticide imports is technical grade material, therefore, formulation locally is relatively minimal. Approximately 95% of formulated pesticides come mainly from China, India and Germany, and smaller quantities from the USA, the UK, Japan, the Netherlands, and Switzerland, among others [11].

The most recent lists of various registered PCPs have been placed on the PCPB database [11], which are available as an open-source to the public on their website, and provide comprehensive information about all products registered for use in Kenya (www.pcpb.go.ke). The first comprehensive list contains information on names of 1447 various products and active ingredients registered for *use in crops*, their trade names, types of formulations, active ingredients, names of international and local manufacturers, local distributors, specified crops, the maximum residues limits (MRLs), the postharvest intervals (PHI) and the WHO toxicity data and any restrictions (e.g. if the product is restricted), respectively. Fungicides, pyrethroids, neonicotinoids, OPs insecticides and herbicides dominating the list, and much fewer numbers of carbamates (mainly *methomyl and propoxur*), petroleum oil, biopesticides (*e.g. Bt, abamectin and azadirachtin*), biological control products in form of predatory mites (including parasitic wasps) and entomopathogenic fungal spores including *Metarhizium anisopliae,* respectively [11], are given. The list also includes adjuvants and surfactants. Natural pyrethrum extracts are manufactured, formulated and distributed by Pyrethrum Board of Kenya; and other local companies are actively involved in manufacturing and distribution of biological control products including entomopathogenic fungal products for use against thrips and mites in Flowers and French beans. Almost all active ingredients, such as glyphosate, are registered in numerous different formulations, manufactured by different companies (more than 50 international and local companies) and distributed by different local companies; making the list very long.

The second list comprises of 157 pesticidal products registered for use in *public health* [11], consisting mainly of various active ingredients of pyrethroids, OPs (*temephos, pirimiphos methyl, chlorpyrifos and fenthion*), carbamates (*propoxur and carbaryl*), rodenticides (*zinc phosphide, brodifacoum, bromadiolone and flocoumafen*), neem oil, boric acid (specified for cockroach control), plant extracts, and neonicotinoids (*imidacloprid*), respectively, in various formulations (liquids, solids, vaporizing liquids with electrical heaters, baits, sticky tapes); and sold by various companies. The registrations are for specified uses, including *pyriproxyfen* as a mosquito larvicide, *deltamethrin* for indoor residual spraying (IRS), *alpha cypermethrin* for Long Lasting Insecticide-treated Nets (LLIN) and *bifenthrin* as a grain storage dust. The third list contains products, which are registered as technical grade materials for formulation purposes only, where information on technical mixtures (a.i.), mostly >95% pure, are given and the formulations for which they are imported are stated. The active ingredients of pyrethroids, OPs, carbamates, fungicides, rodenticides and neonicotinoids, as well as adjuvants such as PPO and plant oils, are given. The last two lists (**4th and 5th**) in the database include the 4th one containing information on *temporarily registered products* with their specified uses; and 5th one for *banned pesticides* including *monocrotophos, alachlor and endosulfan*, which the farmers in Kenya are still using illegally [4], as well as *restricted products* such as DDT for malaria vector control only.

### 3.2 Pesticide manufacturing in Kenya

The PCPB also regulates the manufacture, distribution and sale of PCPs. According to the information in the database, pesticide manufacturing/formulation and trade, respectively, in Kenya involve several multinational companies (e.g. Bayer, BASF, Monsanto, Syngenta and DuPont) with branches in Kenya, as well as numerous local companies. The world's six largest pesticide manufacturers including Syngenta (and ChemChina), Bayer Crop Science, BASF, Dow Agrosciences, FMC and Adama, control nearly 75% of the global pesticide market, with products ranging from insecticides such as DDT, organophosphates, carbamates, herbicides, fungicides, neonicotinoids, and biopesticides [70]. Weed killers (herbicides) account for about one-third of the global pesticide market.

Manufacturing of pesticides involves formulation, packaging and labeling of the product to make it ready for sale. A pesticide formulation is defined as a combination of active ingredients with compatible inert ingredients of chemicals, which ultimately control a pest. Formulating a pesticide involves processing it to improve its storage, handling, safety, application or effectiveness [71, 72]. A pesticide product, which is ready for use, therefore, contains two parts, the active and inert ingredients. Active ingredients (or technical mixtures >95% purity, usually) are chemicals, which actually control the pest. Inert ingredients are solvents, solids and other adjuvants that help present the active ingredients to the target pest. *Adjuvants* assist in the mixing of some formulations during formulation and dilution just before field application and include surfactants, thickeners, baits, buffers, abrasives and *synergists*, which lack any direct pesticidal activity, but they are added to pesticide formulations to optimize product performance while using the minimum amount of it. The inert ingredients serve to enhance the utility of the product by diluting and reducing costs and field effectiveness [73], because an active ingredient in a fairly pure form is not suitable for field application. The formulation process also improves pesticide safety features and enhances handling qualities.

Examples of specific inert materials include *diatomaceous earth, petrolatum, crop oil, biodiesel, surfactants*, etc. Carrier materials can allow the pesticide to be dispersed effectively, e.g. *a talc in a dust formulation*, the *water for mixing a wettable powder* before a spray application, or the *aerosol that disperses the pesticide* in an air blast application. Inert means the carrier or diluent cannot interfere in the toxicity of the active compound. However, inert does not imply that the chemical, say a surfactant, is nontoxic, as some of the inert diluents or carriers can be toxic e.g. to the plant weeds or other non-target plants, and need to be tested alongside the formulation in a performance field trial [56, 74] as well as in a non-target active ingredient toxicity test. Therefore, pest control products exist in different formulations which are manufactured bearing in mind the nature of a.i.'s (solids and liquids), their solubility, ability to control the pest, storage, ease of application and transportation. A review of materials used as carriers in the pesticide industry can be found in other texts [75–77]. The principles involved in formulation are determined by end-use and behavior of the pesticide, and important factors to consider include, chemical and physical properties of a.i. (e.g. bp, mp, specific gravity, vapor pressure, water solubility, rate of hydrolysis, toxicity ($LD_{50}$ or $EC_{50}$), biodegradability and UV-degradability) and inert materials, type of application equipment, nature of target surfaces, containers, marketing and transportation needs. For inert ingredients, there is need to know compatibility with containers, and therefore their physico-chemical properties, as well as the physical properties of the ultimate mixture. The formulation must then be tested to document various characteristics including homogeneity, particle size, storage stability, retention on target surfaces, wetting properties, penetration and translocation in plants, residual nature on target or in soil, nature of deposit, efficiency and potential hazards to users.

### *3.2.1 Manufacturing of synthetic pesticides*

In general, there are approximately twelve (12) types of formulations, which are commonly used and these types of formulations which are discussed briefly below, are included in the manufacturing of various PCPs in Kenya and are listed in the PCPB database.

#### *3.2.1.1 Dusts*

Dusts contain 2 ingredients, i.e. an *inert diluent* and a *toxicant*, with a toxicant accounting for 1–10% by weight of the mixture. Inert diluents here must be relatively non-adsorptive material to avoid inactivating the pesticides, e.g. *talc*, *pyrophyllite* or other clay. The diluents are finely ground for ease of application and coverage. The advantages of dust formulations include simplicity to manufacture and application. However, dust is least effective and least economical since it tends to drift during application resulting in poor deposition on target surfaces. To reduce importation costs, dust can be formulated as dust concentrates, containing say 90% of a.i. by weight for further dilution with local diluents in Kenya, or by mixing or blending at the farm before application.

#### *3.2.1.2 Wettable powders (WPs)*

Wettable powders (WPs) are the most widely used in agriculture, and consist of a toxicant + inert diluent + wetting agent. Inert diluents are usually adsorptive clay, e.g. *attapulgite* (Mg, Al, Si clay), and the wetting agents may be a blend of 2 or more *surfactants*, with the *toxicant* in the range of 25–75% (wt/wt) of the mixture; therefore, highly effective due to high concentrations of the a.i.'s. WPs can be prepared by (1) (i) first spraying the *toxicant* (if liquid) onto the clay material at a controlled temperature or, (ii) mixing the clay with a solution containing the *toxicant* (if solid) and (iii) then allowing the solvent to evaporate or, (2) by direct grinding of *crystalline toxicant* mixed with diluents, to get a homogeneous mixture, which can be ground to powder. Packaged WPs are bought and diluted at the farm by mixing a specified quantity (as on the label) with water, before spraying.

#### *3.2.1.3 Emulsifiable concentrates (ECs)*

Emulsifiable concentrates (ECs) are formulations which consist of a toxicant + a solvent (e.g. water or other types such as *petroleum distillates, kerosene* ($C_9$-$C_{10}$ *fraction*), *Aromax, Solvesso and biodiesel* (e.g. vegetable oil, Neem oil and xylene)) for the toxicant + emulsifier (usually a *surfactant* e.g. calcium alkyl dodecyl benzene sulfonate, or alkyl phenolic polyexthoxylates), which are also imported. The toxicant content of ECs is expressed as weight/volume and not as wt/wt as in dust or WPs. ECs, which are very common in Kenya, typically contain approximately 25–50% by weight of a.i. On mixing at the farm (usually with water) before spraying, the product gives a stable milky emulsion, which can remain stable for up to 24 hours. ECs are more easily absorbed by the skin and plants than WPs and dust and are, therefore, more hazardous, but more effective than WPs since there is no masking effect of diluents.

#### *3.2.1.4 Suspendable concentrates (SCs) or flowables*

Suspendable Concentrates (SCs) or Flowables are used for pesticides, which are just too sparingly soluble to be made inform of ECs but can be formulated to become water-based mixtures, which are handled and applied in the same manner as ECs. SC or flowable is a liquid formulation containing a stable suspension of a *solid* pesticide active ingredient in a solvent, usually intended for dilution with water at the farm before use. They contain 50–90% by weight of toxicants and are basically WP's of much smaller particle sizes (1–5 μm), which remain in suspension for long periods. Suspendability and storage stability are improved by inclusion of adjuvants (such as *surfacta*nts, e.g. emulsifiers, penetrants etc.) which improve the physical and chemical properties and enhance the control effect. Oils such as kerosene can be added when penetration into plant parts such as leaves is desired. A *water-miscible organic liquid*, or a mixture of multiple fluids, is usually used to dissolve water-insoluble or partially soluble pesticide a.i.

#### *3.2.1.5 Water-soluble powders (SP's)*

Water-Soluble Powders (SP's) are water-soluble formulations of solids/powders, in which the a.i.'s (and solid diluents) are easily dissolved in water. The technical grade material is formulated into a finely ground solid, packed and sold, for adding to a spray tank with water, where it dissolves quickly without any binders or surfactants, and spraying without constant agitation. Some examples of common SPs include *acephate* formulations with trade names *Acephate Tur*f and *Acephate PCO Pro 97.4% a.i.*

#### *3.2.1.6 Solutions (S)*

Solutions (S), a Solution formulation is the true solution containing a toxicant in a solvent, which can be used directly without further dilution, for household insect sprays, roadside weed eradication and rangeland spraying. The toxicant is dissolved in low-cost solvents such as *kerosene or aviation fuel*. No surfactants are added because the solvents wet target readily. They are prepared by direct dissolution in the selected solvent where the pesticides are sufficiently soluble in water (subject to hydrolytic stability test); or water-miscible solvents to avoid hydrolysis for mixing with water before application, e.g. Azodrin WMC (*a monocrotophos a.i.*), which is formulated in hexylene glycol, a solvent or coupling agent, giving a water-miscible product.

#### *3.2.1.7 Granules (G)*

Granular formulations are similar to Dust formulations and contain a toxicant (1–10% by weight) and an inert diluent, the major difference being particle size, with *granules* ranging from 20 to 100 mesh and *dusts* passing through 300 mesh screens. Granules are made from active ingredients, in many ways: (i) the toxicant (which is a liquid) may be added so as to impregnate the solid granules, which are then dried and thus completely released only when the granule breaks up after application, or (ii) the toxicant may be surface-coated on to granules using a volatile solvent; i.e. the toxicant, (liquid or solid), dissolves in volatile solvent, which evaporates from the formulation, or (iii) grinding the solid toxicant with inert diluents. The inert diluents for granular formulations can be *clays* or *organic materials* such as corncobs (e.g. Furadan - with *carbofuran* a.i. formulated with *grit material made from corncobs*), pecan shells, tobacco stems and walnut shells, respectively. Granules (mostly for soil and water surfaces) are less likely to drift unlike dust or liquid sprays and have less tendency to adhere to foliage.

#### *3.2.1.8 Water-dispersible granules (WGs)*

Water-dispersible granules (WGs), known as Dry *Flowables,* contain typically a toxicant (50–95%, w/w), a dispersant (e.g. surfactant), a binder and diluents. They are granules intended for application after disintegration and dispersion in water at the farm. They have low dust properties (due to larger particles) and exhibit good flowability. WGs are manufactured by blending and agglomerating a ground solid active ingredient together, with surfactants and other ingredients, mixed with water, then drying step to reduce moisture to a 1–2% range. A binder/an anti-caking agent/carrier such as *inert clay* can be used, e.g. Greensperse® CA-N is an imported binder optimized as granules, which provide enough strength and resistance to abrasion to reduce dusting and maintain the granule's integrity until its application. When powder is agglomerated, particularly by wet granulation flowability is greatly improved. A high MW polyoxyethylene surfactant can be used in agglomeration.

#### *3.2.1.9 Ultra-low volume (ULV)*

Ultra-low volume (ULV) formulations are undiluted technical grade material or the original a.i. dissolved in a minimum amount of solvent, in case of a solid a.i. They are applied, e.g. by helicopter, without further dilution in an extremely fine spray generated by special aerial or ground spray equipment, and are useful for public health vectors and agricultural and forest pests. ULV applications offer several advantages, including high efficacy due to high directed concentrations at the target and absence of masking inert diluents/surfactants, compared with a normal formulation spray [77]. The technique is useful in treating large areas, e.g. a helicopter carrying 100 gallons of ULV-malathion, to spray 400–800 acres of rangeland before reloading. In Kenya, it is used in mosquito control programs, desert locust control and large-scale wheat farms in Narok and Laikipia.

#### *3.2.1.10 Aerosols*

Aerosols are commonly used widely in Kenya for controlling resident flying and crawling insects such as mosquitoes, ants, termites and cockroaches in the domestic sectors. The active ingredient is dissolved in a *volatile solvent*, e.g. a *petroleum solvent,* and the resulting solution is atomized through a jet by means of a propellant. The propellant can be a *gas under pressure* or a *liquid that is gaseous* at atmospheric pressure conditions. Chlorofluorocarbons propellants have now been replaced by other environmentally friendly volatile liquids such as *butane, dimethyl ether, compressed carbon dioxide or nitrogen*.

### *3.2.1.11 Controlled release (CR) formulations*

Controlled release (CR) formulations are recent innovations, where the pesticide is incorporated into a *carrier*, generally, a *polymeric material* and are diffusion-controlled [78]. They are microencapsulated formulations consisting of a solid or liquid inert containing an active ingredient surrounded by a plastic or starch coating [79]. The resulting capsules can be sold as *dispersible granules* (dry flowables) or as *liquid* formulations. Encapsulation enhances applicator safety while providing timed release of the active ingredient. Liquid forms of microencapsulates are further diluted with water and applied as sprays, forming suspensions in the spray tank and having many similar properties as liquid flowables. The rate of release of the pesticide is determined by the properties of the polymer as well as environmental factors. Polymers, e.g. *proteins, synthesized vinyl 2, 4-dichlorophenoxyacetate and vinyl 2-(2, 4, 5-trichlorophenoxy) propionate,* respectively, can be used. An example is the *2-Acryloxyethyl ester of 2, 4-D copolymerized with triethylamine methacrylamide*. There are mainly two types of CR formulations, i.e. the *Reservoir Devices* and the *Monolithic Devices*, respectively. The reservoir devices are made when the toxicant is enclosed in the capsules of a thin polymeric material to become microcapsules of 1–100 μm in diameter; e.g. *Penn Cap-M microcapsules* (methyl parathion product). The *Monolithic Devices* are made when the toxicant is uniformly dissolved or dispersed within the polymer matrix to become microparticles of 1–100 μm diameter or strips; e.g. *Alco No-Pest Strip* (with dichlorvos is the active ingredient), which is used widely in pet flea collars [80]. The manufacturing of these devices is highly mechanized and expensive and, therefore, they are not yet popular in Kenya. There are also other disadvantages including longer-lasting residues and potential toxicity to beneficial insects such as bees.

### *3.2.1.12 Baits*

Baits can be very useful for achieving selective toxicity of insecticides against some species of insects. Spot application, i.e. where the bait is placed in selected places, which are accessible only to target species, permits use of insecticides in a safe manner with no environmental disruption and less human exposure. A bait formulation consists of a *carrier, toxicant and feeding stimulants*. Carriers are made from *laying mash, cracked corn, wheat bran, corn cob grits and peanut hull*, while feeding stimulants include *crude cotton seed oil, refined soybean oil, sucrose, coax brewers concentrate, malt extract, glucose, maltose, honey and wheat germ*. Several types of baits are available in the market, e.g. malathion 4% w/w bait formulated from crude cotton seed oil (5%) and sucrose (10%) on a laying mash carrier, among others [81, 82].

*Formulation labels*-after a formulation has been manufactured, a suitable package is used and labeling is done. What goes into the labels is important for trade purposes and includes sufficient information to inform the buyer about the quality, concentrations and safety of the product as well as any notable special features of the pesticide product. The concentration of the pesticide on the label is very important because it gives guidance on further dilution with water at the site of application. For dry formulations such as WPs, dust and granules, respectively, the insecticide concentration is expressed *as percentage of active ingredient* (a.i.) by weight in the formulations e.g. Diazinon® 50 W or 50WP means it is a formulation containing 50% diazinon (by wt) as the a.i and is a wettable powder formulation. For liquid formulations such as Solutions (S) and emulsifiable concentrates (EC), the concentration of the insecticide is expressed in pounds of a.i. /gallon or grams of a.i./Liter of formulation etc. Diazinon® 4E or Diazinon 4EC means it is an emulsifiable concentrate formulation containing 4 lb. of diazinon (or 4 g/L depending on units used in a particular country) in each gallon or liter of the formulation, respectively.

### *3.2.2 Manufacturing of biopesticides in Kenya*

Biopesticides are derived from micro-organisms (bacteria, fungi, viruses, etc), plants (neem, pyrethrum, etc) and natural enemies of pests (parasitoids, predatory mites and pathogens). Additional substances under biopesticides are semiochemicals (e.g. insect sex pheromones), enzymes (proteins) and insect growth regulators. The biocontrol products presented in **Table 3** include twenty-seven (27) different products/formulations which are manufactured from nineteen (19) different active ingredients by nine (9) different companies in Kenya [11]. Biopesticides have become very popular in the horticultural sector, led by flowers, French beans, peas and avocadoes, which are grown mainly for export. The driving force behind this new shift

| Active ingredients | Species | Number of formulations | Uses |
|---|---|---|---|
| Metarhizium anisopliae | Fungus | 5 | Spider mites (roses); fall army worm (maize); mealy bugs (roses). |
| Atoxigenic fungi | Fungus | 1 | Toxigenic Aspergillus falvus (maize). |
| Amblyseius (strains: *californicus, cucumeris, andersoni, swirski*). | Predatory mite | 5 | Spider mites & whiteflies (French beans, roses); thrips (in green houses). |
| Amphibious transcaspinus | Predatory mite | 1 | Aphids (French beans). |
| Bacillus thuringiensis (var: *aizawa, subtulis, amyloliquefaciens*). | Bacterium | 3 | Caterpillars (roses, chives, French beans, snow peas); mildew (roses); rice blast; coffee leaf rust; black rot (cabbages). |
| Paecilomyces lilacinus | Fungus | 3 | Nematodes (roses, French beans); aphids & white flies (tomatoes). |
| Lecanicillium (strains: *Verticillium, muscarium*) | Fungus | 2 | Aphids (roses, French beans); white flies (tomatoes, roses). |
| Beauveria bassiana | Fungus | 3 | Aphids (cabbages); bollworms; cutworms; caterpillars; thrips (French beans). |
| Steinernema carpocapsae | Nematode | 1 | Caterpillars (roses). |
| Trichoderma harzianum rifai | Fungus | 1 | Soil borne fungi (French beans, roses). |
| Macrocheles robustulus | Predatory mite | 1 | Thrips (roses). |
| Cryptophlebia leucotreta granulovirus | Moth isolate | 1 | Coddling moths (roses, avocadoes, capsicum). |

**Table 3.**
*Some biopesticides manufactured in Kenya.*

towards biopesticides is mainly the need to avoid the strict residue limits imposed by the importing countries in Europe.

## 4. Pesticide use and impacts in Kenya

### 4.1 Pesticide use in agriculture and its impacts

Although developed countries consume 75% of the global pesticide in the market, it is the developing countries that will bear the heaviest burden of pesticides impact despite consuming just 22% [83–85], because of weaknesses in the regulatory mechanism and lack of education and awareness, especially among farmers. Apart from the long-term effects of pesticides, which are already known, several cases of severe impacts have been seen in Kenya including, (i) high mortalities caused by poisoning through suicides [86], (ii) high incidences of occupational exposure among farmers [4, 84, 86, 87], (iii) environmental degradation [26, 83, 84, 80–90], (iv) consumption of contaminated foods and water [26, 89, 91–95 ] and (v) misuse causing threats to wildlife, insects and other species [26, 83, 84, 89–90]. In Kenya, the increased amounts of pesticides being used and the reported potential human impacts, for example, cancer, which is now a major killer [83], seem to coincide, and this has raised concerns among the population. Many cases of acute pesticide poisoning, sometimes fatal, have been detected in people in the agricultural sector, where exposure to pesticides is highest [4, 84, 86, 87]. Even with low exposure, pesticides can cause serious consequences such as acute male infertility, cancers, abortions and other birth defects, and fetal malformations [84].

The rural population in Kenya constitutes approximately 80% of the total population; therefore, human and environmental health risks associated with pesticides are heavily experienced among this population since agriculture is mostly practiced in the rural areas [96]. The government has over the years put strategies to mitigate environmental impacts of pesticides, for example by the ratification of the Stockholm Convention in May 2004 and a national implementation plan in 2007 [97], which led to banning of most of the persistent OCs [11]. Banning of OCs can lead to recovery of affected species in nature [98]. However, a number of highly toxic OPs, carbamates, pyrethroids, neonicotinoids and fungicides, belonging to WHO I and II class, which have been banned in other countries such as the EU, have not been restricted or banned in Kenya [84]. There is, therefore, a dire need for risk assessment of all pesticides on the PCPB database for possible withdrawal or banning of the highly toxic ones.

Several studies on pesticide usage and impacts have been conducted following international best practices, in various agricultural regions in the country, which have revealed that the current group of pesticides used in Kenya are mostly highly toxic and pose threat to humans and the environment [84, 99–102]. Most of the pesticides are used intensively in certain regions that are traditionally agricultural zones, in the *North rift*, the *Central highlands*, the *South Rift*, *Eastern province*, as well as *Western* and *Nyanza provinces*, respectively, where specific types of crops are grown for local consumption and export; and it is in these regions where most studies have been done. In fact, several cases of pesticide misuse by farmers, occupational exposure and potential risks to human and drinking water have been documented since the year 2000. Two recent surveys were conducted in Trans-Nzoia County in Western Kenya which is the largest producer of maize in the country, producing at least 5 million bags of the grains annually from approximately 107,000 acres of cultivated land; which highlighted the toxicities of pesticides used and their impacts on human and the environment. The first study was done in 2018 involved prioritization of the pesticide active ingredients by ranking them according to the *Quantity index* (quantity used) (QI), the Toxicity Potential (TP) and Toxicodynamic Potential (TDP) with regard to *carcinogenicity, mutagenicity, teratogenicity and endocrine disruption*, as described in Dabrowski et al. [99] and Gunier et al. [101], as well as hazard potential (HP), the groundwater ubiquity score (GUS), surface water mobility index (SWMI) to indicate their environmental hazards [99, 100, 102, 103]; and intrinsic toxic potential (ITP) for bioavailability, environmental persistence and bioaccumulation. The **Table 4** shows the criteria for scoring of toxicity potential for specific pesticides, in which a ranking of highest value (8) was given in cases where there was definitive toxic effect

| No. | Pesticide | Quantity kg (a.i) | %use | ITP | HP | WHP | Mobility |
|---|---|---|---|---|---|---|---|
| 1 | Metalaxyl+ mancozeb | 6678 | 19.6 | 3 | 6 | 1.2 | Low |
| 2 | Glyphosate | 5140 | 15.1 | 21 | 21 | 3.2 | Low |
| 3 | Mancozeb | 4443 | 13.0 | 15 | 15 | 2.0 | Low |
| 4 | Terbuthylazine | 4125 | 12.1 | 11 | 14 | 1.7 | Medium |
| 5 | s-metolachlor + Atrazine | 3561 | 10.4 | 7 | 14 | 0.1 | High |
| 6 | Paraquat dichloride | 1774 | 5.2 | 10 | 10 | 0.5 | Low |
| 7 | Tebuconazole | 1244 | 3.6 | 11 | 22 | 0.8 | Medium |
| 8 | Lambda-cyhalothrin | 1230 | 3.6 | 5 | 5 | 0.2 | Extremely low |
| 9 | Imidacloprid | 846 | 2.5 | 6 | 6 | 0.2 | High |
| 10 | Atrazine | 507 | 1.7 | 7 | 14 | 0.2 | High |
| 11 | Carbendazim | 465 | 1.4 | 11 | 22 | 0.3 | Medium |
| 12 | Hexazinone | 436 | 1.3 | 3 | 6 | 0.1 | High |
| 13 | Carbosulfan | 376 | 1.1 | 1 | 1 | 0.01 | Low |
| 14 | Abamectin | 367 | 1.1 | 10 | 10 | 0.1 | Low |
| 15 | Deltamethrin | 362 | 1.1 | 20 | 20 | 0.2 | Low |
| 16 | Topramezone + dicamba | 354 | 1.0 | 10 | 40 | 0.4 | High |
| 17 | s-metolachlor | 345 | 1.0 | 15 | 30 | 0.3 | Low |
| 18 | Alpha-cypermethrin | 344 | 1.0 | 14 | 14 | 0.1 | Extremely low |
| 19 | Cymoxanil + propineb | 344 | 1.0 | 15 | 23 | 0.2 | Low |
| 20 | Chlorpyrifos | 278 | 0.8 | 9 | 9 | 0.1 | Low |
| 21 | Thiamethoxam | 223 | 0.7 | 3 | 4 | 0.03 | High |
| 22 | Cyhalothrin | 221 | 0.3 | 9 | 9 | 0.1 | Extremely low |
| 23 | 2,4-D-Amine | 181 | 0.5 | 9 | 36 | 0.2 | Medium |
| 24 | Profenofos + cypermethrin | 167 | 0.5 | 15 | 40 | 0.1 | Extremely low |
| 25 | Diazinon | 122 | 0.4 | 15 | 23 | 0.1 | Low |

**Table 4.**
*Levels of pesticides (kg a.i.) used in Trans Nzoia, their Intrinsic Toxicity Potential (ITP), Hazard Potential (HP), Weighted Hazard Potential (WHP) and Mobility.*

and zero was awarded to endpoints where there was no evidence of toxic effect, and the toxicity potential (TP) was obtained by adding the scores attributed to each of the five toxic effects (carcinogenicity, endocrine disrupter potential, mutagenicity, teratogenicity and neurotoxicity, respectively) for each active ingredient [101].

The GUS index was applied in a logarithmic scale where those pesticides with a GUS index below 1.8 had lower leaching potential while those with a GUS index higher than 2.8 were classified to have high leaching potential [99]. The potential of a pesticide to contaminate surface water resources was determined from surface water mobility index (SWMI), with pesticides having a SWMI tending towards 1 (one) having higher potential to be carried by surface run-off (**Table 4**) [99]. The criteria for scoring human and wildlife (bees and fish) toxicity potential and environmental impacts reported here are discussed in detail by Odira et al. [56] and Otieno et al. [55], respectively.

From this study, a total of 25 pesticides/active ingredients (**Table 4**) were considered significant in terms of their impacts on the environment and human health. The results showed that *glyphosate, mancozeb, terbuthylazine, metalaxyl-M + mancozeb, paraquat dichloride and carbendazim,* were among the most commonly used active ingredients with far-reaching environmental and health impacts. Although there were some pesticides that were not heavily used (e.g. diazinon), they still had significant toxicity from the evaluation scores and, therefore, presented substantial risk to human and environmental health in the area (**Table 4**). It was observed that the fungicide combination *metalaxy-M and mancozeb* was the most commonly used pesticide in Trans Nzoia County accounting for about 19% of all the active ingredients used (**Table 4**), while *diazinon* was the least used pesticide (0.4%) in the county. Trans Nzoia is generally damp and cold most times of the year, a condition that promotes occurrence of fungal diseases which perhaps explains the heavy usage of metalaxyl+mancozeb fungicide combination in the county. The amounts of herbicides such as *glyphosate, terbuthylazine, paraquat, metolachlor and atrazines*, were also high as expected, because of large farm sizes (5–30 acres) (**Table 4**). In addition, *topramezone + dicamba, 2,4 D-Amine, S-Metolachlor, atrazine, cymoxanil + propineb, diazinon, carbendazim, tebuconazole, glyphosphate and deltamethrin* were prioritized as active ingredients with higher potential to contaminate surface and groundwater, in the area. *Glyphosate, mancozeb, S-Metolachlor, terbuthylazine tebuconazole, paraquat dichloride and topramezone + dicamba* presented enormous risk according the weighted hazard potential (WHP) evaluation, but had low potential to contaminate surface water and groundwater due to their low GUS index, and as a result, they could present minimal risk to aquatic organisms and human through consumption of drinking water. Pesticides with high $K_{oc}$ (as well as high water solubility and low soil half-lives) (data not presented here) have low potential to contaminate water resources and, therefore, present minimal risk to humans. *Thiamethoxam* with low WHP (**Table 4**), had very high GUS index (ranking of 4) and very high SWMI score (4), and also had the highest potential to contaminate the environment and highest potential toxicity scores (4) to birds, mammals, aquatic invertebrates and bees, respectively. Whereas there were also pesticides with high potential to present risks to humans and the environment due to the high WHP, like the top 5 in **Table 4**, including *glyphosate etc,* such risks may not be via water because of their low mobility. The environmental exposure potentials (EEP), and non-target toxicity data of commonly used pesticides in the area were compiled (full data not included here).

In a similar study done in 2019 in the same county, involving different farmers a full range of pesticides (45 a.i.'s in all) was used over 1 year (full data not shown), including their physical-chemical properties and toxicity indices were reported. The toxicity indices, i.e. TP, EEP, GUS, and SWMI, were used to evaluate potential toxicity to humans and the environment. Most of the farmers (99.4%) involved in the survey applied pesticides, consisting of 10 different fungicides in various formulations, 5 OP's, 5 neonicotinoids, 6 pyrethroids, 2 carbamate insecticides, 4 herbicides, heptachlor (which is banned) and Abamectin, respectively; and most of them falling in the WHO Class I and II. The used pesticides in that year included *carbendazim* (32.9%), *epoxiconazole* (17.6%), *diazinon* (20.4%), *imidacloprid* (23.6%), *metolachlor* (28.2%), *amitraz* (56.3%), *chlorpyrifos* (10.6%) and *acetochlor* (9.1%), with smaller amounts of *cypermethrin* (5.5%) and *heptachlor* (1.2%). The most applied pesticide class was the OPs (34.8%). It was found that 18.4% of the pesticides applied in the study area were persistent in soil sub-systems, 31.6% were persistent in water, and 10.5% and 13.2% had the potential of contaminating ground and surface water resources, respectively [55, 56]. The ranked order of human toxicity potential associated with the used pesticides in the area in 2019 was *teratogenicity (31.6%), neurotoxicity (29.0%), endocrine disruption (7.9%), carcinogenicity (7.9%), mutagenicity (2.6%) and multiple toxicity potentials (10.5%)*. In addition, 18.4% of the used pesticides, including *acetamiprid, heptachlor, amitraz, chlorimuron ethyl, azoxystrobin, lufenuron and copper oxychloride*, had higher potential for bioconcentration in the living tissues, while most of the pesticides, (39.5%) and (18.8%), respectively, were highly toxic to aquatic invertebrates and earthworms. All the pesticides applied in the study area in 2019 were potentially harmful to human health, if not properly handled. Round up which is restricted in the EU, as well as carbofuran, carbosulfan and heptachlor, which are restricted and banned, respectively, in Kenya, were also used.

In horticultural farming, where farms are often smaller (1–2 acres), the amounts of insecticides and fungicides used are often higher in comparison to the amounts of herbicides [4, 90]. In a survey done in 2015 and 2016 in Meru in Central Kenya, which is famous for horticultural farming of fruits and vegetables for local consumption in major cities such as Nairobi and for export, respectively, high quantities of insecticides such as *deltamethrin, dimethoate, chlorpyrifos, carbaryl, methoxychlor, λ-cyhalothrin, endosulfan sulfate, cypermethrin, zeta-cypermethrin, malathion, diazinon and propoxur*, and even the banned pesticides including *parathion, carbofuran, heptachlor, dieldrin and endrin*, were used over the 2 year period, compared with only smaller quantities of only two herbicides, *glyphosate and paraquat*, and one fungicide a.i., *mancozeb* [4]. Fungicides such as *carbendazim* and several *neonicotinoids* were also reported in French beans, tomatoes and kales, bought during harvesting on the farms, confirming their usage on the farms [4, 104]. The farmers (26%) reported health effects after using pesticides, with most effects (>12 respondents out of 173) experienced when *dimethoate, malathion, heptachlor, endrin, dursban (chlorpyrifos), parathion and dieldrin*, were used. Nine (9) of the pesticides used in Meru county, including *parathion, methomyl, endosulfan, endrin, dieldrin, methoxychlor, heptachlor epoxide, carbofuran and endosulfan sulfate* were very toxic (WHO class I), 12 were toxic (WHO class II) and 5 were moderately toxic (WHO class III) [4].

In Muranga in Central Kenya, where small-scale farming of avocados, tea and coffee are the main cash crops, and maize, beans and bananas are the main food crops, various categories of pesticides, including neonicotinoids, acaricides, fungicides, insecticides and herbicides, were found to be used in 2021 [90], although the quantities of these products were not reported. Using honey bee pollen as an indicator of used pesticides in county, eleven (11) pesticides were confirmed to be present in the honey, including *carbendazim, carbofuran, Spinosyn A, spinosyn D, acetamiprid, chlorpyrifos, thiamethoxam, imidacloprid, acephate, trifloxystrobin and indoxacarb* [90].

A national survey done in 2020 covering 32 counties located in all agricultural regions in Kenya established that mostly subsistence farming is practiced in the counties, and the major pests affecting crops were insects and rodents, where farmers used various synthetic pesticides (80%) as well as home products (68%), with 84% of the most common pests being caterpillar-related pests such as *stalk borers, white flies, worms, army worms and cut worms, aphids, termites, weevils, rodents and fungi*. A large variety of pesticides, including mainly pyrethroids, organophosphates (*e.g. diazinon, dimethoate, pirimiphos methyl, chlorpyrifos*), fungicides (*metalaxyl-M+ mancozeb*), carbamates (*carbaryl*), neonicotinoids (*thiomethoxam*), IGRs (*pyriproxyfen*), rodenticides (*zinc phosphide*) and unspecified herbicides, were used [105]. Some of the homemade products included *lemon grass, aloe vera, ashes, cloves, marigold extracts, pepper, salt and solanum apple*; for example, *ashes and chillies* were used to control insects such as *aphids in vegetables* [105]. The large variety of pesticides used by farmers, in this study, corroborates those pesticides used in regions such as Trans Nzoia, Muranga and Meru counties [4, 55, 56], and similar types of pests reported here have also been reported and discussed extensively in a government and other reports [106, 107]. The need for pesticides sometimes is absolute because frequent unexpected attacks by pests sometimes occur, for example in 2017, 40% of farms were reported to be infested with the fall armyworm [108].

Other researchers have also reported on pesticide use and impacts in other regions of the country, including vegetable farming districts of Kiambu, Kirinyaga, Nyandarua, Muranga, Meru and Makueni [109, 110], where mostly WHO Class I and II pesticides were used and acute poisoning cases were reported; Lake Victoria basin including Nyando, Kericho and Nandi districts [54], where tea, coffee, sugarcane, maize and vegetables were cultivated, and in which 14 different active ingredients were used against maize stock borer (86% of farmers), aphids (70%), cutworms (60%), diamond black moth (50%), thrips (28%), termites (20%) and weeds (4%), and 4 of the active ingredients were known to be highly toxic to bees and birds [54]. Herbicides were used in tea, coffee and sugarcane and insecticides and fungicides, respectively, largely on vegetables [54], with frequent cases of misuse, including application of banned OCs, and declines in pollinating insects and Red-billed Oxpecker bird species being reported [54]. Mburu et al. [111] found that 141 different pesticides were used in 20 horticultural farms along the small Lake Naivasha shore catchment alone, six of them (4.3%) belonging to WHO Class I, including carbamates (*oxamyl and methomyl*), *bipyridylium, strobilurin, tetranortriterpenoids, azole* and OPs (*fenamiphos*), and 20 of them (14.3%) in the WHO Class II. The farmers also used 4 species of natural predators (*Trichoderma spp*, *Paecilomyces spp*, *phytoseiulus persimilis spp* and *Amblyseius spp*), and entomopathogenic fungi, which are registered by PCPB, as biopesticides [111]. Some of the impacts of pesticides on Lake Naivasha water have been highlighted and residues of carbamates and organophosphates have been detected in the water [92, 95]. However, much less work on pesticides and their impacts than expected in global terms is being done in Sub-Saharan Africa [112].

Recently, however, some impacts on pesticide policy in Kenya have started being felt [43]. The route for food initiative (RTFI) (an NGO) in 2019 conducted a study and found that 77% of the 230 ingredients registered in Kenya have been at least withdrawn from the EU market or are heavily restricted due to their chronic human toxicity and environmental effects (based on fish and bees), and additional 19 of them are not listed in the EU database [43]. The RTFI report further highlighted the carcinogenic, mutagenic, endocrine disruptive, neurotoxic and male infertility effects of most of them [43]. Following these concerns, the Kenya Organic Agriculture

Network (KOAN) in collaboration with Eco-Trac Consulting did a survey in 2020 and produced a comprehensive report giving detailed accounts of pesticide usage and impacts in Kirinyaga and Muranga counties in Central Kenya, where intensive horticulture is practiced for subsistence and export purposes [84]. The aim of the study was to provide the evidence needed to advocate and promote a transition from harmful pesticides, to safer alternatives such as GAPs and bio-pesticides [84]. The risk assessment was done according to the EU protocols. Apart from the information on toxicity of 64 active ingredients and 142 formulations being used to control 30 insect pests, 24 weeds and 11 plant diseases, respectively, in the two counties, they also highlighted issues such as misinformation, misuse, mishandling of pesticides and lack of education

and awareness as some of the main challenges the two counties faced [84]; a good example being *methamidophos,* which was not registered for use by the PCPB and likewise, no product containing *acephate* was registered for use on tomatoes, but these two pesticides were being used by farmers, illegally and incorrectly, resulting in residues of *acephate* and *methamidophos*, which are both very toxic [113, 114] exceeding the MRLs set by KEBS and EU in some samples of tomatoes [84]. They concluded that many of the pesticides used in Kenya are highly toxic, belonging to WHO Class I and II, and have already been banned or heavily restricted in other countries such as China, India and Europe, where most of them are imported from Kenya [43, 84]; and their risks need to be assessed with the aim of withdrawal or banning.

An Expert Taskforce [115], in 2021, was appointed by the NGOs to conduct an evaluation of selected pesticide active ingredients (from the PCPB database), including 20 insecticides, 5 fungicides and 4 herbicides, respectively, which are widely used in agriculture in Kenya [115]. The toxicity scores were obtained according to the methods of Dabrowski et al. [99]. Based on their evaluation, they recommended that seventeen (17) of the active ingredients should be withdrawn immediately, five (5) should go through phased withdrawal and only three (3), *clodinatop, flubendiamide and flufenoxuron*, should be retained [115]. The NGOs used the report and successfully pushed for a 'Pesticide Bill' to be introduced in parliament in 2020 aiming at the withdrawal/banning of pesticides considered harmful, from the Kenyan market. The PCPB is currently in the process of conducting a regulatory review of a priority list of highly active ingredients from the PCPB database, including those recommended by the expert taskforce, in support of the bill [115].

### 4.2 Pesticide use in malaria vector control in Kenya

Malaria remains the major cause of morbidity and mortality globally with 219 million cases reported in 2017, resulting in 435,000 estimated deaths, 61% of them being children under the age of 5 years [116–118]. The integrated malaria vector management program recommended by the WHO outlines a multipronged approach involving five methods, which include, (i) spraying with recommended insecticides against adult mosquitoes in their habitats, (ii) using insecticide-treated mosquito nets (ITNs), (iii) indoor residual spraying (IRS), whenever necessary, (iv) larval source management, and (v) early diagnosis and treatment; but only ITNs, IRS and early diagnosis and treatment, are implemented in Kenya. It is believed that lack of implementation of sustained larval control has reduced the positive gains made in combating malaria in Kenya, and it still remains a major killer accounting for 10,500 deaths annually [7].

Malaria vector control using long-lasting insecticidal nets (LLIN), has gradually increased in Western Kenya from the year 2000 [7], with about 11 million LLINs distributed freely by 2011; still far from reaching the universal coverage of all vulnerable populations. In these LLIN interventions, *permethrin*-treated LLIN has been used in various endemic zones such as Bondo, Teso, Rachuonyo and Nyando [119]. Pyrethroids such as *fenitrothion, lambda-cyhalothrin and alpha-cypermethrin*, and DDT have been used in IRS. However, with the resistance of the mosquito vectors to pyrethroids widely reported, spraying with *pirimiphos-methyl* on walls in Migori county in Nyanza [7] has been done. Although use of biopesticides such as entomopathogenic fungi has not been embraced, several small-scale trials with biolarvicides such as *B. thuringiensis israelensis (Bti)* and *Bacillus sphaericus (B.s)* in form of water-dispersible granules have reported positive results against various species of mosquito larvae along Lake Victoria shores [7]. The low residual activity makes larval control using the two interventions costly since repeated applications of the bacterial strains to the breeding sites would be necessary, and suitable formulations such as slow-release methods have to be considered. Biorational pesticides such as *Wolbachia*, *Metarhizium anisopliae, methoprene*, *hydropene,* pyriproxyfen, *B. thuringiensis and Spinosad*, which have very low human toxicity and are biodegradable, have not been significantly adopted for larval and adult mosquito control in Kenya [7].

## 5. Conclusions

The chemistry, manufacturing, importation and regulatory processes regarding pesticides in Kenya as well as their usage and impacts on humans and the environment have been discussed. All the various categories of pesticides, i.e. organochlorine, organophosphate, carbamate, pyrethroid and neonicotinoid insecticides, as well as fungicides, herbicides and biopesticides, which are used in the country, have been considered. Important information on a total of 1447 formulations and 157 active ingredients, respectively, for use in agriculture and public health sectors, are listed on the Pest Control Products Board database and is available freely to the public. A significant number of biopesticides are manufactured in the country and are used in horticulture. A number of studies have been conducted in major agricultural regions, which have characterized pesticides, their toxicities, types of crops and pests, usage and human and environmental health risk indices, since the 2000, but the reports have not made any impact on pesticide regulation, and very toxic active ingredients belonging to the WHO Class I and II, some of them already banned or removed from the EU, seem to dominate the market in Kenya. However, recent pressure from NGOs made an impact on government and parliament and a bill was introduced in 2020, aiming at more strict enforcement and banning of some of the very toxic pesticides, which have already been banned in the EU market. The PCPB which is the government institution charged with the responsibility of regulating pesticides in the country is currently reviewing some of the products, which can be replaced by safer alternatives, for banning.

## Acknowledgements

The authors wish to thank the National Research Fund of Kenya (NRF) for their support under the Multidisciplinary Research Grants.

## Author details

Joseph O. Lalah[1]*, Peter O. Otieno[2], Zedekiah Odira[3] and Joanne A. Ogunah[4]

1 School of Chemistry and Material Science, Technical University of Kenya, Nairobi, Kenya

2 Pest Control Products Board, Kisumu, Kenya

3 Department of Biological Sciences, Masinde Muliro University of Science and Technology, Kakamega, Kenya

4 Department of Physical Sciences, University of Embu, Embu, Kenya

*Address all correspondence to: joseph.lalah@tukenya.ac.ke

## References


[1] Jones MM, Netterville JT, Johnston DO, Wood JL, Joesten MD. Chemistry and Society. 5th ed. New York: Saunders College Publishing; 1987. pp. 491-519

[2] World bank. World Development Report 2020. 2020. Available from: Worldbank.org/en/publication/wdr.2020

[3] Yu JS. The Toxicology and Biochemistry of Pesticides. Boca Raton, USA: CRC Press; 2008. ISBN 978-1-4200-5975-5

[4] Marete GM, Lalah JO, Mputhia J, Wekesa VW. Pesticide usage practices as sources of occupational exposure and health impacts on horticultural farmers in meru county, Kenya. Heliyon. 2021;7:1-13. DOI: 10.1016/j.heliyon.2021.e06118

[5] Nauen R. Insecticide resistance in disease vector of public health importance. Pest Management Science. 2007;**63**:628

[6] Ware GW, Whitcare DM. The Pesticide Book. 6th ed. Willoughby, OH: Meister Pro Information Resource; 2004

[7] Ogunah JO, Lalah JO, Schramm K-W. Malaria vector control strategies: What is appropriate towards sustainable global eradication? Sustainable Chemistry and Pharmacy. 2020;**18**:100339. DOI: 10.1016/j.scp.2020.100339

[8] Helm R, Cockrell G, Stanley JS, Brenner RJ, Burks W, Bannon GA. Isolation and characterization of a clone encoding a major allergen (Bla Bd90k) involved in IgE-mediated cockroach hypersensitivity. The Journal of Allergy and Clinical Immunology. 1996;**98**:172

[9] Donaldson T, Kiely T, Grube A. Pesticide Industry Sales and Usage: 1998-1999 Market Estimates. Wahsington DC: USEPA; 1996

[10] Zilberman D, Schmitz A, Lichtenberg E, Siebert JB. The economics of pesticide use and regulation. Science. 1991;**253**:518

[11] PCPB. Pest control products board. Fully registered pest control products Version_2018. 2018. Available from: www.pcpb.go.ke

[12] Sande D, Mullen T, Wetzstein M, Houston J. Environmental Impacts from pesticide use: A case study of soil fumigants in Florida tomato production. International Journal of Environmental Research and Public Health. 2011;**8**: 4649-4661

[13] USEPA. Pesticide Assessment Guideines, Subdivision F: Hazard evaluation –Human and Domestic Animals Series. Washington DC: USEPA; 1991. p. PB-158394

[14] Lalah JO, Kaigwara PN, Getenga ZM, Mghenyi JM, Wandiga SO. The major environmental factors that influence rapid disappearance of pesticides from tropical soils in Kenya. Toxicological Environmental Chemistry. 2001;**81**:167-197

[15] Mnif W, Hassine AIH, Bounaziz A, Bartegi A, Thomas O, Roig B. Effect of endocrine disrupting pesticides: A review. International Journal of Environmental Research and Public Health. 2011;**8**(6):2265-2303

[16] Shen L, Wania F. Compilation, evaluation and selection of physical chemical property data for OCs. Journal of Chemical & Engineering Data. 2005;**50**(3):742-766

[17] Freed H, Schmedding D, Kohnert R, Haque R. Physical Chemical properties of several organophosphates: Some implication in environmental and biological behaviour. Pesticide Biochemistry and Physiology. 1979;**10**(2):203-211

[18] Keadtisuk S, Dheranetra W, Nakatsugawa T, Fukuto TR. Liver damage induced in rats by malathion impurities. Toxicology Letters. 1990;**52**:35

[19] Umetsu N, Grose FH, Allahyari R, Abu-el-Haj S, Fukuto TR. Effect of impurities on the mammalian toxicity of technical malathion and acephate. Journal of Agricultural and Food Chemistry. 1977;**25**:946

[20] Kuhr RJ, Dorough HW. Carbamate Insecticides, in: Chemistry, Biochemistry and Toxicology. Cleveland USA: CRC Press; 1976

[21] Mineau P, Tucker KR. Improving Detection of Pesticide Poisoning in Birds Part II. Journal of Wildlife Rehabilitation. Summer: Wildlfe Conservation; 2002. p. 2002

[22] Mineau P. Estimating the probability of bird mortality from pesticide sprays on the basis of the field study record. Environmental Toxicology and Chemistry. 2002;**24**(7):1497-1506

[23] Sherwani SI, Arif IA, Khan HA. In: Herbicides AP, Kelton J, Sarunaite L, editors. Modes of Action of Different Classes of Herbicides. London, UK: Intech Open; 2015. DOI: 10.5772/61779

[24] Grossman K. Mediation of herbicide effects by hormone interaction. Journal of Plant Growth Regulators. 2003;**22**:109-122

[25] Singh S, Kumar V, Chauhan A, Datta S, Wani AB, Singh N, et al. Toxicity, degradation and analysis of the herbicide atrazine. Environmental Chemistry Letters. 2018;**16**:211-237

[26] Lalah JO, Omwoma S, Osano FO, Omukunda E, Wafubwa G, Muyekho D, et al. Assessment of potential risks and effectiveness of agrochemical usage in a catchment: A case study of the Nzoia Nucleus Estate sugarcane farms in western Kenya. In: Maestroni B, Cannavan A, editors. Chapter 16 in: Integrated Analytical Approaches for Pesticide Management. Academic Press Publishers, Elsevier; 2018 eBook: 9780128161562, Paperback: ISBN: 9780128161555 (338 p)

[27] Bigley WS, Plapp FW. Metabolism of cis and trans-14C-permethrin by tobacco budworm and bollworm. Journal of Agricultural and Food Chemistry. 1978;**26**:1128

[28] Bryant R, Bite MG, Hopkins WI. Global Insecticide Directory. 2nd ed. Kent, UK: Agranova; 1999

[29] Isman MB. Botanical insecticides, deterrents, and repellants in modern agriculture and an increasingly regulated world. Annual Review of Entomology. 2006;**51**:45-66

[30] Chermenskaya TD,Stepanycheva EA, Shchenikova AV, Chakaeva AS. Insecticid acaricidal and deterrent activities of extracts of Kyrgyzstan plants against three agricultural pests. Industrial Crops and Products. 2010;**32**:157-163

[31] El-Wakeil. Botanical Pesticides and their modes of action. Gesunde Pflanzen. 2013;**65**:125-149

[32] Fusetto R, O'Hair RAJ. Nicotine as an Insecticide in Australia. J Royal Australian Institute, October: A short history; 2015. p. 2015

[33] Boomquist JR. Insecticides: Chemistries and Characteristics. 2nd ed. Radcliffe World Text book: Ipmworld. umn.edu/bloomquist-insecticides; 1996

[34] Singha A, Thareja V, Singla AK. Application of neem seed kernel extrats result in mouthpart deformites and subsequent mortality in *Nezara viricula* (L.) (hem: Pentatomidae). Journal of Applied Entomology. 2007;**131**:197-201

[35] Silva JCT, Iham GN, RDL O, Brown L. Purification of the seven tetranortriterpenoids in neem (*Azadirachta indica*) seed by counter-current chromatography sequentially followed by isocratic preparative reversed-phase high-performance liquid chromatography. Journal of Chromatography. 2007;**1151**:203-210

[36] Ihara M, Matsuda K. Neonicotinoids: Molecular mechanisms of action, insights into resistance and impact on pollinators. Current Opinion Insect Science. 2018;**30**:86-92. DOI: 10.1016/j.cois.2018.09.009

[37] Tailebois E, Carlereau A, Jones AK, Thany SH. Neonicotinoid insecticides mode of action on insect nicotinic acetylcholine receptors using binding studies. Pesticide Biochemistry and Physiology. 2018;**151**:59-66

[38] Luetan L, Merwin I, Kovach J. Assessing the relative environmental impacts of agricultural pesticides: The quest for a holistic method. Agriculture, Ecosystem & Environment. 1995;**55**(3):153-168

[39] Odinokov VN, Ishmuranov GY, Kharisov RY, Serebreykov EP, Tostikov GA. Synthsis of s-(+)methoprene. Bulletin of the Academy of USSR, Division of Chemical Sciences. 1993;**42**(1):98-99

[40] Ravon A. In: Charles JF, Delecluse A, Nielsen-LeRoux C, editors. *Bacillus thuringiensis* Application in Agriculture, in Entomopathogenic Bacteria: From Laboratory to Field Application. London: Kluer; 2000. p. 255

[41] Bravo A, Soberon M, Gill SS. In: Gilbert LI, Iatrou K, Gill SS, editors. *Bacillus thuringiensis*: Mechanisms and use., in: Comprehensive Molecular Insect Science. Vol. 6. London: Elsevier; 2005. p. 175

[42] Mcguire AV, Northfield TD. Tropical occurrence and agricultural importance of Beauveria bassiana and Metarhizium anisopliae. Frontiers in Sustainable Food Systems. 2020;**4**:6. DOI: 10.3389/fsufs.2020.00006

[43] RTFI. Pesticides in Kenya. Why our Health, Environment and Food Security is at Stake. 2019 Available from: https://routetofood.org/pesticides-in-kenya-whats-at-stake/

[44] Werren JH, Guo L, Windsor DW. Distribution of *Wolbachia* among neotropical arthropods. Proceedings of the Royal Society B. 1995;**262**(1364):197-204. DOI: 10.1098/rspb.1995.0196

[45] Aliota MT, Walker EC, Yepes AU, Velez ID, Christensen BM, Osorio JE. The mel strain of *Wolbachia* Reduces transmission of chikungunya virus in aedes aegypti. PLOS Neglected Tropical Diseases. 2016, 2016;**10**(4):e0004677

[46] Dutra HLC, Rocha MN, Dias FBS, Mansur SB, Caragata EP, Moreira LA. Wolbachia blocks currently circulating zika virus isolates in Brazilian *Aedes aegypti* Mosquitoes. Cell Host Microbe. 2016, 2016;**19**(6):771-774

[47] Zhang D, Lees RS, Xi Z, Gilles JRL, Bourtzis K. Combining the sterile insect technique with wolbachia-based approaches: II- A safer approach to Aedes albopictus population suppression programmes, designed to minimize the consequences of inadvertent female release. PLoS One. 2015, 2015;**10**(8):e0135194

[48] Hoffmann AA, Montgomery BL, Popovici J, Iturbe-Ormaetxe I, Johnson PH, Muzzi F, et al. Successful establishment of *Wolbachia* in *Aedes* populations to suppress dengue transmission. Nature. 2011;**476**(7361):454-457

[49] Bian G, Joshi D, Dong Y, Lu P, Zhou G, Pan X, et al. Wolbachia invades Anopheles stephensi populations and induces refractoriness to Plasmodium infection. Science. 2013;**340**(6133):748-751

[50] McGrath MT. Fungicides and Other Chemical Approaches for Use in Plant Disease Control. Encyclopedia of Microbiology. 3rd ed. Amsterdam: Science Direct; 2009

[51] Zubroid JP, Brundschuh M, Arts G, Bruehl CA, et al. Fungicides: An overlooked pesticide class? Environmental Science & Technology. 2019;**53**(7):3347-3365

[52] Yang C, Hamel C, Vujanovic V, Gan Y. Fungicides: Modes of action and possible impacts on non-target organisms. International Scholarly Research Network. 2011;**2011**:1-8. DOI: 10.5402/2011/130289

[53] Sunday AO, Alafara BA, Oladel OG. Toxicity and speciation analysis of organotin. Chemistry, Speciation and Bioavailability. 2012;**24**(4):216-226

[54] Abong'o DA, Wandiga SO, Jumba IO, Madadi VO, Kylin H. Impacts of pesticides on human health and environment in the River Nyando catchment, Kenya. International Journal of Humanities, Arts, Medicine and Sciences. 2014;**2**(3):1-14

[55] Otieno PO, Lalah JO. Prioritization and Assessment of Large Scale Pesticide use in Trans Nzoia County and their Potential Impacts on Human and Environmental Health. Unpublished Report; 2022

[56] Odira Z, Lalah JO, Otieno PO, Omwoma S, Osano O, Omukunda E. Screening and Prioritization of Pesticides Application for Potential Human and Environmental Health Risk Assessment in Large Scale Farms in Trans Nzoia County. Kenya: Unpublished Report; 2022

[57] Oparaeke AM, Dike MC, Amatobi CJ. Insecticidal efficacy of Sabruka formulations as protectants of cowpeas against field pests. Journal of Entomology. 2006;**3**:130-135

[58] D'Silva C, Krishna B. Rodenticide poisoning. Indian Journal of Critical Care Medicine. 2019;**33**(4):S272-S277

[59] Rumbeiha WK, Snider DB. Veterinary Toxicology, In: Encyclopedia of Toxicology. 3rd ed. Amsterdam: Science Direct; 2014

[60] Lalah JO, Wandiga SO. Persistence and Fate of Malathion residues in stored beans (Phaseoulus Vulgaris) and maize (Zea Mays). Pesticide Science. 1996;**46**:215-220

[61] USEPA. Epa.gov/ozone-layer-protection/ozone depleting substances. 2022. [Accessed: April 26, 2022]

[62] Cheke RA, Sidatt ME. A review of alternatives to fenthion for quela bird control. Crop Protection. 2019;**116**:15-23

[63] Bustinduy AL, King CH. Schistosomiasis in Manson's Tropical Intectious Diseases. 23rd ed. Amsterdam: Science Direct; 2014

[64] Hodgson E, Levi PE. In: Krieger RI, editor. Metabolism of Pesticides, in: Handbook of Pesticide Toxicology. 2nd ed. Vol. 2. San Diego: Academic Press; 2001

[65] Usmani K, Karoly ED, Hodgson E, Ros RL. In vitro sulfoxidation of thioether compounds by cytochrome

P450 and flavin-containing monooxygenase isoforms with particular reference to the CYP2C subfamily. Drug Metabolism and Disposition: The Biological Fate of Chemicals. 2004;**32**(3):333-339

[66] Jing T-X, Tan Y, Ding B-Y, Dou W, Wei D-D, Wang J-J. NADPH–cytochrome P450 reductase mediates the resistance of *Aphis* (*Toxoptera*) citricidus (Kirkaldy) to abamectin. Frontiers in Physiology. 2018;**9**:986

[67] PCPB. The Pest Control Products Act, Chapter 346, Laws of Kenya, Revised Edition, 1985. Nairobi: Government Printer; 1985

[68] Ngaruiya. Overview of registration of pesticides in Kenya. In: Wabule MN, Ngaruiya PN, Kimmins FK, Silverside PJ, editors. Registration for Biocontrol Agents in Kenya. Proceedings of the PCPB/KARI/DFID CPP Workshop; Nakuru, Kenya; 14-16 May 2003. pp. 79-85

[69] Retkute R, Hinton RGK, Kressman K, Gilligan. Regional differences in control operations during 2019-2021 Desert locust upsurge. Agronomy. 2019;**221**(11):2529

[70] Sharma A, Kumar V, Shahzad B, et al. Worldwide pesticide usage and its impacts on ecosystem. S. N. Applied Sciences, Article number 1446, Springer. 2019. DOI: 10.1007/542452-019-1485-1

[71] Label Review Manual. US Environmental Protection Agency, Office of Pesticide Programs. Washington, DC: U.S. Government Printing Office; 1998. Available from: https://www.epa.gov/pesticide-registration/label-review-manual

[72] UNEP. Globally Harmonized System of Classification and Labelling of Chemicals (GHS): Seventh. Revised ed. New York: UN; 2017. DOI: 10.18356/e9e7b6dc-en

[73] Fishel FM. What are inert ingredients. Gainsville. University of Florida Institute of Food and Agriculture Sciences. 2010 https://edis.ifas.ufl.edu/pi231

[74] Kwiatkowska M, Huras B, Bukowska B. The effect of metabolites and impurities of glyphosate on human erythrocytes (in vitro). Pesticide Biochem Physiology. 2014;**109**:34-43

[75] Yusoff SNM, Kamari A, Aljafree NFA. A review of materials used as carrier agents in pesticide formulations. International Journal of Environmental Science and Technology. 2016;**13**:2977-2994

[76] Fishel FM. Pesticide formulations. Gainsville: University of Florida Institute of Food and Agriculture Sciences. Available from:; 2010 https://edis.ifas.ufl.edu/pi231

[77] Roberts DM, Buckley NA, Mohamed F, Eddleston M, Goldstein DA, Mehrsheikh A, et al. A prospective observational study of the clinical toxicology of glyphosate-containing herbicides in adults with acute self-poisoning. Clinical Toxicology (Phila.). 2010;**48**:129-136. DOI: 10.3109/15563650903476491

[78] Bahadir M, Pfister G. Controlled release formulations of pesticides. In: Haug G, Hoffmann H, editors. Controlled Release Biochemistry Effects of Pesticides, Inhibition of Plant Pathogenic Fungi, Chemistry of Plant Protection. Vol. 6. Berlin: Springer; 1990. DOI: 10.1007/978-3-642-46674-8_1

[79] Tsuji K. Microencapsulation of pesticides and their improved handling safety. Journal of Microencapsulation. 2001;**18**(2):137-147

[80] Nyangi M. Efficacy of Crude Onion Extracts and Yellow Sticky Trap in the Control of Leafminers Liriomyza Trifolii in Capsicum. BTech Industrial Chemistry Project, Nairobi: Technical University of Kenya; 2015

[81] Lancaster. https://lancaster.unl.edu/pest/roach.Chapter 7: Insecticide Baits. Visited 26th April 2022

[82] Jordan BW, Bayer BE, Koehler PG, Pereira RO. Bait evaluation methods for urban pest management. London, UK: Intechopen.; 2013. DOI: 10.5772/5342

[83] Oesterlund A, Thomson JF, Sekimpi DK, Maziina J, Apio R, Joers E. Pesticide knowledge, practice and attitude and how it affects the health of small scale farmers in UgandaL a cross sectional study. African Health Sciences. 2014;**14**(2):420-433

[84] KOAN. Kenya Organic Agriculture Network: Pesticide use in Kirinyaga and Muranga Counties. 2020.Available from: https://www.koan.co.ke

[85] Hayes TB, Khoury V, Narayan A, et al. Atrazine induces complete femninization and chemical castration in male African clawed frogs (Xenopus Laevis). Proceedings of National Academy of Science of USA. 2010;**107**(10):4612-4617. DOI: Doi.org/10.1073/pnas.0909519107

[86] Nyamu D, Maitai CK, Mecca LW, Mwangangi EM. Trends of acute poisoning cases occurring at the kenyatta national Hospital, Nairobi, Kenya. East and Central African Journal of Pharmaceutical Sciences. 2012;**15**(2012):29-34

[87] Pamela F, Tsimbiri PF, Moturi WN, Sawe J, Henley P, Bend JR. Health impact of pesticides on residents and horticultural workers in the lake naivasha Region Kenya. Occupational Diseases and Environmental Medicine. 2015;**2015**(3):24-34. DOI: 10.4236/odem.2015.32004

[88] Gyawali K. Pesticide uses and its effects on public health and environment. Journal of Health Promotion. 2018;**6**:28-36

[89] Otieno PO, Lalah JO, Virani M, Jondiko IO, Schramm K-W. Carbofuran and its toxic metabolites provide forensic evidence for Furadan exposure in Vultures (Gyps africanus) in Kenya. Bulletin Environmental Contamination Toxicology. 2010;**84**(5):536-544

[90] Koech SJ, Karanja RHN, Kurgat J.K, Mokaya HO, Dubois T, Lattorff MG. Pesticide Contamination and their botanical sources in pollen loads collected by honey bees in Kenya: A spatio-temporal context (in press). 2022

[91] Kenya Plant Health Inspectorate Service (KEPHIS). Annual Report and Financial Statement. Kenya: Nairobi; 2018

[92] Otieno PO, Okinda PO, Lalah JO, Pfister G, Schramm K-W. Monitoring the Occurrence and Distribution of Selected Organophosphates and Carbamate Pesticide Residues in the Ecosystem of Lake Naivasha. Kenya: Toxicological & Environmental Chemistry; 2014. DOI: 10.1080/02772248.2014.942309

[93] Getenga ZM, Keng'ara FO, Wandiga SO. Determination of organochlorine pesticides in soil and water from river Nyando Drainage System within Lake Victoria Basin, Kenya. Bulletin of Environmental Contamination. 2004;**72**(2):335-343

[94] Abong'o DA, Wandiga SO, Jumba IO, Van den Brink PJ, Nazariwo BB, Madadi VO, et al. Organochlorine pesticide residue levels in

soil from the Nyando River Catchment, Kenya. Africa Journal of Physical Sciences. 2015;**2**(1):18-32

[95] Omwenga I, Kanja L, Zomer P, Louisse J, Rietjens IMCM, Mol H. Organophosphate and carbamate pesticide residues and accompanying risks in commonly consumed vegetables in Kenya. Food Additives & Contaminants Surveillance: Part B. 2021;**14**:48-58

[96] FAO/WHO. International Code of Conduct on Pesticide Mangement. Managing pesticides in agriculture and public health: A compendium of FAO and WHO guidelines and other resources. 2nd ed. FAO/WHO 2021. 2021

[97] GoK. Government of Kenya, Ministry of Agriculture, Livestock and Fisheries report: Capacity Building Strategy for Agruclture Sector, November, 2017. 2017

[98] Chen Y, Yan C, Sun Z, Wang Y, et al. Organochlorine pesticide ban facilitated reproductive recovery of Chinese striped Hamsters. Environmental Science & Technology. 2021;**55**(9):6140-6149

[99] Dabrowski DM, Shadung JM, Wepener V. Prioritizing agricultural pesticides used in South Africa based on their environmental mobility and potential human health effects. Environment International. 2014;**62**:31-40

[100] Valcke M, Chaverri F, Monge P, Bravo V, Mergler D, Partanen T, et al. Pesticide prioritization for a case–control study on childhood leukemia in Costa Rica: A simple stepwise approach. Environmental Research. 2005;**97**(3):335-347

[101] Gunier RB, Harnly ME, Reynolds P, Hertz, Behren V. Agricultural pesticide use in california: Pesticide prioritization, use densities, and population distributions for a childhood cancer study. Environmental Health Perspectives. 2001;**109**:1071-1078. Avilable from: http://ehpnet1.niehs.nih.gov/docs/2001/109p1071-1078gunier/abstract.htm

[102] Gustafson DI. Groundwater ubiquity score: A simple method for assessing pesticide leachability. Enviromental Toxicology and Chemistry. 1989;**8**:339-357. DOI: 10.1002/etc.5620080411

[103] Chen H et al. Estrogenicity of organophosphorus and pyrethroid pesticides. Journal of Toxicology and Environmental Health Part A. 2002;**65**:1419-1435. DOI: 10.1080/00984100290071243

[104] Marete GM, Shikuku VO, Lalah JO, Mputhia J, Wekesa VW. Occurrence of pesticides residues in French beans, tomatoes, and kale in Kenya, and their human health risk indicators. Environmental Monitoring and Assessment. 2020;**192**:692. DOI: 10.1007/s10661-020-08662-y

[105] Soko JJ. Agricultural Pesticides used by small-scale farmers in Kenya: What are the experiences of the farmers? ORCiD: https://orcid.org/0000-0001-7427-9188. 2020. DOI: 10.21203/rs.3.rs-20784/v1

[106] MOA. Ministry of Agriculture, Livestock, Fisheries and Irrigation National Agricultural and Rural inclusive Growth Project (NARIGP), "Integrated Pest Management Plan,". Nairobi: Government of Kenya; 2018. p. 2018

[107] Loha KM, Lamoree M, Weiss JM, de Boer J. Import, disposal, and health impacts of pesticides in the East Africa Rift (EAR) zone: A review on management and policy analysis. Crop Protection. 2018;**112**(2018):322

[108] World Vision International. Fall armyworm invasion in Kenya. World Vision International report 2017

[109] Macharia I, Mithofer D, Waibel H. Potential environmental impacts of pesticide use in the vegetables. Africa Journal of Horticultural Science. 2009, 2009:1998-9326

[110] Macharia I. Pesticides and health in vegetable production in Kenya. BioMed Research International. 2015;**2015**(241516):10. DOI: 10.1155/2015/241516

[111] Mburu NS, Matuku T, Osano O, Gichuho CM. Pesticide Preferences and patterns of use along the shore of Lake Naivasha, Kenya. Greener Journal of Environmental Management and Public Safety. 2013;**2**(3):115-120

[112] Fuhrimann S, Wan C, Blouzard E, Veludo A, Holtman Z, Chetty-Mhlanga S, et al. Pesticide research on environmental and human exposure and risks in sub-saharan africa: A systematic literature review. International Journal of Environmental Research and Public Health. 2022;**19**:259. DOI: 10.3390/ijerph19010259

[113] Farag AT, Eweidah MH, El-Okazy AM. Reproductive toxicology of acephate in male mice. Reproductive Toxicology. 2000;**14**(5):457-462. DOI: 10.1016/S0890-6238(00)00094-0

[114] FOOTPRINT, 2006. The Footprint Pesticide Properties Database. Database collated by the University of Hertfordshire as part of the EU-funded Footprint project (FP6-SSP-022704). http://www.eu-footprint.org/ppdb.html.

[115] Expert Taskforce. Scientific Report on Pesticides in the Kenyan Market. Contributers: Biodiversity and Biosafety Association of Kenya, Kenya Organic Agricultural Network, Resources Oriented Development Initiatives and Route for Food Initiative, September 2021. 2021

[116] WHO (2018). WHO | World malaria report 2018 [Internet]. WHO. 2018 [cited 2019 Oct 14]. Available from: http://www.who.int/malaria/publications/world-malaria-report-2018/en/

[117] WHO (2003). WHO | Guidelines on the management of public health pesticides. Available from: http://www.who.int/whopes/resources/who_cds_whopes_2003.7/en/. [Accessed: October 22, 2019]

[118] WHO (2006). Guidelines for the treatment of malaria [Internet]. 2006 [cited 2019 Oct 22]. Available from: http://archives.who.int/publications/2006/9241546948_eng.pd

[119] Bayoh MN, Mathias DK, Odiere MR, Mutuku FM, Kamau L, Gimnig JE, et al. Anopheles gambiae: Historical population decline associated with regional distribution of insecticide-treated bed nets in western Nyanza province, Kenya. Malaria Journal. 2010;**9**(1):62. DOI: 10.1186/1475-2875-9-62

Chapter 12

# Extraction and Identification Techniques for Quantification of Carbamate Pesticides in Fruits and Vegetables

*Nasir Md Nur 'Aqilah, Kana Husna Erna, Joseph Merillyn Vonnie and Kobun Rovina*


## Abstract

The usage of carbamate pesticides in agriculture is increasing year by year. Carbamate pesticides are thioesters and esters, which are derived from aminocarboxylic acid. Carbamates are commonly utilized to improve agricultural production and protect humans and animals from disease. They were also used to control and prevent agricultural pests. However, carbamate can be highly toxic if not applied properly. Therefore, carbamate pesticides need to be monitored in fruits and vegetables. Sensitive and selective detection of carbamate pesticides using nanotechnology helps overcome the drawback of conventional methods of detecting carbamates. Nowadays, the demand for rapid, highly sensitive, and selective pesticide detection techniques is expanding to facilitate detection without complicated equipment. Due to this, this chapter focuses on nanotechnology and current detection methods for detecting residual carbamate pesticides in fruits and vegetables more precisely and faster.


**Keywords:** carbaryl, carbofuran, toxicology, recent approaches, traditional techniques, fresh produce

## 1. Introduction

A pesticide is a hazardous chemical compound or a mixture of biological agents or chemicals that are deliberately presented into the environment to prevent, dissuade, eliminate, or control populations of insects, rodents, weeds, fungus, or other unwanted pests. Pesticides play an important role in attracting, enticing, and killing or repelling organisms. Generally, pesticides are widely applied and reported at approximately 5.2 billion pounds per year to reduce various harmful species such as microscopic fungi, weeds, rodents, and insects. Pesticide is highly applicable for pest control in agricultural areas and households to control mosquitoes, ticks, cockroaches, rats, fleas, and other dangerous creatures [1]. Using pesticides improves crop yields by controlling pathogenic microorganisms, resulting in better consumption of fresh fruits and vegetables [2]. There are four types of pesticides, namely organochlorines, carbamates, organophosphates, and pyrethroids, illustrated in **Figure 1** with their chemical structures.

Carbamate pesticides are known as esters of carbamic acid ($R^1$-S-CO-$NR^2R^3$), which are not structurally complex. They are commonly employed in farming to protect many crops, including fruits, cotton, rice, and vegetables, due to their broad biological activity, less mammalian toxicity, and minimal bioaccumulation potential [3]. Besides, it was applied as a therapeutic drug in human medicine and veterinary medicine. Carbamate has a high polarity, is water-soluble and thermodynamically unstable, which contains insecticides like carbaryl, acaricides, and fungicides [4]. Previous research found that carbamate pesticides are capable absorb in the food source's tissues such as fish, poultry, and meat, in processed foods such as vegetables, nuts, dehydrated fruits, and vegetable oils [5]. Based on FAO and WHO, in 2016, Codex Alimentarius Commission for carbamate maximum residue levels was set up to 4844 but required the presence of different combinations of pesticides. However, in European Union, carbaryl was banned in most countries [6]. This is because the carbamate residual in foods functions as acetylcholinesterase inhibitors, which can damage the brain, nervous systems, liver, muscles, and pancreas over the long term [7, 8].

It is essential to track and measure the carbamate amounts in fresh products and improve the sensitivity of the detection methods that have been developed.

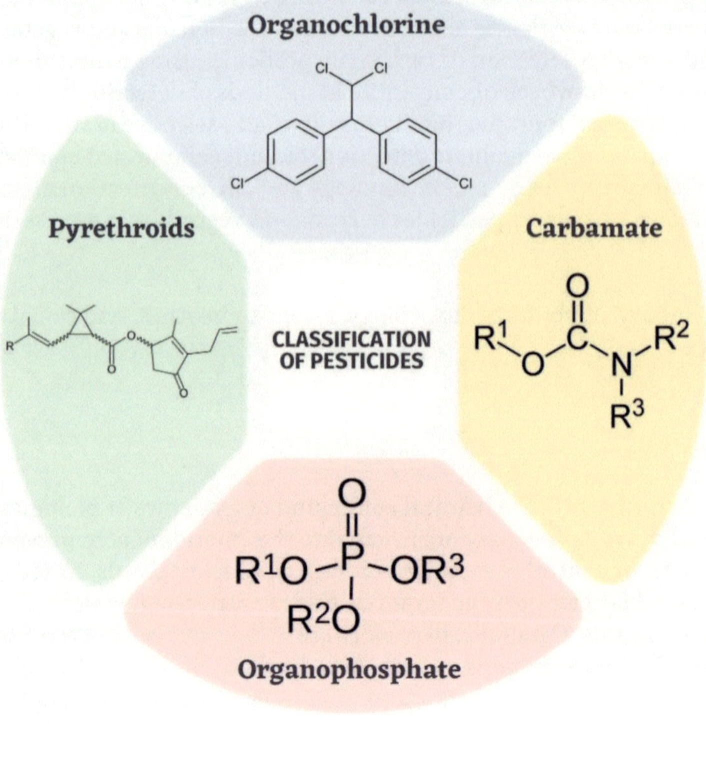


**Figure 1.**
*Classification of pesticides with chemical structure.*

Previously, conventional methods such as chromatography, immunoassay, and surface-enhanced Raman spectroscopy (SERS), have been applied and show reliability and sensitivity to determine the presence of carbamate. However, these techniques are typically insufficient for real-time and on-site detection, which necessitates advancements in terms of preparation time and cost of machinery and highly skilled workers [9]. Hence, the development of advanced nanotechnology is one of the alternative methods that show rapid, low-cost, easy to use, and capable of detecting low concentrations of carbamate in food samples. This chapter focuses on the latest information on sample pretreatment and analytical detection strategies available from 2000 to 2021. Also, we highlight the reader with an understanding of some innovative ways to increase carbamate pesticides detection in food products.

## 2. Types of carbamate pesticides

Carbamate is an N-methyl produced from carbonic acid, responsible for the carbamylation of acetylcholinesterase at neuromuscular junctions in the brain and spinal cord and at neuronal synapses. Carbamate is classified as an insecticide that is physically and mechanistically comparable to organophosphate (OP) insecticides in both structure and mechanism of action. Carbamates have a reversible binding to acetylcholinesterase and do not cause the irreversible phosphorylation of the enzyme that occurs when organophosphates interact with it [10]. Consequently, carbamates are toxicologically similar to OP poisoning, with a toxic period of fewer than 24 hours [11]. Aldicarb, carbaryl, carbofuran, bendiocarb, fenobucarb, methomyl, oxamyl, propoxur, and methiocarb are the most common agents that lead to dangerous exposure. **Figure 2** below illustrates the chemical structure of carbamate pesticides available in agriculture applications.

Carbaryl is a member of the chemical family N-methyl carbamate and was discovered in 1959 for use as a carbamate pesticide in cotton in the United States. Carbaryl is a popular insecticide in agriculture, specialist turf control, ornamental production, and residential settings. Carbaryl is mildly toxic when taken orally and has low toxicity when applied topically or inhaled [12]. In outdoor conditions, carbaryl has a low persistence rate. Human exposure occurs by ingestion of residues in food, skin contact, and inhalation of airborne particles. Carbaryl blocks acetylcholinesterase in the neurological system, causing acetylcholine buildup and cholinergic hyperstimulation. In contrast to adults, immature organisms are more sensitive to the inhibition of cholinesterase (ChE). In addition to reproductive and developmental toxicity, carbaryl can also alter the immune system. It may also cause cancer in humans and be highly harmful to non-target organisms [13, 14].

Aldicarb is a carbamate insecticide active against insects, mites, and nematodes belonging to the chemical family of N-methyl carbamates. Aldicarb is water-soluble at pH 7 and a colorless crystalline substance that acts as a cholinesterase inhibitor, soil contaminant, carcinogen, and a possible endocrine disruptor. Aldicarb is acutely toxic and causes cholinergic symptoms by inhibiting acetylcholinesterase (AChE), neither genotoxic nor cancer-causing. Much information about toxicity includes developmental, long-term, short-term, reproductive, and neurotoxic studies. They are dose-dependent, rapidly reversible, and do not manifest at levels of human exposure predicted [15]. The toxicity of aldicarb is evident in even small doses with stomach cramping, dizziness, nausea, diarrhea, and convulsions [16, 17].

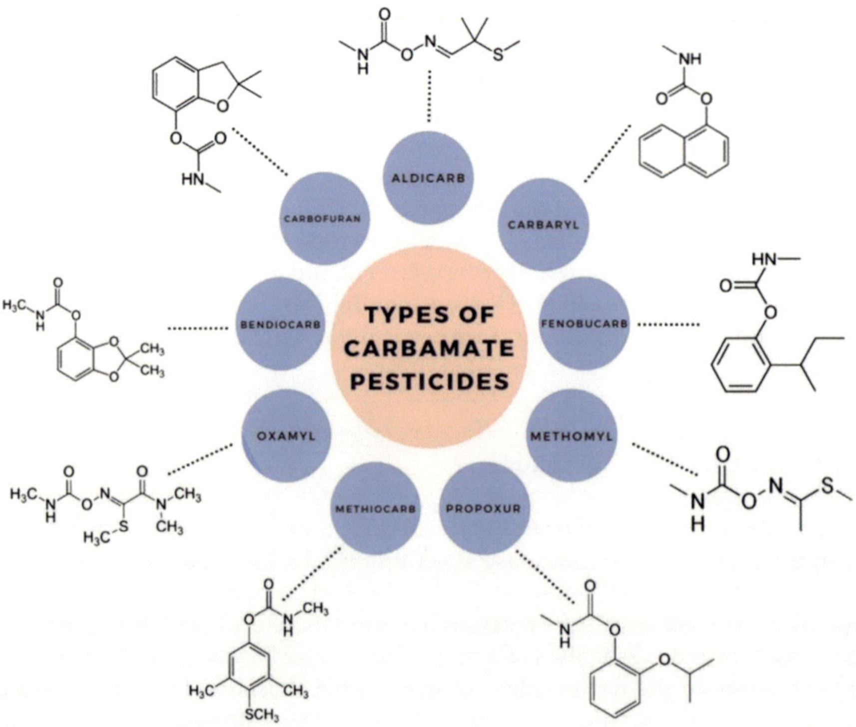


**Figure 2.**
*Chemical structure of types of carbamate pesticides.*

Carbofuran is a wide-spectrum of N-methyl carbamate insecticide commonly used in farming to combat insects, nematodes, and mites in soil or protect forest crops, fruit, and vegetables. It is incredibly toxic to birds, mammals, fish, and wildlife due to its anticholinesterase action that inhibits acetylcholinesterase and butyrylcholinesterase. Carbofuran can disrupt the neuroendocrine system, cause reproductive disorders, and be genotoxic and cytotoxic to humans [18]. However, it did not affect a humoral immune response [15]. Besides, it is a relatively unstable chemical that degrades in weeks or months. Recently, Amatatongchai et al. [19] found carbofuran in potatoes, corn, soybean, fruits, and vegetables. Similarly, Lan et al. [20] detected carbofuran in watermelon, long bean, mango, and chives samples.

Methomyl is known as metomil or mesomile, commonly used to treat crops. It is a colorless crystalline structure soluble in organic solvents and water, which may pollute the environment. It has a wide application in biological activities and is efficient against insects [21]. Methomyl is categorized as a harmful and dangerous pesticide by the World Health Organization and the European Union [22]. Acetylcholinesterase (AChE) is inhibited by methomyl lead in a reduction of the ability of the enzyme to hydrolyze acetylcholine that buildup in the body. The most common signs of methanol include tearing of the eyes, vomiting, nausea, stomach pain, diarrhea, loss of consciousness (coma), and death due to respiratory failure [23–25]. The endocrine system is also affected by methomyl because of its capability to influence estrogen production and reproductive capabilities [26]. Presently, Guo et al. [27] identified methomyl residue in barley, millet, wheat, and rice grains. Besides, Rasolonjatovo et al. [28] found methomyl residues in tomatoes.

Methiocarb is a carbamate pesticide that colorless, crystalline substance sparingly soluble in water and xylenes. However, it is unstable in alkaline media (pH 9). Methiocarb is a contact wide-spectrum, a residual insecticide which acts as a molluscicide, acaricide, and bird repellent since the 1960s [13]. Methiocarb is used on fruit crops and orchids to control snails and rice insects [29]. Sivaperumal et al. [30] found the methiocarb residues in mango fruits. The molecule is oxidized sequentially to sulfoxide and sulfone in the vertebrate liver. Methiocarb sulfoxide is also available in methiocarb sulfone in the form of iocarb sulfone and the combination known as methiocarb [31].

The chemical name for propoxur is 2-isopropoxyphenyl-N-methylcarbamate with a molecular weight of 209.24, which is hydrolyzed by strong alkali. Propoxur is unstable in alkaline media and has a half-life at a pH of 10 for 40 minutes. It is a non-systemic insecticide primarily used against household insect pests and domestic animal pests [32]. However, propoxur causes neurotoxicity by inhibiting acetylcholinesterase in a significant reversible manner [33]. Based on Borahan et al. [34], propoxur has been detected in raisins by gas chromatography-mass spectrometry (GC-MS). Besides, Xiao-Xue et al. [35] found propoxur in fruit samples such as plum, pear, and loquat by employing the molecularly imprinted photoelectrochemical sensor.

Through the oral pathway, oxamyl is highly toxic. Like other carbamates, exposure to oxamyl can result in cholinesterase inhibition over a short period [36]. The pure compound has a slightly sulfurous odor and is a white crystalline solid, which melts at 100–102°C and shifts to a different crystalline structure between 108 and 110°C [32]. Yaseen et al. [37] found oxamyl in peach fruit using a surface-enhanced Raman scattering. Bendiocarb is a carbamate insecticide efficient against a broad spectrum of agricultural pests. Bendiocarb is poisonous to fish, birds, and bees, and research has demonstrated that bendiocarb is unable to bioaccumulate in animals [38]. Kowalska et al. [39] stated that terbucarb residues were found in plants, and HP-LC detected it with tandem mass spectrometry (HPLC-MS/MS). Liquid fenobucarb pesticides are pale yellow or pale red. Pelle et al. [40] found fenobucarb residues in grain samples.

## 3. Physical and chemical properties of carbamate pesticide

The straightforward technique to identify carbamate pesticides is to look at their carbamic acid N- or S-substitutions. The carbamates are classified into nine major groups: dithiocarbamates, thiocarbamates, benzimidazole carbamates, N-phenyl carbamates, ethylenebisdithiocarbamate, N,N-methyl carbamates, N-methyl carbamates, aminophenyl N-methylcarbamates, and oxime N-methylcarbamates [41]. Carbamates are typically insoluble in water molecules because it has low solubility in polar organic solvents, ethanol, or acetone. Carbamate is a polar molecule soluble in solvents with a medium polarity, including benzene, chloroform, toluene, xylene, dichloromethane, or 1,2-dichloromethanebut are insoluble in nonpolar organic solvents [42, 43]. Pure carbamate pesticides are crystalline, white, practically odorless solids with low vapor pressure and high melting point. Carbamate pesticide features include physical form, melting point, vapor pressure, and solubility [41].

## 4. Toxicology of carbamate pesticide

Carbamates are carbamic acid esters substituted for N-methyl carbamic acid that act as AChE inhibitors to catalyze acetylcholine (ACh). The reaction enhanced the

ACh level at a nerve synapse or neuromuscular junction, raising nerve-ending stimulation by reversible cholinesterase inhibition [44]. In contrast to organophosphates, the cholinesterase-inhibiting action of carbamates is reversible. Carbamates are toxic to rodents in doses ranging between LD50 > 200 mg/kg and LD50 > 50 mg/kg [45]. According to the classification system, the US Environmental Protection Agency and the World Health Organization (WHO) have classified carbamate as class II (moderate). Several additional factors, such as route and frequency of exposure, interactions with other impurities, and compromised physiological conditions, such as liver impairment, may all impact the level of toxicity [25, 46]. Besides, WHO includes carbamates on its endocrine-disrupting chemicals (EDCs), potentially harmful to animals and human health [47]. They discovered that EDCs might disrupt hormone production, transport, metabolism, and elimination, with developmental, behavioral, and reproductive effects resulting from these hormone-active compounds. De Coster and Van Larebeke [48] examined the endocrine-disrupting properties of chlorpropham, carbaryl, benomyl, methiocarb, pirimicarb, and propamocarb by highlighting various pathways, including nuclear receptor activation, estrogen-associated receptor activation, and membrane-bound estrogen-receptor activation, among others.

High-potential AChE-inhibitors have been utilized as toxicants, but low-potential AChE-inhibitors have been used as prevention agents against nerve agents or as therapeutic agents in treating illnesses such as glaucoma, Alzheimer's disease, and myasthenia gravis, among other things [49]. The primary benefits of carbamate are its intense insecticidal action and poor durability since it degrades swiftly within weeks or months after being applied to crops. Carbamates are effective against a wide range of pests by blocking the enzyme cholinesterase, causing neurotoxicity, and interfering with the nervous system of the pests [50]. These chemicals also exhibit a range of neurotoxic effects not mediated by a cholinergic mechanism. Carboxylated acetylcholinesterase enzyme is a volatile version of the enzyme, and regeneration of this enzyme is comparatively quick when contrasted with the regeneration of a phosphorylated form of the enzyme [51]. Carbamates produce mild eye irritation and moderate skin irritation, depending on the specific vehicle employed, the duration of contact, and the substance applied directly to the skin that has been harmed or is in good condition, according to the manufacturer [51, 52].

## 5. Extraction techniques of carbamates pesticides

The separation of pesticides is necessary from the sample before introducing into the instrument. This approach is expected to limit measuring interferences while enhancing the analyte concentration for research. Besides, the extraction method is a standard procedure that begins with releasing a preferred analyte from matrices and ends with a purification procedure, which directs to a series of stages via the analytical approach wherein a high proportion of potential interference co-extracts is eliminated using chemical or physical means [53]. Liquid-liquid extraction (LLE), solid-phase extraction (SPE), solid-phase microextraction (SPME), quick, easy, cheap, effective, rugged, and safe, microwave-assisted extraction (QuEChERS), and microwave accelerated selective Soxhlet extraction are among the extraction technologies available.

### 5.1 Liquid-liquid extraction (LLE)

Liquid-liquid extraction (LLE) has become a standard procedure in sample preparation due to its convenience and efficacy for insecticide contamination of food [54]. However, LLE requires a lot of solvents, which is terrible for the environment compared to solventless extraction technologies like solid-phase microextraction. On the other hand, the LLE approach is poor in yield analyte concentration, laborious, and requires a significant volume of toxic organic solvents [55]. Previously, liquid-liquid extraction/low-temperature purification incorporated with HPLC-UV was applied for determining aldicarb, carbofuran, and carbaryl in water samples. The separation for the carbamates aldicarb, carbofuran, and carbaryl show a high recovery rate. Although in small amounts of material and solvent, the extraction method was selective, with a limit of detection was found 5.0 and 10.0 g $L^{-1}$ [56].

### 5.2 Solid-phase extraction (SPE)

Solid-phase extraction (SPE) was initially presented during the 1970s, then widely accessible in 1978. At the moment, the most often used widely is SPE procedures for the pretreatment of environmental materials [55]. SPE is simpler, acceptable, and convenient than traditional LLE. Wang et al. [57] recently published an SPE technique utilizing porous organic polymers as an absorbent to extract isoprocarb, metolcarb, bassa, carbaryl, and lastly, diethofencarb, from white wine, milk, and juice before HPLC-diode array detection. The findings showed that milk and white wine samples have excellent linearity, with low detection limits for milk, white wine, and juice samples.

Earlier, Li et al. [58] used a simple one-step synthesis technique to make graphene-based magnetic nanoparticles by using MSPE to detect trace carbamate insecticides in tomatoes. Under ideal conditions, this technique has high enrichment factors, good linearities, low detection, and satisfactory spiking recoveries. The findings show that this approach was an adequate preparation and enhancement approach that may be used to extract and determine trace carbamate pesticides in complicated matrices. Besides that, Shi et al. [59] used graphene-based solid-phase extraction with ultra-HPLC-tandem mass spectrometry to analyze carbamate in ambient water samples. The LOD ranged from 0.5 to 6.9 ng $L^{-1}$, with relative standard deviations of 5.54%. The graphene-packed SPE cartridge may be reused over 100 times for a typical solution after proper regeneration with no appreciable performance degradation. The target analytes' has good enrichment values, which indicate that the developed approach successfully determined carbamate pesticide residues in ambient water samples.

### 5.3 Solid-phase microextraction (SPME)

Solid-phase microextraction (SPME) is a technology that is a highly selective, sensitive, and solvent-free sample and is frequently used to extract volatile and semi-volatile chemicals by its absorption fibers. The range of SPME coatings available, dependent on the analytes' polarity, results in high sensitivity and selectivity because of the strong coating affinity for particular analytes that build up in the environment until they reach equilibrium [60]. Zhou and Fang [61] developed a graphene-modified $TiO_2$ nanotube array by electrodeposition utilizing a cyclic voltammetric reduction approach to detect carbamate. When utilized in $TiO_2$ nanotube arrays for MSPE, the combination of graphene's adsorptive solid properties and its higher extraction capabilities results in remarkable sample preconcentration performance. These results indicate that graphene-modified $TiO_2$ nanotube arrays have a high capacity for adsorption of contaminants. The technique demonstrates a quick and efficient alternative analytical solution for detecting and quantifying carbamate in fruits and vegetables.

### 5.4 QuEChERS (quick, easy, cheap, effective, rugged, safe) extraction

Quick, easy, cheap, effective, rugged, safe (QuEChERS) is a sensitive food analysis technology that has undergone numerous revisions and advancements. QuEChERS is a two-stage technology employed to detect carbamate residues in foods that includes salting-out partitioning, which involves the transition between an aqueous and an organic layer. This technique necessitates further cleaning to remove interfering chemicals by combining magnesium sulfate with various sorbents like C18, graphitized carbon black (GCB), or primary-secondary amines (PSA). It may be used to clean a variety of complex substances like food products while also allowing for a less organic solvent [62]. Due to its numerous advantages, the QuEChERS technique has gained massive attention and is widely utilized and regarded as a preferable approach for measuring toxic contaminants in foods.

Previously, Anastassiades et al. [63] introduced the QuEChERS technique to extract carbamate from food matrices by using a small quantity of acetonitrile, followed by a clean-up step employing DSPE. This method was first used to examine fruits and vegetables. Nonetheless, recent research adapted QuEChERS and used dried samples, animal-based food, cereal, milk-based products, and soil-sediment analysis [64]. The approach is based on analyte extraction in buffered acetonitrile (MeCN) and subsequent separation by salting out and d-SPE. The primary disadvantage of this technique is that the natural elements of the sample must be removed. Based on a study by Zhang et al. [65], they adopted LC-MS/MS to assess 60 different insecticide contaminants in cinnamon bark using a repeated dispersive SPE with QuEChERS.

Some studies reported that almost 54 pesticides residues were extracted and analyzed by acetonitrile. Furthermore, Reddy and Reddy [66] employed QuEChERS to extract pesticides from sunflower oil using modified charcoal to reduce fat and pigment thermal deterioration during analysis. Furthermore, according to Neufeld et al. [67], QuEChERS extraction has a high sensitivity to organophosphates and carbamates. Besides, the QuEChERS technique combined with magnetic SPE and DLLME was developed to remove pesticides from high-solid vegetable, fruit, and nectar samples [68].

### 5.5 Microwave-assisted extraction (MAE)

Environmental Canada pioneered microwave-assisted extraction (MAE), which is currently used in research applications and industrial settings. This approach employs microwave radiation to induce polar molecules and ions to migrate and dipoles to spin to heat solvents and assist the transfer of the target from the food matrix to the extractant [69]. According to Wang et al. [70], the significant edges of adopting MAE are reducing the time extraction, which could be assigned to the differences in the microwave and traditional heating performance. MAE also allows for on-the-fly connection to different analytical processes and the simultaneous execution of several samples. A quick and straightforward analytical method based on LC-MS/MS has been established to measure carbamate residues and mycotoxins in apples using MAE simultaneously. In the recovery rate range of 70–116%, the technique displayed strong linearity with high acceptable accuracy and a lower limit of detection [71].

### 5.6 Microwave accelerated selective Soxhlet extraction (MA-SSE)

Microwave accelerated selective Soxhlet extraction (MA-SSE) is a technique similar to traditional Soxhlet extraction but employs microwaves to improve the procedure [72]. Although MA-SSE is fast and effective, its poor selectivity requires additional cleaning operations. Besides, a selective MA-SE approach is required due to its time-consuming and labor-intensive nature. Zhou et al. [72] employed MA-SSE as a selective extraction strategy in their investigation to detect the carbamate contaminants in ginseng. The MA-SSE extracts the sample's target analytes and interfering components using microwave-irradiated extraction solvent. After the solvent passed through the extraction container, the sorbent adsorbed the interfering elements in the solvent and collected the target analytes. Because of the effect of microwave irradiation, MA-SSE outperformed conventional extraction processes significantly. According to the findings, MA-SSE has much potential as a fast and reliable method for preparing samples to detect pesticide residue in complex matrices.

## 6. Conventional techniques for detection of carbamate pesticides

Various techniques for identifying carbamate residues are summarized in **Table 1**.

### 6.1 Capillary electrophoresis (CE)

Capillary electrophoresis (CE) is a proper analytical method that could also be applied in various situations and is expected to offer several advantages, including fewer chemicals and samples, higher removal efficiency, and time efficiency. The capillary's inner diameter (50–75 m) is tiny, allowing only a limited sample volume to be injected into the system, thus limiting sensitivity detection. Due to the small volume of sample that can be injected into such a capillary system, CE has been combined with sensitive detection [81] and combined with internet-based-concentration methods. Attig et al. [82] described a microextraction technique for selective preconcentration of N-methyl carbamate in water prior to CE analysis using temperature-controlled IL-DLPME in an alkaline buffer. Microextraction with ionic liquid and elution with a trace amount of dichloromethane was used to obtain the samples. MMWCNTs enhanced ionic liquid-analyte binding and recovery compared to using simple nanomaterials as a sorbent. Cheng et al. [83] developed a CE with amperometric detection based on a polyamide-modified carbon paste electrode to determine carbamate in alkaline water solutions. According to Zhang et al. [84], an efficient method for simultaneous determination of carbamate pesticides in vegetables included solid-phase microextraction for purification and enrichment, followed by CE separation. Standard addition recoveries of 86.1–115.8% for vegetable samples are quick and accurate. The presence of carbamates has been determined using nanomaterials such as graphene and gold nanoparticles in pesticide biosensors [85]. Direct electrodeposition of electrochemically reduced graphene oxide-gold

| Detection method | Carbamate pesticides | Food products | Reference(s) |
|---|---|---|---|
| Surface-enhanced Raman spectroscopy (SERS) | Carbaryl | Orange juice, grapefruit, milk | [73] |
| ELISA immunoassay (IA) | Carbofuran | Cucumbers, apples, leek, sweet potato, potato | [74, 75] |
| Terahertz time-domain spectroscopy (THz-TDS) | Methomyl | Wheat, rice flour | [76] |
| Gas chromatography-mass spectroscopy (GC-MS/MS) | Methiocarb | Cabbage | [77] |
| High performance liquid chromatography (HPLC) | Propoxur | Lemonade, grape juice | [78] |
| Liquid chromatography-tandem mass spectrum (LC-MS/MS) | Aldicarb | Vegetable | [79] |
| Surface-enhanced Raman spectroscopy (SERS) | Oxamyl | Peach, milk | [37, 80] |
| High performance liquid chromatography with tandem mass spectrometry (HPLC-MS/MS) | Terbucarb | Plants | [39] |
| High performance liquid chromatography (HPLC) | Fenobucarb | Lemonade, grape juice | [57] |

**Table 1.**
*Detection techniques of carbamate pesticides.*

nanoparticles-cyclodextrin and Prussian blue-Chitosan modified glass carbon electrodes was used to identify pesticides. Carbamate pesticides inhibit AChE activity, with malathion having a LOD of 4.14 pg $mL^{-1}$ and carbaryl having a LOD of 1.15 pg $mL^{-1}$.

## 6.2 Micellar electrokinetic capillary chromatography (MEKC)

Micellar electrokinetic capillary chromatography (MEKC), a hybrid methodology incorporating chromatographic and electrophoretic extraction principles, extends the usability of capillary electrophoretic procedures to neutral analytes. Surfactants are added to the buffer solution at quantities remarkably different from their essential micellar concentrations, producing micelles that move electrophoretically like any other charged particle. The separation is based on the differential partitioning of an analyte between two-phase systems: the moving aqueous phase and the micellar pseudo stationary phase [86]. Using MEKC with a UV-Vis detector, the best separation conditions were 20 mM phosphate buffer (pH 8.0) and 15 mM sodium dodecyl sulfate. The detecting wavelength was set at 200 nm, with a voltage of 12.5 kV supplied. Baseline separation of five pesticides took 15 minutes under these circumstances with low detection limits. This method produced high repeatability, reproducibility, separation efficiency, and a reasonable recovery rate in rice samples [87]. MEKC has evolved into an effective separation technology for neutral and ionic chemicals in complex mixtures, including a broad spectrum of analytes. MEKC is based on the separation of the micellar and aqueous phases. See et al. [88] originally described a technique for determining glyphosate and aminomethylphosphonic acid in tap and river water using a dynamic supported liquid membrane tip extraction approach followed by MEKC with capacitively linked contactless conductivity detection. Besides, Sung et al. [89] used in-line LLE surface analysis with CE to detect pesticides on solid surfaces of apples. Other research used the SPE-MEKC approach to identify trifloxystrobin, tebufenozide, and halofenozide in foods with detection limits ranging from 0.088 to 0.094 mg/kg [90]. Moreover, Santalad et al. [91] described an SPE-MEKC approach for determining the presence of six carbamate pesticides with low detection limits. Water-soluble CdTe/CdSe core-shell quantum dots were employed to enhance pesticides selective fluorescence enhancement [92]. The baseline separation took 12 minutes, and the detection limits obtained varied from 50 to 180 μg/kg [93]. DLLME coupled with sweeping in MEKC, a quick, easy, and sensitive approach for detecting certain neonicotinoid pesticides in cucumber samples has been devised. Under optimal circumstances, enrichment factors ranging from 4000 to 10,000 were obtained. The method's linearity ranged from 2.7 to 200 ng $g^{-1}$ for thiacloprid, acetamiprid, and imidacloprid in cucumber samples and from 4.0 to 200 ng $g^{-1}$ for imidaclothiz, with the limit of detection varied from 0.8 to 1.2 ng $g^{-1}$. The new approach successfully analyzed neonicotinoid pesticides in cucumbers, promising outcomes [94].

### 6.3 Enzyme-linked immunosorbent assay (ELISA)

Immunochemical techniques, such as enzyme-linked immunosorbent assay (ELISA), have recently gained interest and recognition as rapid and low-cost extraction and detection procedures for pesticide compounds. Based on the antigen-antibody interaction, this analytical technique can give high sensitivity and specificity (selectivity) for particular kinds of pesticides. Additionally, since it can load many samples concurrently, it enables rapid and precise assessment of pesticide residues in agricultural items prior to shipping. Indeed, the primary advantage of ELISA for identifying pesticide residues is the convenience of sample preparation methods [95]. Bellemjid et al. [96] created a rapid ELISA to detect carbamates such as carbendazim and carbofuran using synthetic compounds with acid functions linked with BSA protein and injected into rabbits with antibodies collected for the immunoanalytical test. Zhang et al. [97] used nanobody Nb316 to develop an indirect competitive enzyme-linked immunosorbent test (ELISA) to detect carbofuran in vegetable and fruit samples. A phage display platform was used to extract and characterize unique nanobodies against the pesticide carbofuran from an immunized library. The average recovery rate of spiked samples was 82.3–103.9%, comparable to the conventional UPLC-MS/MS approach.

### 6.4 Gas chromatography-mass spectroscopy (GC-MS)

James and Martin [98] devised the gas chromatography (GC) technology in 1952. The fundamental working concept of gas chromatography is the volatilization of the sample in the input or injector of the gas chromatograph, followed by the separation of the mixture's components in a specially designed column. Pesticide residues were recently found in Chinese liquor using gas chromatography-mass spectrometry [99]. In general, Chinese liquor is an extraction of fermented food. They are a trendy alcoholic beverage in China. In Chinese liquor, ethyl carbamate was found at a detection limit of 0.56 μg/L and a limit of quantification of 1.87 μg/L. Ethyl carbamate was also discovered in Chinese rice wine using gas chromatography-mass spectrometry [100].

According to Yao et al. [101], GC-MS detected ethyl carbamate in grain co-products. A gas chromatography-mass spectrometry assay with the limit of detection of 0.7 ng/g was developed to measure ethyl carbamate extracted from different distillers grains co-products. It was identified in all of the co-products of distillers grains examined in this investigation. The greatest concentration of ethyl carbamate was found in corn condensed distillers solubles, ranging from 1618 to 2956 ng/g. Other kinds of distillers grains co-products exhibited ethyl carbamate concentrations ranging from 17 to 917 ng/g.

## 7. Advanced techniques for detection of carbamate pesticides

In pesticide analysis, advanced technologies are presented as an alternative to the conventional chromatographic methods combined with selective sensors. The chromatographic procedures yielded sensitive, specific, and dependable analytical findings. However, they are time-consuming, complicated, and costly, with a high organic solvent usage, which is unsuitable for analyzing large samples [102]. New approaches are challenging to implement in most developing countries. The advancement of improved methodologies has resulted in promising instruments for easy and fast operation, affordable cost, and suitable for in-situ evaluation. Furthermore, they perform well in terms of pesticide detection accuracy and precision.

### 7.1 Molecular imprinted polymer (MIP) biosensor

Biosensors based on molecularly imprinted polymers (MIP) are widely used as sensitive sensing materials because they detect molecules with many biological weights. MIP has effectively created artificial materials that behave similarly to biological receptors; however, it has limited stability. MIP has also been indicated as a biosensing breakthrough due to its ability to overcome the drawbacks of current specific molecular elements such as antibodies, peptides, and enzymes [103]. MIP is used to detect pesticides by imitating biological receptors, polymerizing a functional monomer in the analyte, and finally removing the template using a polymer matrix [104]. Hence, this approach can detect pesticide residues in food since they are inexpensive, simple to use, and have excellent chemical and physical stability. Recently, Li et al. [105] published a work that demonstrated the construction of a MIPs biosensor to detect pesticides utilizing a carbon paste electrode modified with surface MIP microspheres and evaluated using cyclic voltammetry. The approach used on vegetable samples showed high sensitivity, with significant recoveries ranging from 97.2 to 101%. Additionally, Wang et al. [106] used a MIP sensor modified with polyquercetin(Qu)-polyresorcinol(Re)-AuNPs to assess methyl parathion in waters, juice drinks, and vegetable juice. Nevertheless, the analytical performance of sensors created to detect methyl parathion was lower. Xie et al. [107] detected pesticides in brown rice using MIP sensors and linear sweep voltammetry. Additionally, the MIPs sensor was produced via free-radical polymerization of p-vinylbenzoic acid on the surface of a modified glassy carbon electrode. The study demonstrated that the approach could detect thiamethoxam residues with an 88.7–94.0% recovery range. Li et al. [108] used differential pulse voltammetry to build a MIP-based sensor to analyze paraoxon and exhibited excellent stability after 3 months.

### 7.2 Optical biosensors

Optical biosensors have attracted considerable interest and are being applied in various fields, including food safety and security, biological sciences, environmental sensing, and medical science. The optical characteristics of the optical transducers, including absorption, reflectance, and fluorescence emission, will change in response to the analyte. In many instances, optical biosensors have been used to detect pesticides, especially enzyme-based biomolecules Yotova and Medhat [109] developed an optical biosensor to identify pesticides contaminants based on the parallel immobilization of AChE and choline oxidase enzymes in silicon dioxide hybrid membranes. The bioactive component of the sensor is a multi-enzyme system that includes AChE and choline oxidase covalently immobilized on new hybrid membranes. It demonstrates a constant value of acetylcholine at concentrations ranging from 2.5 to 30 mM. Previously, Xavier et al. [110] studied an optical fiber biosensor for assessing propoxur and carbaryl in vegetable crops, employing chlorophenol red as an optical transducer of the analyte's inhibitory impact on the AChE enzyme. The linear dynamic ranges of carbaryl and propoxur are 0.8–3.0 mg $L^{-1}$ and 0.03–0.50 mg $L^{-1}$, respectively. However, propoxur has a lower detection limit (0.4 ng) than carbaryl in the biosensor (25 ng). Ultrasonic extraction was utilized to detect propoxur in spiked onion and lettuce, with recovery rates ranging from 93 to 95% for onion samples at the different concentration levels studied.

### 7.3 Electrochemical biosensor

Electrochemical biosensors are gaining traction as a novel detection principle, increasing sensitivity, specificity, and repeatability [111]. Biosensors, in theory, are made up of two or three-electrode systems, comprising auxiliary, reference, and working electrodes, that create electrical signals when a target biomolecule interacts with a recognition element [112, 113]. For example, Chauhan and Pundir [114] used iron oxide nanoparticles and carboxylated multi-walled carbon nanotubes nanocomposite-based AChE enzymes. The enzyme AChE was isolated from maize seedlings and covalently attached to a modified gold electrode as a working electrode. The modified gold electrode was developed to measure the presence of different pesticides, including malathion, chlorpyrifos, monocrotophos, and endosulfan in water and milk samples with LODs as low as 0.1 nmol $L^{-1}$.

Similarly, Zhao et al. [84] established direct electrodeposition of electrochemically based reduced graphene oxide-gold nanoparticles-cyclodextrin and Prussian blue-Chitosan modified glass carbon electrodes for pesticide determination. The AChE enzyme was immobilized via adsorption with a low detection limit for carbaryl. An AChE enzyme-based biosensor based on rGO-coated GCE was also created to detect carbamate herbicides in tomatoes with a detection limit of 1.9 nmol $L^{-1}$ [115]. Additionally, Sun et al. [116] have created an amperometric AChE biosensor-based poly (diallyldimethyl-ammonium chloride)-multi-walled carbon nanotubes-graphene hybrid film to evaluate carbaryl in vegetables. Besides, Cesarino et al. [117] used polyaniline and multi-walled carbon nanotubes core-shell modified glassy carbon electrode to construct electrochemical AChE biosensors to measure carbamate pesticides in apple, broccoli, and cabbage. The detection limits for carbaryl and methomyl were 1.4 and 0.95 mol $L^{-1}$, which shows lower than the allowed concentrations indicated by Brazilian regulatory regulations for the pesticides tested in the samples. Besides that, Song et al. [118] detected the carbamate pesticides using citrate-capped gold nanoparticles. The biosensor was made by first creating 3D MPS networks on an Au electrode and then adding citrate-capped AuNPs via an Au–S bond. Based on the inhibitory effect of carbamate insecticides on AChE activity, the pesticide's action may be evaluated at a shallow potential. It was also demonstrated that the method could detect carbamate pesticides in real-world samples.

## 8. Conclusion

Pesticides and other environmental pollutants are being extensively monitored due to their potential threat to humans and agriculture. As a result, multiple methods for assessing pesticide residues in various matrices have arisen. Scientifically, capillary electrophoresis, immunoassay, GC, HPLC, and fluorescence detectors have high sensitivity. On the other hand, the earlier approaches are time-consuming, costly, and need highly skilled personnel. As a result, newer technologies have developed as a feasible choice for determining insecticide contaminant levels. Pesticides are increasingly analyzed using enzyme-based biosensors instead of analytical methods. Experts seek to build low-cost, ecologically friendly technologies as pesticide residues become increasingly urgent. The established enzymatic biosensor methods must be used to detect pesticide residuals below the approved safety level. Nanobiosensors allows for simultaneous monitoring of food products such as packaged food components, fruits, vegetables, juices, and the environment. A single and miniature biosensor that employs nanomaterials has a bright future in pesticides detection.

## Acknowledgements

The authors would like to thank all the researchers involved in the project. This work was supported by the Grant from Universiti Malaysia Sabah (Grant No: PHD0024-2019).

## Conflict of interest

The authors declare no conflict of interest.

## Author details

Nasir Md Nur 'Aqilah, Kana Husna Erna, Joseph Merillyn Vonnie and Kobun Rovina*
Faculty of Food Science and Nutrition, Universiti Malaysia Sabah,
Kota Kinabalu, Sabah, Malaysia

*Address all correspondence to: rovinaruby@ums.edu.my

## References

[1] Kaur R, Mavi GK, Raghav S, Khan I. Pesticides classification and its impact on environment. International Journal of Current Microbiology and Applied Sciences. 2019;**8**(3):1889-1897. DOI: 10.20546/ijcmas.2019.803.224

[2] Duke SO. Interaction of chemical pesticides and their formulation ingredients with microbes associated with plants and plant pests. Journal of Agricultural and Food Chemistry. 2018;**66**(29):7553-7561. DOI: 10.1021/acs.jafc.8b02316

[3] Lushchak VI, Matviishyn TM, Husak VV, Storey JM, Storey KB. Pesticide toxicity: A mechanistic approach. EXCLI Journal. 2018;**17**:1101. DOI: 10.17179%2Fexcli2018-1710

[4] Mustapha MU, Halimoon N, Johar WL, Abd Shukor MY. An overview on biodegradation of carbamate pesticides by soil bacteria. Pertanika Journal of Science and Technology. 2019;**27**(2):547-563

[5] Chung SW, Chen BL. Determination of organochlorine pesticide residues in fatty foods: A critical review on the analytical methods and their testing capabilities. Journal of Chromatography. A. 2011;**1218**(33):5555-5567. DOI: 10.1016/j.chroma.2011.06.066

[6] Fan Y, Lai K, Rasco BA, Huang Y. Determination of carbaryl pesticide in Fuji apples using surface-enhanced Raman spectroscopy coupled with multivariate analysis. LWT- Food Science and Technology. 2015;**60**(1):352-357. DOI: 10.1016/j.lwt.2014.08.011

[7] Karami-Mohajeri S, Abdollahi M. Toxic influence of organophosphate, carbamate, and organochlorine pesticides on cellular metabolism of lipids, proteins, and carbohydrates: A systematic review. Human & Experimental Toxicology. 2011;**30**(9):1119-1140. DOI: 10.1177/0960327110388959

[8] Oliveira TM, Ribeiro FW, Sousa CP, Salazar-Banda GR, de Lima-Neto P, Correia AN, et al. Current overview and perspectives on carbon-based (bio) sensors for carbamate pesticides electroanalysis. TrAC, Trends in Analytical Chemistry. 2020;**124**:115779. DOI: 10.1016/j.trac.2019.115779

[9] Songa EA, Okonkwo JO. Recent approaches to improving selectivity and sensitivity of enzyme-based biosensors for organophosphorus pesticides: A review. Talanta. 2016;**155**:289-304. DOI: 10.1016/j.talanta.2016.04.046

[10] Silberman J, Taylor A. Carbamate toxicity. In: Bookshelf National Center for Biotechnology Information. Treasure Island, FL: StatPearls Publishing LLC; 2018

[11] Vale A, Lotti M. Organophosphorus and carbamate insecticide poisoning. Handbook of Clinical Neurology. 2015;**131**:149-168. DOI: 10.1016/B978-0-444-62627-1.00010-X

[12] Krieger R, editor. Handbook of Pesticide Toxicology, Two-Volume Set. London, United Kingdom: Academic Press; 2001

[13] Koshlukova SE, Reed NR. Carbaryl. Amsterdam: Elsevier; 2014. pp. 668-672. DOI:10.1016/B978-0-12-386454-3.00107-X

[14] Jing X, Wang H, Huang X, Chen Z, Zhu J, Wang X. Digital image colorimetry detection of carbaryl in food samples based on liquid phase microextraction coupled with a microfluidic thread-based

analytical device. Food Chemistry. 2021;**337**:127971. DOI: 10.1016/j.foodchem.2020.127971

[15] Blacker AM, Kelly ID, Lantz JL, Mihlan GJ, Jones RL, Young BM. Aldicarb: Toxicity, exposure and risks to humans. In: Hayes' Handbook of Pesticide Toxicology. 2010. pp. 1619-1632. DOI: 10.1016/B978-0-12-374367-1.00075-6

[16] Paranjape K, Gowariker V, Krishnamurthy VN, Gowariker S. The Pesticide Encyclopedia. United Kingdom: CABI; 2014

[17] Liu L, Suryoprabowo S, Zheng Q, Song S, Kuang H. Rapid detection of aldicarb in cucumber with an immunochromatographic test strip. Food and Agricultural Immunology. 2017;**28**(3):427-438. DOI: 10.1080/09540105.2017.1293015

[18] Mishra S, Zhang W, Lin Z, Pang S, Huang Y, Bhatt P, et al. Carbofuran toxicity and its microbial degradation in contaminated environments. Chemosphere. 2020;**259**:127419. DOI: 10.1016/j.chemosphere.2020.127419

[19] Amatatongchai M, Thimoonnee S, Jarujamrus P, Nacapricha D, Lieberzeit PA. Novel amino-containing molecularly-imprinted polymer coating on magnetite-gold core for sensitive and selective carbofuran detection in food. Microchemical Journal. 2020;**158**:105298. DOI: 10.1016/j.microc.2020.105298

[20] Lan J, Wang M, Ding S, Fan Y, Diao X, Li QX, et al. Simultaneous detection of carbofuran and 3-hydroxy-carbofuran in vegetables and fruits by broad-specific monoclonal antibody-based ELISA. Food and Agricultural Immunology. 2019;**30**(1):1085-1096. DOI: 10.1080/09540105.2019.1664997

[21] Shen S, Chen M, Wang T, Fei T, Yang D, Cao M, et al. Determination of methomyl residue in tobacco samples by heart-cutting two-dimensional liquid chromatography with tandem mass spectrometry. International Journal of Analytical Chemistry. 2020;**2020**: 8813142. DOI: 10.1155/2020/8813142

[22] Derbalah A, Massoud A, El-Mehasseb I, Allah MS, Ahmed MS, Al-Brakati A, et al. Microbial detoxification of dimethoate and methomyl residues in aqueous media. Water. 2021;**13**(8):1117. DOI: 10.3390/w13081117

[23] Fan J, Shao X, Xu H, Feng S. A novel inhibitory kinetic fluorimetric method for the determination of trace methomyl in environmental samples. Luminescence: The Journal of Biological and Chemical Luminescence. 2009;**24**(4):266-270. DOI: 10.1002/bio.1120

[24] Kinoshita H, Tanaka N, Jamal M, Kumihashi M, Okuzono R, Tsutsui K, et al. Application of energy dispersive X-ray fluorescence spectrometry (EDX) in a case of methomyl ingestion. Forensic Science International. 2013;**227**(1-3):103-105. DOI: 10.1016/j.forsciint.2012.08.026

[25] Kim KH, Kwon IH, Lee JY, Yeo WH, Park HY, Park KH, et al. Clinical significance of national patients sample analysis: Factors affecting mortality and length of stay of organophosphate and carbamate poisoned patients. Healthcare Informatics Research. 2013;**19**(4):278-285. DOI: 10.4258/hir.2013.19.4.278

[26] Silva CC, De Souza D. Polished silver solid amalgam electrode and cationic surfactant as tool in electroanalytical determination of methomyl pesticide. Talanta. 2018;**189**:389-396. DOI: 10.1016/j.talanta.2018.07.029

[27] Guo Y, Wang H, Chen Z, Jing X, Wang X. Determination of methomyl in grain using deep eutectic solvent-based extraction combined with fluorescence-based enzyme inhibition assays. Spectrochimica Acta, Part A: Molecular and Biomolecular Spectroscopy. 2022;**266**:120412. DOI: 10.1016/j.saa.2021.120412

[28] Rasolonjatovo MA, Cemek M, Cengiz MF, Ortaç D, Konuk HB, Karaman E, et al. Reduction of methomyl and acetamiprid residues from tomatoes after various household washing solutions. International Journal of Food Properties. 2017;**20**(11):2748-2759. DOI: 10.1080/10942912.2016.1250099

[29] Pietrzak D, Kania J, Malina G, Kmiecik E, Wątor K. Pesticides from the EU first and second watch lists in the water environment. Clean: Soil, Air, Water. 2019;**47**(7):1800376. DOI: 10.1002/clen.201800376

[30] Sivaperumal P, Anand P, Riddhi L. Rapid determination of pesticide residues in fruits and vegetables, using ultra-high-performance liquid chromatography/time-of-flight mass spectrometry. Food Chemistry. 2015;**168**:356-365. DOI: 10.1016/j.foodchem.2014.07.072

[31] European Food Safety Authority. The 2010 European Union report on pesticide residues in food. EFSA Journal. 2013;**11**(3):3130

[32] Hayes WJ, Laws ER. Handbook of Pesticide Toxicology. San Diego: Academic Press; 1991. DOI: 10.1016/j.microc.2020.105298

[33] Karimi Z, Alizadeh AM, Dolatabadi JE, Dehghan P. Nigella sativa and its derivatives as food toxicity protectant agents. Advanced Pharmaceutical Bulletin. 2019;**9**(1):22. DOI: 10.15171%2Fapb.2019.004

[34] Borahan T, Tekin Z, Zaman BT, Chormey DS, Bakirdere S. An analytical strategy for propoxur determination in raisin samples with matrix matching method after dispersive liquid-liquid microextraction. Journal of Food Composition and Analysis. 2019;**84**:103315. DOI: 10.1016/j.jfca.2019.103315

[35] Xiao-Xue SH, Xiu-Qi LI, Xiao-Ping WE, Jian-Ping LI. Molecularly imprinted photoelectrochemical sensor based on AgBiS2/Bi2S3 for determination of propoxur. Chinese Journal of Analytical Chemistry. 2020;**48**(3):396-404. DOI: 10.1016/S1872-2040(20)60004-4

[36] Fan AM. Biomarkers in toxicology, risk assessment, and environmental chemical regulations. In: Biomarkers in Toxicology. London, United Kingdom: Academic Press; 2014. pp. 1057-1080. DOI: 10.1016/B978-0-12-404630-6.00064-6

[37] Yaseen T, Pu H, Sun DW. Fabrication of silver-coated gold nanoparticles to simultaneously detect multi-class insecticide residues in peach with SERS technique. Talanta. 2019;**196**:537-545. DOI: 10.1016/j.talanta.2018.12.030

[38] Ghoraba Z, Aibaghi B, Soleymanpour A. Ultrasound-assisted dispersive liquid-liquid microextraction followed by ion mobility spectrometry for the simultaneous determination of bendiocarb and azinphos-ethyl in water, soil, food and beverage samples. Ecotoxicology and Environmental Safety. 2018;**165**:459-466. DOI: 10.1016/j.ecoenv.2018.09.021

[39] Kowalska G, Pankiewicz U, Kowalski R. Estimation of pesticide residues in selected products of plant origin from Poland with the use of the HPLC-MS/MS technique. Agriculture.

2020;**10**(6):192. DOI: 10.3390/agriculture10060192

[40] Pelle FD, Angelini C, Sergi M, Del Carlo M, Pepe A, Compagnone D. Nano carbon black-based screen printed sensor for carbofuran, isoprocarb, carbaryl and fenobucarb detection: Application to grain samples. Talanta. 2018;**186**:389-396. DOI: 10.1016/j.talanta.2018.04.082

[41] Toldra F. Handbook of Food Analysis-Two Volume Set. Boca Raton: CRC Press; 2015. DOI: 10.1201/b18668

[42] Nollet LM, Rathore HS, editors. Handbook of Pesticides: Methods of Pesticide Residues Analysis. Boca Raton: CRC Press; 2016

[43] Sabljic A, editor. Environmental and Ecological Chemistry—Volume II. Oxford, United Kingdom: EOLSS Publications; 2009

[44] Jokanović M. Medical treatment of acute poisoning with organophosphorus and carbamate pesticides. Toxicology Letters. 2009;**190**(2):107-115. DOI: 10.1016/j.toxlet.2009.07.025

[45] Saber M, Parsaeyan E, Vojoudi S, Bagheri M, Mehrvar A, Kamita SG. Acute toxicity and sublethal effects of methoxyfenozide and thiodicarb on survival, development and reproduction of *Helicoverpa armigera* (Lepidoptera: Noctuidae). Crop Protection. 2013;**43**:14-17. DOI: 10.1016/j.cropro.2012.09.011

[46] Gil HW, Jeong MH, Park JS, Choi HW, Kim SY, Hong SY. An outbreak of food borne illness due to methomyl pesticide intoxication in Korea. Journal of Korean Medical Science. 2013;**28**(11):1677-1681. DOI: 10.3346/jkms.2013.28.11.1677

[47] WHO. State of the Science of Endocrine Disrupting Chemicals 2012: An Assessment of the State of the Science of Endocrine Disruptors Prepared by a Group of Experts for the United Nations Environment Programme and World Health Organization. Geneva: WHO; 2013

[48] De Coster S, Van Larebeke N. Endocrine-disrupting chemicals: associated disorders and mechanisms of action. Journal of Environmental and Public Health. 2012;**99**(8):34-38. DOI: 10.1155/2012/713696

[49] Masondo NA, Stafford GI, Aremu AO, Makunga NP. Acetylcholinesterase inhibitors from southern African plants: An overview of ethnobotanical, pharmacological potential and phytochemical research including and beyond Alzheimer's disease treatment. South African Journal of Botany. 2019;**120**:39-64. DOI: 10.1016/j.sajb.2018.09.011

[50] Abubakar Y, Tijjani H, Egbuna C, Adetunji CO, Kala S, Kryeziu TL, et al. Pesticides, history, and classification. In: Natural Remedies for Pest, Disease and Weed Control. London, United Kingdom: Academic Press; 2020. pp. 29-42. DOI: 10.1016/B978-0-12-819304-4.00003-8

[51] Nollet L. Handbook of Food Analysis. New York: Marcel Dekker; 2004

[52] Roberts JR, Reigart JR. Recognition and Management of Pesticide Poisonings. United States: Govt Printing; 2013

[53] Narenderan ST, Meyyanathan SN, Babu B. Review of pesticide residue analysis in fruits and vegetables. Pre-treatment, extraction and detection techniques. Food Research International. 2020;**133**:109141. DOI: 10.1016/j.foodres.2020.109141

[54] Musarurwa H, Tavengwa NT. Deep eutectic solvent-based dispersive

liquid-liquid micro-extraction of pesticides in food samples. Food Chemistry. 2021;**342**:127943. DOI: 10.1016/j.foodchem.2020.127943

[55] Nasiri M, Ahmadzadeh H, Amiri A. Sample preparation and extraction methods for pesticides in aquatic environments: A review. TrAC, Trends in Analytical Chemistry. 2020;**123**:115772. DOI: 10.1016/j.trac.2019.115772

[56] Goulart SM, Alves RD, Neves AA, De Queiroz JH, De Assis TC, De Queiroz ME. Optimization and validation of liquid–liquid extraction with low temperature partitioning for determination of carbamates in water. Analytica Chimica Acta. 2010;**671**(1-2):41-47. DOI: 10.1016/j.aca.2010.05.003

[57] Wang X, Meng X, Wu Q, Wang C, Wang Z. Solid phase extraction of carbamate pesticides with porous organic polymer as adsorbent followed by high performance liquid chromatography-diode array detection. Journal of Chromatography. A. 2019a;**1600**:9-16. DOI: 10.1016/j.chroma.2019.04.031

[58] Li N, Chen J, Shi YP. Magnetic graphene solid-phase extraction for the determination of carbamate pesticides in tomatoes coupled with high performance liquid chromatography. Talanta. 2015;**141**:212-219. DOI: 10.1016/j.talanta.2015.04.018

[59] Shi Z, Hu J, Li Q, Zhang S, Liang Y, Zhang H. Graphene based solid phase extraction combined with ultra high performance liquid chromatography–tandem mass spectrometry for carbamate pesticides analysis in environmental water samples. Journal of Chromatography. A. 2014;**1355**:219-227. DOI: 10.1016/j.chroma.2014.05.085dq

[60] Sýkora M, Vítová E, Jeleń HH. Application of vacuum solid-phase microextraction for the analysis of semi-hard cheese volatiles. European Food Research and Technology. 2020;**246**(3):573-580. DOI: 10.1007/s00217-020-03426-x

[61] Zhou Q, Fang Z. Graphene-modified $TiO_2$ nanotube arrays as an adsorbent in micro-solid phase extraction for determination of carbamate pesticides in water samples. Analytica Chimica Acta. 2015;**869**:43-49. DOI: 10.1016/j.aca.2015.02.019

[62] Nantia EA, Moreno-González D, Manfo FP, Gámiz-Gracia L, García-Campaña AM. QuEChERS-based method for the determination of carbamate residues in aromatic herbs by UHPLC-MS/MS. Food Chemistry. 2017;**216**:334-341. DOI: 10.1016/j.foodchem.2016.08.038

[63] Anastassiades M, Lehotay SJ, Štajnbaher D, Schenck FJ. Fast and easy multiresidue method employing acetonitrile extraction/partitioning and "dispersive solid-phase extraction" for the determination of pesticide residues in produce. Journal of AOAC International. 2003;**86**(2):412-431. DOI: 10.1093/jaoac/86.2.412

[64] Manav ÖG, Dinç-Zor Ş, Alpdoğan G. Optimization of a modified QuEChERS method by means of experimental design for multiresidue determination of pesticides in milk and dairy products by GC–MS. Microchemical Journal. 2019;**144**:124-129. DOI: 10.1016/j.microc.2018.08.056

[65] Zhang Z, Dong M, Hao X, Han L, Song S, Yao W. Evaluation of cleanup procedures in pesticide multi-residue analysis with QuEChERS in cinnamon bark. Food Chemistry. 2019a;**276**:140-146. DOI: 10.1016/j.foodchem.2018.10.019

[66] Reddy CS, Reddy GV. Pesticide residues recovery studies by QuEChERS sample technique using gas chromatography. The Madras Agricultural Journal. 2017;**104**(1-3):64-67

[67] Neufeld DS, Åkerson N, Barahona D. Pesticide analysis in vegetables using QuEChERS extraction and colorimetric detection. Virginia Journal of Science. 2015;**66**(4):4. DOI: 10.25778/8rzd-1z63

[68] Farajzadeh MA, Safi R, Yadeghari A. Combination of QuEChERS extraction with magnetic solid phase extraction followed by dispersive liquid–liquid microextraction as an efficient procedure for the extraction of pesticides from vegetable, fruit, and nectar samples having high content of solids. Microchemical Journal. 2019;**147**:571-581. DOI: 10.1016/j.microc.2019.03.074

[69] Luque D, Castillo P. Microwave-assisted extraction of food components. In: Innovative Food Processing Technologies. Amsterdam: Elsevier; 2016. pp. 57-110. DOI: 10.1016/B978-0-08-100294-0.00003-1

[70] Wang H, Ding J, Ren N. Recent advances in microwave-assisted extraction of trace organic pollutants from food and environmental samples. TrAC, Trends in Analytical Chemistry. 2016;**75**:197-208. DOI: 10.1016/j.trac.2015.05.005

[71] Kalogiouri NP, Papadakis EN, Maggalou MG, Karaoglanidis GS, Samanidou VF, Menkissoglu-Spiroudi U. Development of a microwave-assisted extraction protocol for the simultaneous determination of mycotoxins and pesticide residues in apples by LC-MS/MS. Applied Sciences. 2021;**11**(22):10931. DOI: 10.3390/app112210931

[72] Zhou T, Xiao X, Li G. Microwave accelerated selective soxhlet extraction for the determination of organophosphorus and carbamate pesticides in ginseng with gas chromatography/mass spectrometry. Analytical Chemistry. 2012 Jul 3;**84**(13):5816-5822. DOI: 10.1021/ac301274r

[73] Alsammarraie F, Lin M. Using standing gold nanorod arrays as surface-enhanced Raman spectroscopy (SERS) substrates for detection of carbaryl residues in fruit juice and milk. Journal of Agricultural and Food Chemistry. 2017;**65**(3):666-674. DOI: 10.1021/acs.jafc.6b04774

[74] Yao L, Liu L, Song S, Kuang H, Xu C. Development of indirect competitive enzyme-linked immunosorbent and immunochromatographic strip assays for carbofuran detection in fruits and vegetables. Food and Agricultural Immunology. 2017;**28**(4):639-651. DOI: 10.1080/09540105.2017.1309359

[75] Zhao Y, Tan G, Wang M, Lin H, He L, Li L, et al. Application of immunoassays for rapid monitor of carbofuran residue in vegetables. Journal of Food Science. 2019;**84**(11):3296-3302. DOI: 10.1111/1750-3841.14826

[76] Baek S, Kang J, Hwang Y, Ok K, Kwak K, Chun H. Detection of methomyl, a carbamate insecticide, in food matrices using terahertz time-domain spectroscopy. Journal of Infrared, Millimeter, and Terahertz Waves. 2015;**37**(5):486-497. DOI: 10.1007/s10762-015-0234-9

[77] Wanwimolruk S, Duangsuwan W, Phopin K, Boonpangrak S. Food safety in Thailand 5: The effect of washing pesticide residues found in cabbages and tomatoes. Journal of Consumer Protection and Food Safety. 2017;**12**(3):209-221. DOI: 10.1007/s00003-017-1116-y

[78] Wang X, Feng T, Wang J, Hao L, Wang C, Wu Q, et al. Preparation of magnetic porous covalent triazine-based organic polymer for the extraction of carbamates prior to high performance liquid chromatography-mass spectrometric detection. Journal of Chromatography. A. 2019b;**1602**:178-187. DOI: 10.1016/j.chroma.2019.06.046

[79] Da J, Wang G, Cao J, Zhang Q. Determination of 30 carbamate pesticide residues in vegetative foods by QuEChERS-liquid chromatography-tandem mass spectrometry. Chinese Journal of Chromatography. 2015;**33**(8):830. DOI: 10.3724/sp.j.1123.2015.04013

[80] Hussain A, Pu H, Sun D. Cysteamine modified core-shell nanoparticles for rapid assessment of oxamyl and thiacloprid pesticides in milk using SERS. Journal of Food Measurement and Characterization. 2020;**14**(4):2021-2029. DOI: 10.1007/s11694-020-00448-7

[81] Sánchez-Hernández L, Hernández-Domínguez D, Bernal J, Neusüß C, Martín MT, Bernal JL. Capillary electrophoresis–mass spectrometry as a new approach to analyze neonicotinoid insecticides. Journal of Chromatography. A. 2014;**1359**:317-324. DOI: 10.1016/j.chroma.2014.07.028

[82] Attig JB, Latrous L, Zougagh M, Ríos Á. Ionic liquid and magnetic multiwalled carbon nanotubes for extraction of N-methylcarbamate pesticides from water samples prior their determination by capillary electrophoresis. Talanta. 2021;**226**:122106. DOI: 10.1016/j.talanta.2021.122106

[83] Cheng X, Wang Q, Zhang S, Zhang W, He P, Fang Y. Determination of four kinds of carbamate pesticides by capillary zone electrophoresis with amperometric detection at a polyamide-modified carbon paste electrode. Talanta. 2007;**71**(3):1083-1087. DOI: 10.1016/j.talanta.2006.06.001

[84] Zhang W, Yang F, Zhang Y, Zhou K. Simultaneous determination of seven carbamate pesticide residues in vegetable by capillary electrophoresis with solid phase microextraction. International Journal of Electrochemical Science. 2021;**16**:1-4. DOI: 10.20964/2021.06.06

[85] Zhao H, Ji X, Wang B, Wang N, Li X, Ni R, et al. An ultra-sensitive acetylcholinesterase biosensor based on reduced graphene oxide-Au nanoparticles-β-cyclodextrin/Prussian blue-chitosan nanocomposites for organophosphorus pesticides detection. Biosensors & Bioelectronics. 2015;**65**:23-30. DOI: 10.1016/j.bios.2014.10.007

[86] Hancu G, Simon B, Rusu A, Mircia E, Gyéresi Á. Principles of micellar electrokinetic capillary chromatography applied in pharmaceutical analysis. Advanced Pharmaceutical Bulletin. 2013;**3**(1):1. DOI: 10.5681/apb.2013.001

[87] Hsu CH, Hu CC, Chiu TC. Analysis of carbofuran, carbosulfan, isoprocarb, 3-hydroxycarbofuran, and 3-ketocarbofuran by micellar electrokinetic chromatography. Journal of Separation Science. 2012;**35**(10-11):1359-1364. DOI: 10.1002/jssc.201101108

[88] See HH, Hauser PC, Sanagi MM, Ibrahim WA. Dynamic supported liquid membrane tip extraction of glyphosate and aminomethylphosphonic acid followed by capillary electrophoresis with contactless conductivity detection. Journal of Chromatography. A. 2010;**1217**(37):5832-5838. DOI: 10.1016/j.chroma.2010.07.054

[89] Sung IH, Lee YW, Chung DS. Liquid extraction surface analysis in-line coupled with capillary electrophoresis for direct analysis of a solid surface sample. Analytica Chimica Acta. 2014;**838**:45-50. DOI: 10.1016/j.aca.2014.05.013

[90] Araujo L, Rojas C, Cubillan D, Villa N, Mercado J, Prieto A. Determination of trifloxystrobin, tebufenozide, and halofenozide in foods by micellar electrokinetic capillary chromatography. Analytical Letters. 2010;**43**(15):2340-2348. DOI: 10.1080/00032711003717372

[91] Santalad A, Zhou L, Shang F, Fitzpatrick D, Burakham R, Srijaranai S, et al. Micellar electrokinetic chromatography with amperometric detection and off-line solid-phase extraction for analysis of carbamate insecticides. Journal of Chromatography. A. 2010;**1217**(32):5288-5297. DOI: 10.1016/j.chroma.2010.06.024

[92] Chen Q, Fung Y. Capillary electrophoresis with immobilized quantum dot fluorescence detection for rapid determination of organophosphorus pesticides in vegetables. Electrophoresis. 2010;**31**(18):3107-3114. DOI: 10.1002/elps.201000260

[93] Chang PL, Hsieh MM, Chiu TC. Recent advances in the determination of pesticides in environmental samples by capillary electrophoresis. International Journal of Environmental Research and Public Health. 2016;**13**(4):409. DOI: 10.3390/ijerph13040409

[94] Zhang S, Yang X, Yin X, Wang C, Wang Z. Dispersive liquid–liquid microextraction combined with sweeping micellar electrokinetic chromatography for the determination of some neonicotinoid insecticides in cucumber samples. Food Chemistry. 2012;**133**(2):544-550. DOI: 10.1016/j.foodchem.2012.01.028

[95] Watanabe E. The present state and perspective on simple and rapid immunochemical detection for pesticide residues in crops. Japan Agricultural Research Quarterly: JARQ. 2011;**45**(4):359-370. DOI: 10.6090/jarq.45.359

[96] Bellemjid N, Iddar A, Moussaif A, Abbadi NE, Mesfioui A. Analysis of carbamates pesticides: Immunogical technique by local development of enzyme-linked immuno-sorbent assay. The Journal of Pharmacy and Pharmacology. 2018;**6**:395-402. DOI: 10.17265/2328-2150/2018.04.010

[97] Zhang JR, Wang Y, Dong JX, Yang JY, Zhang YQ, Wang F, et al. Development of a simple pretreatment immunoassay based on an organic solvent-tolerant nanobody for the detection of carbofuran in vegetable and fruit samples. Biomolecules. 2019b;**9**(10):576. DOI: 10.3390/biom9100576

[98] Sparkman OD, Penton Z, Kitson FG. Gas Chromatography and Mass Spectrometry: A Practical Guide. London, United Kingdom: Academic Press; 2011. DOI: 10.1016/C2009-0-17039-3

[99] Zhang W, Si G, Ye M, Feng S, Cheng F, Li J, et al. An efficient assay for simultaneous quantification of ethyl carbamate and phthalate esters in Chinese liquor by gas chromatography-mass spectrometry. Food Analytical Methods. 2017;**10**(11):3487-3495. DOI: 10.1007/s12161-017-0906-2

[100] Liu J, Xu Y, Zhao GA. Rapid determination of ethyl carbamate in Chinese rice wine using headspace solid-phase microextraction and gas chromatography–mass spectrometry. Journal of the Institute of Brewing. 2012;**118**(2):217-222. DOI: 10.1002/jib.33

[101] Yao L, Maloley K, Broeckling C, Ensley S, Crain S, Coffey R. Quantitative analysis of ethyl carbamate in distillers grains co-products and bovine plasma by gas chromatography–mass spectrometry. Journal of Agricultural and Food Chemistry. 2020;**68**(39):10984-10991. DOI: 10.1021/acs.jafc.0c03535

[102] Van Dyk JS, Pletschke B. Review on the use of enzymes for the detection of organochlorine, organophosphate and carbamate pesticides in the environment. Chemosphere. 2011;**82**(3):291-307. DOI: 10.1016/j.chemosphere.2010.10.033

[103] Lowdon JW, Diliën H, Singla P, Peeters M, Cleij TJ, van Grinsven B, et al. MIPs for commercial application in low-cost sensors and assays—An overview of the current status quo. Sensors and Actuators B: Chemical. 2020;**325**:128973. DOI: 10.1016/j.snb.2020.128973

[104] Verma N, Bhardwaj A. Biosensor technology for pesticides—A review. Applied Biochemistry and Biotechnology. 2015;**175**(6):3093-3119. DOI: 10.1007/s12010-015-1489-2

[105] Li D, Qiao X, Lu J, Xu Z. Synthesis and evaluation of a magnetic molecularly imprinted polymer sorbent for determination of trace trichlorfon residue in vegetables by capillary electrophoresis. Advances in Polymer Technology. 2018;**37**(4):968-976. DOI: 10.1002/adv.21745

[106] Wang FR, Lee GJ, Haridharan N, Wu JJ. Electrochemical sensor using molecular imprinting polymerization modified electrodes to detect methyl parathion in environmental media. Electrocatalysis. 2018;**9**(1):1-9. DOI: 10.1007/s12678-017-0411-9

[107] Xie T, Zhang M, Chen P, Zhao H, Yang X, Yao L, et al. A facile molecularly imprinted electrochemical sensor based on graphene: Application to the selective determination of thiamethoxam in grain. RSC Advances. 2017;**7**(62):38884-38894. DOI: 10.1039/C7RA05167K

[108] Li S, Luo Q, Liu Y, Zhang Z, Shen G, Wu H, et al. Surface molecularly imprinted polymer film with poly (p-aminothiophenol) outer layer coated on gold nanoparticles inner layer for highly sensitive and selective sensing paraoxon. Polymers. 2017;**9**(8):359. DOI: 10.3390/polym9080359

[109] Yotova L, Medhat N. Optical biosensor with multienzyme system immobilized onto hybrid membrane for pesticides determination. International Journal Bioautomation. 2011;**15**(4):267

[110] Xavier MP, Vallejo B, Marazuela MD, Moreno-Bondi MC, Baldini F, Falai A. Fiber optic monitoring of carbamate pesticides using porous glass with covalently bound chlorophenol red. Biosensors & Bioelectronics. 2000;**14**(12):895-905. DOI: 10.1016/S0956-5663(99)00066

[111] Jain U, Saxena K, Hooda V, Balayan S, Singh AP, Tikadar M, et al. Emerging vistas on pesticides detection based on electrochemical biosensors—An update. Food Chemistry. 2021;**371**:131126. DOI: 10.1016/j.foodchem.2021.131126

[112] Moon JM, Thapliyal N, Hussain KK, Goyal RN, Shim YB. Conducting polymer-based electrochemical biosensors for neurotransmitters: A review. Biosensors & Bioelectronics. 2018;**102**:540-552. DOI: 10.1016/j.bios.2017.11.069

[113] Uniyal S, Sharma RK. Technological advancement in electrochemical biosensor based detection of organophosphate pesticide chlorpyrifos in the environment: A review of status and prospects. Biosensors &

Bioelectronics. 2018;**116**:37-50. DOI: 10.1016/j.bios.2018.05.039

[114] Chauhan N, Pundir CS. An amperometric biosensor based on acetylcholinesterase immobilized onto iron oxide nanoparticles/multi-walled carbon nanotubes modified gold electrode for measurement of organophosphorus insecticides. Analytica Chimica Acta. 2011;**701**(1): 66-74. DOI: 10.1016/j.aca.2011.06.014

[115] da Silva MK, Vanzela HC, Defavari LM, Cesarino I. Determination of carbamate pesticide in food using a biosensor based on reduced graphene oxide and acetylcholinesterase enzyme. Sensors and Actuators B: Chemical. 2018;**277**:555-561. DOI: 10.1016/j.snb.2018.09.051

[116] Sun X, Gong Z, Cao Y, Wang X. Acetylcholinesterase biosensor based on poly (diallyldimethylammonium chloride)-multi-walled carbon nanotubes-graphene hybrid film. Nano-Micro Letters. 2013;**5**(1):47-56. DOI: 10.1007/BF03353731

[117] Cesarino I, Moraes FC, Lanza MR, Machado SA. Electrochemical detection of carbamate pesticides in fruit and vegetables with a biosensor based on acetylcholinesterase immobilised on a composite of polyaniline–carbon nanotubes. Food Chemistry. 2012;**135**(3):873-879. DOI: 10.1016/j.foodchem.2012.04.147

[118] Song Y, Chen J, Sun M, Gong C, Shen Y, Song Y, et al. A simple electrochemical biosensor based on AuNPs/MPS/Au electrode sensing layer for monitoring carbamate pesticides in real samples. Journal of Hazardous Materials. 2016;**304**:103-109. DOI:10.1016/j.jhazmat.2015.10.058

Chapter 13

# Pesticides Occurrence in Water Sources and Decontamination Techniques

*Sophia Subhadarsini Pradhan, Gadratagi Basana Gowda, Totan Adak, Govindharaj Guru-Pirasanna-Pandi, Naveenkumar B. Patil, Mahendiran Annamalai and Prakash Chandra Rath*

**Abstract**

Pesticides are essential in crop protection as they keep the plants safe from insects, weeds, fungi, and other pests in order to increase crop production and feed billions of people throughout the world. There are more than 500 pesticide molecules currently in use all around the world. Their non-judicious use has noticeably contaminated the environment and caused negative effects on humans and other life forms. The rainfall or irrigation water takes away the pesticide residues to nearby surface water bodies through runoff or to the groundwater sources through leaching. The occurrence of pesticides in water resources could have multiple consequences. Exposure of pesticides through contaminated water becomes the cause of acute and chronic health problems in people of all ages. Pesticide residues have the potential to disrupt the ecosystem equilibrium in water bodies. Contaminated irrigation water can contaminate other crops as well as their environment. This chapter will discuss the major exposure routes of pesticides in water bodies mainly from agricultural sectors and their effect on the ecosystem. The chapter will also discuss decontamination techniques to eliminate pesticide contaminants from water bodies.

**Keywords:** pesticides, residues, water, leaching, runoff, decontamination, ecosystem

## 1. Introduction

The growth of the population of the world is increasing at an alarming rate, which draws the attention of researchers, scientists, environmentalists, and policymakers across the globe. According to a scientific report, the global human population is likely to increase up to 9 billion by 2050 [1]. To meet the food requirement of this growing population as well as to cover their modified consumption patterns, there is an ultimate requirement of intensification and diversification of agricultural sectors. The current food production of the globe needs to be increased by 60% by 2050 [2].

Thanks to the green revolution that instigated the use of various agrochemicals that effectively increased agricultural productivity by many folds. Besides traditional agrochemicals (fertilizers and pesticides), new ones such as hormones, antibiotics, vaccines, growth promoters, etc., also brought revolutionary changes in different food production sectors. Undoubtedly, the use of agrochemicals has directly or indirectly benefitted millions of people all over the globe by increasing food production, there are instances that the action has put questions toward the well-being of the environment. Among all the environmental compartments, water resources are especially affected to a greater extent as agricultural works mostly depend on water and use about 70 % of total water resources globally [3]. In crop production sectors, some of the most important crops such as rice and wheat generally consume a huge amount of water and the total amount of water used; most part is for irrigation. Production of 1 kg of wheat requires approximately 1 $m^3$ of water and 1 kg of rice requires 1.2 $m^3$ of water [4]. Rice, which is the staple food for most people living in Asia, consumes about 80% of freshwater resources for irrigation. Apart from crop production, a huge amount of water is also used indirectly in livestock sectors through the production of fodder crops and forage. These amounts of water are directly or indirectly recycled back to surface water as well as groundwater sources carrying the pollutants such as pesticides, fertilizers, salts, sediments, hormones, antibiotics, etc., from crop fields. Now agriculture has become a major source of freshwater pollution in rivers and lakes, the second major source for wetland pollution, and the third major source for estuaries and groundwater pollution [5].

## 2. Pesticides and their groups

A pesticide is a chemical substance or combination of different chemical substances used to eliminate pests to protect crops. Food and Agriculture Organization (FAO) defined a pesticide as "a substance intended for preventing, destroying, repelling or mitigating any pest in crops either before or after harvest to prevent deterioration during storage and transport." Pesticides are designed to control pests of the standing crop in the crop fields as well as to protect the stored crops after harvest, thus finely ensuring food security. Pesticides are classified according to their chemical nature, their target, modes of action, period of activity, mode of formulation, activity spectrum, toxicity level, etc [6].

### 2.1 Mode of action

After application, pesticides either remain on the part of the plant to which those are applied or enter into the vascular system of the plant body and get transported to different organs. According to this principle, pesticides are categorized as systemic and non-systemic ones. In the case of systemic pesticides, the compound penetrates the plant body, gets into the vascular tissue system, and spreads to different parts of the plant showing its effects uniformly. In contrast to this, non-systemic pesticides do not effectively penetrate the plant tissue and remain at the applied area on the plant body.

### 2.2 Target of the pesticides

This classification of pesticides is the most common and familiar as the categorization is based on the effectiveness of the pesticide on different types of pests. For an instance, pesticides those act on insects are called insecticides, those acts on fungi are called fungicides, and those acts on herbs are called herbicides and so on. Likewise, there are rodenticides, molluscicides, nematicides, plant growth regulators, etc., used to protect plants.

### 2.3 Chemical composition of pesticide compounds

This type of classification of pesticides is done based on their chemical composition and the active ingredients they contain. This classification of pesticides is actually the most useful one as it helps in studying the occurrence of pesticides in the field, which implies their physical and chemical properties, helps to know their persistency in the environment etc. Based on their chemical nature, pesticides are categorized mainly into seven groups; those are organochlorines (OC), organophosphates (OP), carbamates, pyrethroids, amides, anilines, and azotic heterocyclic compounds. Of these seven classes, organochlorines are highly toxic pesticides. In their chemical structure, they contain five or more chlorine atoms. The chemical structure of this group of pesticides makes them highly persistent in the environment. However, the use of these pesticides is now banned in many countries due to certain problems such as their toxicity toward humans and persistency in ecosystems. Other groups of toxic pesticides are organophosphates and carbamates. Organophosphates have a chemical structure that makes them easily degraded in nature, and hence, these constitute a group of most commonly used pesticides in almost all countries. These pesticides are comparatively less toxic but effective pest controlling chemicals nowadays. However, their widespread use has now become a serious problem for ecosystems due to the occurrence of residues in different environmental compartments including water resources. The groups of pesticides—anilines, pyrethroids, amides, azotic heterocyclic compounds—constitute comparatively less toxic groups. Pyrethroid pesticides derive from a plant-based product and are made from flowers of Pyrethrum *(Chrysanthemum cinerariaefolium)*. These are used for their quick action against insect pests, easy biodegradability, and low toxicity toward mammals [7]. However, these pesticides are found to be toxic to aquatic organisms. Amide pesticides are also less persistent, and in many studies, they have been found to be completely degraded after 10 weeks of their field application. Though aniline pesticides are found to be very effective against insect pests, their toxicity toward mammals and aquatic animals made them banned in many European countries.

### 2.4 Mode of formulation

Pesticides constitute mainly of two parts—active ingredient (AI) and inert ingredient. The active ingredient is the pure form of the chemical, and this gives a pesticide its actual pesticidal property. However, for improving its activities, long-term storage, safe handling, and enhanced effectiveness, the active ingredients are usually mixed with some inert ingredients. This is called pesticide formulation, and it is of different types such as emulsifiable concentrates (EC), wettable powder (WP), soluble concentrate (SL), soluble powder (SP), suspension concentrate (SC), capsule suspensions (CS), water-dispersible granules (WG), granules (GR), dusts (Dp), etc [8].

### 2.5 Active spectrum

Pesticides that are active against a wide range of crop pests are included under broad-spectrum pesticides and those which act only on a selective group of pests are called selective pesticides.

## 2.6 Toxicity

Pesticides are categorized into five groups according to the potential risk they exert on humans and based on that pesticides are extremely hazardous, highly hazardous, moderately hazardous, slightly hazardous, products unlikely to present acute hazards in normal use [7].

Organochlorine insecticides were the first group of pesticides that were used successfully in eliminating crop pests. However, due to the reported toxicity toward humans and other mammals and persistency in different ecosystems, the use of organochlorines is now withdrawn. New groups of pesticides developed later, such as organophospahtes in 1960s, carbamates in 1970s, and pyrethroids in 1980s, herbicides in 1970s–1980s brought revolutionary changes in the field of crop pest regulation. Today pesticide production is a large industry with an annual turnover worth USD 35 billion. Currently, about 4.6 million tonnes of chemical pesticides are applied to crop plants, thereby put into the environment each year. In 2004, this amount included 47.5% of herbicides, 29.5% of insecticides, 17.5% of fungicides, and other group of pesticides account for 5.5% [9]. The overall usage of pesticides from 1990–2019 is depicted in **Figure 1**. The trend of use of different groups of pesticides is now changed. For example, the use of herbicides has been increased and the use of insecticides, fungicides, and bactericides has decreased largely in the last few decades [10].

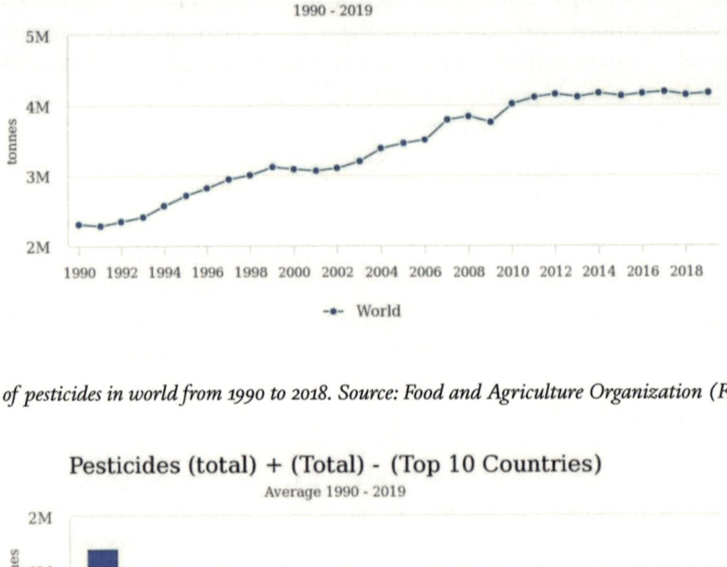


**Figure 1.**
*Consumption of pesticides in world from 1990 to 2018. Source: Food and Agriculture Organization (FAO).*

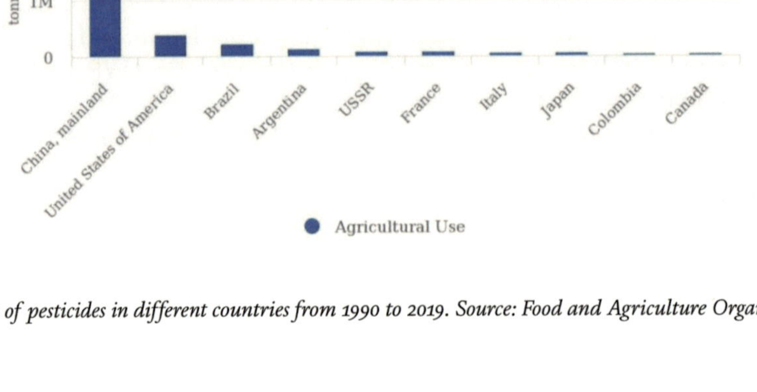


**Figure 2.**
*Consumption of pesticides in different countries from 1990 to 2019. Source: Food and Agriculture Organization (FAO).*

China tops the list of the highest amount of pesticide user in the world followed by the United States [11]. India ranks fourth in pesticide manufacturing and 12th in the list of highest pesticide-consuming countries (**Figure 2**) [12].

## 3. Exposure routes of pesticides in water bodies through crop production

To cope with the growing world's population, crop production has also been increasing. Till 2015, cereal production has increased threefold, production of vegetables increased fourfold, production of tomatoes increased fivefold, and production of soybean increased eightfold as compared with 1970 [13]. This huge increase in crop production has been achieved through the expansion of crop lands, cultivation of high-yielding crop varieties, and most importantly through the use of pesticides. In India, cotton is at the top of the list consuming the highest amount of pesticides (45%) followed by rice (22%), vegetable (9%), plantation crops (7%), wheat (4%), and other crops (9%). Among the vegetables, cabbage consumes the highest amount of pesticides. On an overall basis, pesticide consumption is the highest in fruit and vegetable cropping. In developing countries such as India, about 600g/ha of pesticides are used, whereas the amount is 6000g/ha in developed nations. According to estimation, about 4.6 million tonnes of pesticides are being integrated into the environment each year through crop production of which 51.3% was consumed in Asia, 33.3% in the Americas, 11.8% in Africa, and 1.4% in Oceania in 2016.

Pesticides are usually directly applied on plant parts or plant parts are subject to pesticide pretreatment. However, only 1% of the applied pesticide reaches the target pest, and the rest amount gets incorporated into different environmental compartments exerting its harmful effects on biodiversity, and nontarget organisms. The aerial application of pesticides may pollute surrounding areas with macro-droplets or micro-droplets of pesticides. Several studies showed that pesticide spraying enhances the distribution of pesticides in areas far from the spraying site. For an instance, spraying of pesticides caused health-related issues in children living within 1000 m of a greenhouse [14].

Depending on the chemical composition of pesticides, they show different degrees of solubility, according to which they follow different pathways to reach the water bodies after their application to crop fields (**Figure 3**). The common pathway through which pesticides enter the surface water sources such as ponds, pools, ditches, lakes, streams, rivers, etc., is through irrigation or when immediate rainfall occurs after pesticide application. Small water bodies situated adjacent to agricultural fields are more prone to pesticide pollution as the pesticides applied to the crop fields directly washed away into those water sources. These water bodies receive considerably higher amounts of pesticide as compared with farther or larger water bodies [15]. In the case of groundwater systems, the common pathway for the entry of pesticides is through leaching. Also, other routes of exposure of pesticide molecules include soil erosion, direct disposal, or sedimentation, etc.

Some major pathways through which pesticides reach water sources are as follows.

### 3.1 Leaching

Leaching of pesticides is the vertically downward movement of pesticide molecules through the minute capillaries formed by soil particles or channels formed by roots and root hairs to the groundwater table and deeper aquifers. The pesticides with

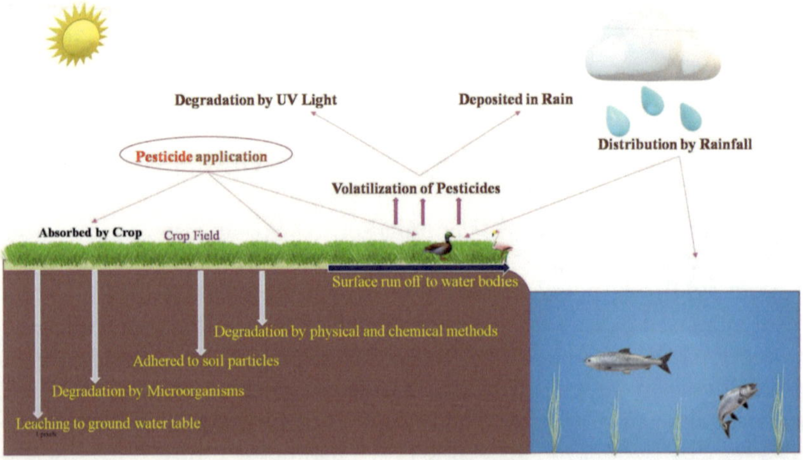


**Figure 3.**
*Pathways of pesticide entry into different ecosystems.*

a lower persistency value tend to degrade within less time posing a comparatively lesser threat to groundwater. There are two types of leaching observed, which may provide pathways for pesticide movement toward the groundwater [16].

- Preferential flow—leaching of pesticides in soil profile through the cracks and crevices, large voids, channels formed by already penetrated roots and root hairs, etc.
- Matrix flow—water moves through capillaries formed by small soil pores.

Among the pesticides used, atrazine, an herbicide, has a high potential to leach into the underground water table due to its high persistency. In contrast to this, cyanazine and methyl parathion show low leaching potential due to their shorter half-life, high rate of adsorption to soil particles, and low persistency. Herbicide, 2, 4-D, is a hydrophilic pesticide and easily gets broken down by the actions of microorganisms, hence less chance of accumulation in soil [17], thereby exerting a lesser chance of water contamination. Pesticide leaching to the groundwater may be enhanced by rainfall or through irrigation only when the concerned pesticide is fairly soluble in water. The pesticide may get dissolved in water or form suspension or emulsion. Water that is moving at a higher speed as in rivers or streams as compared with ponds or ditches is more likely to carry heavy pesticides and to a farther distance. Several factors affect the rate of leaching of pesticides such as physical and chemical characteristics of pesticides, the permeability of pesticides in soil, volatilization of pesticide molecules, crop-root uptake, methods and doses of pesticide application and types of weather conditions, variation in temperature and precipitation pattern [18].

### *3.1.1 Soil organic matter*

The organic matter content of the soil is the most important soil property that affects pesticide breakdown by microorganisms. Organic matter present in the soil helps in better adsorption of pesticide molecules by providing a larger surface area. The presence of organic matter also helps the soil to hold more amount of water, thereby increasing the chance of degradation by microorganisms. This ultimately decreases the rate of leaching of pesticides to the groundwater table.

#### *3.1.2 Soil texture*

Soil is composed mainly of sand, silt, and clay. This composition affects the movement of water through the soil. The coarse-textured soil will have more sand particles and large pores, allowing water to move rapidly carrying pesticides to the water table. Clay-textured soils will have more clay and hence will have less pore size that provides low permeability. This slows the downward movement of pesticides and increased the rate of degradation of pesticides on the soil surface.

#### *3.1.3 Soil structure*

In the soil where particles are loosely packed, pesticides tend to leach faster in the soil. Compact soil holds water back and prevents the free flowing of water through it. the soil in which openings and channels are formed, for an instance, burrows formed by earthworms or crevices due to freezing and thawing allow downward movement of water that may contain pesticides. Plant roots penetrate the soil, thus creating channels that allow water to carry pesticides downward toward the water table.

#### *3.1.4 Soil water content*

The amount of water present in the soil determines the leaching of pesticides into the groundwater sources. Pesticides that are more soluble will have a greater chance of leaching to the water table when the soil is fully saturated. However, in the case of dry soil when water is added, the water molecules just fill the pores in the soil surface, decreasing the chances of carrying the pesticide residues down through the soil profile via water.

#### *3.1.5 Depth of groundwater*

The water table is usually separated from the soil surface through a number of soil layers. The soil layers above the water table determine the pesticide adsorption and degradation. The more the depth of the water table, the more the groundwater is protected decreasing the probability of contamination. The water table is more prone to pesticide contamination when it is present nearer to the soil surface.

#### *3.1.6 Type of bedrock*

Bedrock is the bottommost layer present beneath soil or rock fragments. The types of bedrock determine the leaching of water that may carry pesticides. For example, in the case of limestone bedrocks, the downward water channels are comparatively larger, thereby allowing water to leach quickly. Limestone is highly soluble in water and hence dissolves in water creating underground passages that let water move out of the area rapidly, carrying pesticides to farther distances.

#### *3.1.7 Slope*

The topography of an area affects the rate of movement of water flow across the earth's surface. The areas with steep slopes allow fast surface runoff but reduce the chances for water to leach to the groundwater table. In contrast in valleys and flat areas, a runoff will be slow, but leaching to underground will be comparatively faster.

### 3.2 Runoff

Surface runoff is the movement of water molecules on the earth's surface in case of the availability of excess water on the soil surface that accumulates from different sources. It occurs when the amount of surface water reaches such a quantity that the soil fails to infiltrate or absorb that. It happens when there is irrigation, rainfall, or when the snow melts that add more water to soil surface that eventually flows down toward ponds, pools, ditches, canals, streams, rivers, or lakes. During runoff, the pesticide molecules present in the crop field soils tend to be carried away that get stored in the lentic water systems. Pesticides stored in the standing water systems get a longer period for leaching into the groundwater sources. Several factors such as environmental conditions, pesticide composition, soil characteristics, etc., affect the transfer of pesticides through runoff water.

#### *3.2.1 Soil moisture content*

The water content of the soil in an area will determine the amount of runoff that will occur from the site. Soil that is already saturated with water faces more risk of surface runoff. In the case of dry soils, the addition of water will lead to filling of the pores of soil decreasing the chances of runoff.

#### *3.2.2 Soil texture*

Soils that contain clay are more compact and hence more prone to runoff losses, whereas loose sandy soils possess less chance of surface runoff.

#### *3.2.3 Weather or irrigation*

Climatic conditions such as the temperature of the atmosphere, precipitation, etc., determine largely the rate of surface runoff. Pesticides applied in the crop fields when subjected to immediate rainfall lead to washing off of the applied pesticide molecules. The wasted pesticides along with surface runoff may reach the nearby water bodies. Also, pesticides that are applied where the soil is already saturated with previous rainfall or irrigation may be subjected to runoff if light rainfall or additional irrigation follows. At times when the temperature is very low, i.e., in the case of frozen soils, applied pesticides face the problem of runoff. Therefore, it is usually recommended not to apply pesticides in frozen soils and the pesticide application should not be followed by heavy rainfall or irrigation.

#### *3.2.4 Slope*

The slope is an important deciding factor for the runoff of pesticides with water. The type of landscapes where the ground has a slope will facilitate the runoff of pesticides.

#### *3.2.5 Pesticide characteristics*

The physical and chemical properties of a pesticide are the deciding factors for the surface runoff of those molecules. Pesticides that are hydrophilic or more soluble in water will get the opportunity for easy runoff. The hydrophobic pesticides get adhered to the soil particles and hence get less chance for surface runoff.

#### 3.2.6 *Pesticide persistence*

Some pesticide molecules are easily degraded by the action of microorganisms and hence will not be available for surface runoff.

### 3.3 Soil erosion

Soil is composed of different constituents such as sand, silt, clay, minerals, organic matters, etc. These components of soil facilitate the adsorption of pesticide molecules to soil particles. The adsorption of pesticides determines their persistency in soil ecosystems. Pesticides that are hydrophobic tend to be get adsorbed to soil particles when applied in the crop fields [19]. These pesticides strongly bind in the soil and lose the chance of surface runoff. However, when the weather conditions become dry, that leads to soil erosion leading to the transfer of pesticides from crop fields to other regions and may reach the nearby aquatic systems. Some examples of pesticides that are displaced only when the soil particles are eroded are organochlorines, paraquat, and arsenical pesticides. These pesticides strongly bind to the soil particles and contaminate the water bodies only when erosion occurs in that area.

### 3.4 Irrigation

Pesticide movement on or within the soil surface is greatly determined by the process of irrigation, which is a common practice in crop production systems. Irrigation facilitates the movement of pesticides on the soil surface as runoff or leaching to the ground water table. When the rate of irrigation exceeds the rate of infiltration, soil promotes runoff that will carry pesticides away to nearby water bodies. Irrigation made the water molecules available on the soil surface, which interferes with the physical and chemical properties of pesticides and thereby facilitates their movement.

## 4. Pesticide residues in water bodies

Pesticide use in both developed and developing countries has no doubt enhanced food production and ensured food security, the inappropriate and poorly regulated practices of pesticide handling and application have led to contamination of water bodies. There are several scientific reports those indicate that only 0.1% of the applied pesticides in the field reach the target organisms, and a huge amount is lost into different environmental compartments [20]. Pesticides are chemical substances with harmful chemical properties such as toxicity and persistency. They remain as such in various ecosystems for a long time and are hence called persistent organic pollutants (POPs). "Persistency may be defined as the tendency of a chemical compound to conserve its molecular integrity and chemical, physical, and functional characteristics for a certain time after being released into the soil." Pesticides are grouped into two categories—hydrophobic and hydrophilic based on which the extent of persistency of a pesticide is determined. The persistency of pesticides in the environment depends on several factors such as the type of soil, method of pesticide application, the capacity of soil to adsorb pesticides, organic matter content of the soil, etc. Hydrophobic pesticides are persistent and hence have the properties of bioaccumulation in the environment, e.g., organochlorines (DDT, endosulfan, endrin, heptachlor, lindane). Some pesticides being persistent persist in the soil and in that course of time may experience a variety of fates. Some amount of the pesticides will be taken up by the plants, and some amount will be degraded by the native microorganisms present in that area. The remaining amount of the pesticide active ingredients or their transformed products will be carried away by water at the time of rainfall or irrigation to different sources of water. Pesticides that will percolate vertically downward in the soil horizon finally reach the groundwater table and those that will move in surface water runoff reach nearby water bodies. Some amount of insoluble chemicals that get tightly bound to soil particles on the topsoil layer are subjected to erosion and ultimately reach surface waters. Pesticide residues that remain in the soil are sometimes subjected to volatilization, in the atmosphere that get accumulated in the rain and during rainfall, finally reach different water bodies. However, water source contamination through this pathway is insignificant. Some pesticides such as herbicides, carbamates, fungicides, and some organophosphates are hydrophilic, hence transported through runoff to surface water bodies and may be leached to groundwater sources.

The occurrence of pesticide residues in the ground as well as surface water sources is a widespread issue globally [21]. Some pesticides detected in major water bodies in different countries are presented in **Table 1**. Pesticide molecules are often found more frequently in surface water sources as compared with groundwater tables [33]. The reason is that the pesticides tend to slowly filter down the soil horizon and reach the deep aquifers, whereas the precipitations and frequent irrigations enhance the chances of pesticide transfer to surface water sources. It is hard to decontaminate the water in the groundwater table and the deep aquifers once pesticide residues contaminate the sources.

Surface water source contamination by pesticides is now common case in developing countries such as India. Not only the surface and groundwater sources but also the direct drinking water sources are found to be contaminated with some pesticide residues in almost all countries around the globe. In several reports where drinking water samples were collected from hand pumps or tube wells from one state of India, about 58% of the samples were found contaminated with various pesticide residues, mainly organochlorines above United States Environmental Protection Agency standards [34]. In China, drinking water samples were found contaminated with 42 different organochlorine pesticides at a concentration ranging from 0.001 to 2.65 μg/l [35–37]. Twenty-three OC pesticide residues were detected at a concentration of 0.01–0.34 μg/l in water samples from India [23]. Water samples from Turkey had 18 different types of OCs at a concentration of 0.007–0.159 μg/l [38]. OCs at a concentration of 0.01–0.03 μg/l were found in water samples from South Africa [39]. Fourteen OCs with a concentration of 0.003–0.09 μg/l were found from Mexico water samples [40]. Twelve OCs were found in water samples of the Philippines at a concentration of 0.02–0.74 μg/l [41]. In some studies of water samples from the United States [42] and Ireland [43], two different OCs were found at a concentration of 0.0004–0.22 μg/l. The occurrence of OC pesticides in water sources of the above-said countries may be due to the previous application of pesticides as insecticides in crop fields. For an instance, in China, a pesticide, dicofol, was applied in cotton fields that later became the cause of DDT contamination of water sources [44]. In the United States also organochlorine pesticides were widely applied in cotton farms that later became

| Country | Water sources studied | Detected pesticides | Concentration of pesticides | Reference |
|---|---|---|---|---|
| Japan | Chikuma river, shinano river | Bromobutide | 3 ng/l | [22] |
| | | Isoprothiolane | 8200 ng/l | |
| India | Yamuna river | Hexachloro-cyclohexane | 12.76–593.49 ng/l | [23] |
| | | DDT | 66.17–722.94 ng/l | |
| Nigeria | Lagos Lagoon | Chlordane | 0006–0.950 μg/l | [24] |
| | | Heptachlor | 0.067 μg/l | |
| | | Methoxychlor | 0.123 μg/l | |
| | | Hexachloro-benzene | 0.015–0.774 μg/l | |
| | | Endosulfan | 0.015–0.996 μg/l | |
| | | Dtrichloro-ethane | 0.012–0.910 μg/l | |
| | | Dieldrin | 0.015–0.996 μg/l | |
| | | Aldrin | 0.080–0.790 μg/l | |
| Bangladesh | Surface water samples from paddy and vegetable fields | Diazinon | 0.9 μg/l | [25] |
| | | Carbofuran | 105.2–198.7 μg/l | |
| | | Malathion | 105.2 μg/l | |
| | | Carbaryl | 14.1–18.1 μg/l | |
| Southern Iran | Lake Tashk | DDT | 0.028 ppb | [26] |
| | | DDE | 0.075 ppb | |
| | | Lindane | 0.082 ppb | |
| | | Endosulfan | 0.068 ppb | |
| Nepal | Ansikhola watershed | Endosulfan | 50 μg/l | [27] |
| | | Iprobenfos | 3980 μg/l | |
| | | Monochrotofos | 118 μg/l | |
| | | Mevinphos | 103 μg/l | |
| | | Acephate | 43 μg/l | |
| | | Butamifos | 3980 μg/l | |
| Bangladesh | Fish ponds, Tube wells | Malathion | 42.58–922.8 μg/l | [28] |
| | | Diazinon | 31.5 μg/l | |
| Ecuador | Guayas river basin | Cadusafos | 0.081 μg/l | [29] |
| | | Butachlor | 2.006 μg/l | |
| | | Pendimethalin | 0.557 μg/l | |
| India | Chilika lake | Chlorpyrifos | 0.019–2.73 μg/l | [30] |
| | | Dichlorvos | 0.647 μg/l | |
| China | Taihu lake | Carbendazim | 508 ng/l | [31] |
| | | Imidacloprid | 438 ng/l | |
| Malaysia | Tengi river | Imidacloprid | 57.7 ng/l | [32] |
| | | Tebuconazole | 512.1 ng/l | |
| | | Propiconazole | 4493.1 ng/l | |
| | | Difenoconazole | 1620.3 ng/l | |
| | | Buprofezin | 729.1 ng/l | |

**Table 1.**
*Pesticides detected in major water bodies in different countries.*

a major cause of water pollution [45]. Many agricultural practices sometimes enhance the distribution of pesticides in nearby water sources from crop fields. For example, rice cropping requires flooding of the fields for a long duration, which increases the chances of transfer of pesticide residues from a contaminated site to non-contaminated sites as well as to water sources. In India, the huge application of organochlorine insecticides in crop fields has become the major source of surface soil contamination [46] and water pollution [47, 48] nowadays. Organochlorine pesticides remain for a longer period in the environment and cycle through various routes such as volatilization, runoff, or leaching [49]. As a result of which organochlorine pesticide residues get transported to water sources via environmental components. Organochlorine pesticides have high $K_{ow}$ values and hence persist in soil for a longer duration as they get adsorbed to clay or organic matter present in soil and gradually released into water [50–52]. Sometimes organochlorine pesticides get evaporated from crop field soils into the surrounding atmosphere, get deposited in the rain, and eventually distributed in different water sources during rainfall events [53–55].

Organophosphorus pesticide residue detection in drinking water sources all around the world is noted in several published studies. This may be due to intensive OP application for crop protection. In China, OP pesticides are used at a higher amount that is about 1.5–4-fold higher as compared with other parts of the world [56]. OP pesticides were detected from water sources of Spain [57], Brazil [58], Canada [59], and United States [42] at concentration ranges of 1.01–21.95 μg/l, 0.21–0.57 μg/l, 0.01–2.56 μg/l, 0.001–0.06 μg/l and 0.06–0.22 μg/l, respectively. Compared with organochlorines, organophosphorus pesticides are less frequently detected in water sources due to their susceptibility to water hydrolysis at alkaline pH [60], photochemical degradation [61], and degradation by microbes in water bodies [62].

Carbamate pesticides such as carbofuran, carbaryl, methiocarb, fenobucarb, propoxur were found in water samples in Brazil, Spain, Vietnam, Burkina Faso. Carbofuran, carbaryl, methiocarb, fenobucarb, propoxur were detected at a concentration range of 0.06–2.95 μg/l, 0.17 μg/l, 1.35 μg/l, 0.04–0.074 μg/l, and 0.029–0.023 μg/l, respectively. The occurrence of carbamate pesticides in water bodies may be due to their use in agricultural sectors [63, 64], leaching in the soil profile [65], wash-off from plant surfaces during rainfall [66]. However, detection of a low amount of carbamate residues in water bodies may be due to its susceptibility to water hydrolysis [67], degradation by exposure to UV light [68], and degradation through the action of microbes [69].

Pyrethroids, neonicotinoids, and other pesticides were found in water samples all around the world at a concentration of 0.001–0.041 μg/l [22, 70, 71]. Drinking water samples from Burkina Faso [71], Brazil [58], Spain [57], and China [72] were found to have imidacloprid pesticide with a concentration of 0.01, 1.28, 3.99, and 8.33 μg/l, respectively. The low detected concentration of these pesticides may be due to their sensitivity to photo-degradation [73], and the concentration may be due to their usage in agricultural sectors [74].

Approximately 31 different parent herbicide residues were detected in more than 768 water samples collected from 18 countries around the world. Herbicide residues were detected in water samples from Portugal [75], Brazil [58, 76], Spain [57], Vietnam [77, 78], United States [79], Canada [59], China [80], Germany [81] at concentrations of 0.002–027 μg/l, 0.01–4.90 μg/l, 1.16–32.32 μg/l, 0.0001–0.47 μg/l, 0.03–1.8 μg/l, 0.0001–0.051 μg/l, 0.001–0.021 μg/l, 1.22–79.02 μg/l, respectively.

Herbicide glyphosate is highly water-soluble (10.5 g/l) and has a high dissociation constant and low partitioning coefficient, therefore considered as a nontoxic pesticide to humans; however, it is highly toxic to aquatic organisms. Due to widespread use, glyphosate residues have been found in many water sources, including drinking water, and also detected at a concentration of 1.42 μg/l in the groundwater table [82].

Different fungicides have been detected in water samples from different countries. Water samples collected from different places in Japan showed fungicide residues at concentration ranges of 0.013–0473 μg/l [22]. Fungicides are also detected at a concentration of 4.82–101.03 μg/l in Spain [57], 0.001–0.39 μg/l in Brazil [58], 0.0011–0.077 μg/l in China [83]. Fungicides are found in water samples due to their use in agricultural practices such as to control soil-borne plant diseases, seed dressings, foliar sprays, etc.

## 5. Consequences

Though the application of pesticides provides a range of benefits such as enhancing the quality of food and increasing the quantity of food production by reducing pest-related issues of crop plants; however, the inappropriate use of pesticides has also led to potential negative effects on the environment, mainly water sources. The adverse effects of the pesticides remain in the environment for a long time as the pesticide molecules also remain persistent for a long period. Surface water bodies such as ponds, pools, ditches, streams, lakes, estuaries, and groundwater remain vulnerable to pesticide pollution. Even when the amount of pesticide residues that enter the water bodies is very less, subjected to biomagnifications, and the residues get deposited at a noticeable amount. Pesticides in water bodies have the chance to enter the body of aquatic organisms and then get transferred to others in the food chain. Man occupies the highest trophic position in a food chain, and also man has access to a number of other food chains, hence tends to acquire the highest amount of pesticide residues than other organisms by a process of biomagnification. The accumulated pesticides in the human body interfere with physiological processes, and the consequences are decreased immunity, hormonal balance disruption, reproductive system abnormalities [84], and more importantly carcinogenic effects [49], the occurrence of breast cancers [85], prostate cancers [86], abnormalities in the endocrine system [87], the occurrence of Parkinson's disease [88], and imbalance in cardiovascular system [84]. Pesticides such as organochlorines when reach non-target insects disrupt their nervous systems leading to paralysis and ultimate death. Organochlorine residues in water bodies promote endocrine system disorders in aquatic organisms such as fishes. Hence these toxic pesticides are now banned in many nations worldwide. Organophosphate pesticides inhibit the function of the enzyme-acetylcholine esterase that hydrolyzes acetyl choline [89]. Farmers and field workers sometimes when exposed to pesticides while handling or applying face pesticide poisoning, and this adds to the negative impacts of pesticides with respect to public health problems [90]. Each year about 3 million cases are registered as pesticide poisoning of which the death of 250–370,000 people is reported [91]. This may be due to handling, spraying, and storage of pesticides without improper protection measures. Not only human beings but also plants, birds, and aquatic organisms get affected when exposed to pesticide-contaminated water. Contaminated aquatic organisms such as fishes or shell fishes transfer pesticide residues in their body to humans. Hence humans may acquire pesticide residues through two major pathways—ingestion of food and water. World Health Organization (WHO) and many other health and environmental agencies established the maximum allowable quantities of about 33 pesticides for daily ingestion under the term "acceptable daily intake (ADI)."

## 6. Decontamination techniques

Though the use of pesticides since nineteenth century has brought revolutionary advancements in crop production sectors, the inappropriate usage has now put questions to the sustainability of the environment. The pesticide active ingredients, as well as their transformation products in different ecosystem compartments , and more importantly in drinking water sources, have now drawn the attention of environmentalists to work in the field of removal of pesticides. Pesticides are usually organic compounds, hence put through various physical, chemical, and microbial degradation processes. Microorganisms mineralize the pesticides into final small molecules such as $CO_2$ and water. Sometimes microbes transform the pesticides into a new modified compound by changing their chemical structure, which is called co-metabolism. Photochemical degradation or photolysis is a process where the pesticide molecules are broken down in the presence of ultraviolet rays. Chemical degradation of pesticides occurs via oxidation-reduction reactions as well as by hydrolysis in air and water.

Naturally, pesticides are removed from the environment through the exposure of UV light, sedimentation, adsorption-desorption, and microbial action, but to a smaller extent. On a large scale, the removal of pesticides from the environment may involve both physical and biological processes. The typical physical methods for removal of pesticides in treatment plants include ozonation [92], fluid extraction [93], solid-phase extraction [94], photocatalytic degradation [95, 96], adsorption [97], filtration [98], and sedimentation. These methods of pesticide decontamination of water usually have high operational costs and also may create the chances of the development of secondary pollutants such as sludge. So, now there are requirements of alternative pesticide removal processes, which will be long term and feasible. One of the most promising and clean technologies for decontamination of water is Advanced Oxidation Processes (AOPs). It is now the most accepted technique for water purification as it is thermodynamically feasible and has broad-spectrum applicability. The mechanism of the process involves the production of highly reactive hydroxyl radicals within the system. Highly reactive hydroxyl radicals are formed by different processes such as by using oxidants, catalysts, or UV rays. These in situ generated hydroxyl radicals carry out the oxidation of a wide range of chemical contaminants including pesticides and their transformation products and lead to their complete mineralization to $CO_2$, water, and inorganic elements [99, 100]. In more complex systems, AOPs are recommended as a pretreatment process that converts the pesticides into a more biodegradable form followed by a biological treatment process that converts the pesticides into $CO_2$, water, inorganic minerals, and biomass.

Adsorption of pesticides on activated carbon materials in its different forms such as granular activated carbon [101], powdered activated carbon [102], carbon cloth [103], carbon fibers [103], black carbon [104], activated carbon composites [105], etc., has now become a cheaper and renewable method of pesticide removal from waste water. Researchers are now trying to synthesize activated carbon from cheaper sources such as agricultural wastes such as coconut fibers, sal wood, coconut shells, horseshoe crab shell, corn stillage, oil palm fronds, wood, date stones, and biochar, etc., for effective removal of pesticides.

In the last few decades, membrane technologies such as reverse osmosis and nanofiltration are found to remove pesticides from waste water efficiently. Nanofiltration is the most suitable technology for removing pesticides while reserving the inorganic nutrients in the water. The principle behind the process is the charged surface of the membrane that effectively removes pesticide molecules from treated water [106].

Reverse osmosis (RO) is a process that eliminates impurities from drinking water including pesticides residues. Here water is passed through a membrane having a pore size of 0.0001 micron under high pressure. Only 5–10% of the ions can pass through the membrane [107], and those are included under acceptable levels as per World Health Organization (WHO). RO systems are helpful in the removal of pesticide residues; however, the cost varies depending on the capacity of the plants, level of utilization, level of salinity, presence of other contaminants, and distance from the source of water. Removal of pesticides from water by the process of reverse osmosis through the use of membranes such as aromatic polyamines, cross-linked polyethylenimine membranes, e.g., NS-100, PA300 [107], cross-linked m-phenylenediamine membrane (FT-30) [108], etc., was successfully applied later.

Biotic degradation or biodegradation is defined as the breakdown of complex pesticide molecules into smaller products. The rate at which pesticides biodegrade varies widely. Some pesticides such as DDT and dieldrin are recalcitrant. Pesticides such as organophosphates, which are biodegradable, are nowadays given more preference over recalcitrant ones such as organochlorines. The biodegradation process involves both aerobic and anaerobic methods. Also, biodegradation is divided into three categories based on the location where bioremediation is done, i.e., ex situ and in situ. In in situ treatment, bioremediation is carried out at the contaminated site itself, and it is usually the aerobic process. Some of the in situ bioremediation techniques that can be instigated to eliminate pesticides are attenuation, bioaugmentation, biostimulation, bioventing, and biosparging. In ex situ treatment, the contaminated water is removed from the polluted site, transported to other sites where the pesticides in the water are biodegraded. During biodegradation, microbes use pesticides as co-substrates in their metabolic reactions, mineralizing them and thus eliminating them from the environment. The key microbial enzymes that carried out the process are hydrolases, peroxidases, oxygenases, etc. The process of biodegradation involves three steps. In the first step, through the processes such as oxidation, reduction, and hydrolysis, the pesticides are converted into more water-soluble forms. The transformed products are converted into sugars and amino acids, which are again

| Pesticides | Microorganisms | Reference |
|---|---|---|
| Glyphosate | *Fusarium* | [111] |
| Chloropyrifos | *Ochrobactrum sp. JAS2* | [112] |
| Cypermethrin | *Bacillus subtilis* | [113] |
| Deltamethrin | *Streptomyces rimosus* | [114] |
| Fentopropathrin | *Rhodopseudomonas palustris* | [115] |
| Phorate | *Brevibacterium frigoritolerans*<br>*Bacillus aerophilus*<br>*Pseudomonas fulva* | [116] |
| Acetachlor | *Tolypocladium geodes*<br>*Cordyceps* | [117] |
| Tebuconazole | *Serratia mercescens* | [118] |
| DDT | *Fomitopsis pinicola*<br>*Ralstonia pickettii* | [119] |

**Table 2.**
*Microorganisms capable of degrading several pesticides.*

more water-soluble and less toxic in the second step and finally converted into CO2, salts, minerals, and water in the final step. The availability of pesticides for microbes depends on their solubility, pH of water, temperature, microbial diversity, etc. The microorganisms that can carry out the degradation of pesticides are bacteria, fungi. In some cases, it is easier when a group of microorganisms called microbial consortium is used as compared with the pure culture. Among fungi, molds, yeast, and filamentous fungi are more useful for the biodegradation of pesticides [109]. Fungi are better degraders of pesticides than bacteria due to characteristics such as specific bioactivity, growth morphology, and high resistance even at high concentrations of pesticides [110] (**Table 2**).

## 7. Conclusion

Clean water is an important part of human life and plays a major role in the sustainability of life on earth. Access to clean water is a fundamental human right and vital to sustaining a healthy life. However, the occurrence of pesticide residues in different water sources including drinking water has now become a universal problem. Nowadays, the increasing demand for food has resulted in intensive agricultural practices that resulted in contamination of water sources with pesticide residues; degrade the water quality in both developed and developing nations. Freshwater is a scarce and vulnerable resource that can be easily contaminated and whose original quality is hard and expensive to be restored. Water pollution through pesticides is posing deleterious effects on many types of organisms, including useful microorganisms, insects, birds, fishes, and humans.

Briefly, it can be said that agriculture has no beneficial effects on water resources. As agriculture is a primary requirement for human society, it cannot be disregarded. So only we can minimize or regulate the activities in agricultural sectors to keep down the extent of water pollution. Although pesticides are considered as easy, cheap, quick methods for eliminating pests and weeds from crop fields, pesticide users should be recommended to completely eliminate chemical pesticides and replace that with bio-pesticides that will minimize the risks of environmental hazards. Also, there are reports that showed that cheaper pesticides sustain in the environment for a long time as they are resistant to natural degradation processes. In some developed countries, the use of such pesticides is banned already but due to their low cost, these are still in use in many developing nations. Integrated pest management (IPM) is another clean way for the management of insects and pests where the growth of healthy crops is emphasized that will discourage pest attack. The areas where pesticide occurrence in water bodies became more common should undergo constant observations. The water bodies where residues have been detected should be subjected to various treatment processes for decontamination and the potable water sources should undergo advanced decontamination processes. Finally to reduce the pesticide load in water sources as well as in other ecosystem compartments is the duty for all of us to do our part through the use of non-chemical pest control methods.

## Acknowledgements

Authors gratefully acknowledge Director, ICAR-National Rice Research Institute, Cuttack, India, for constant support and providing all the facilities.

## Conflict of interest

The authors declare that they have no known competing interests or personal relationships that could have appeared to influence the work reported in this paper.

## Author details

Sophia Subhadarsini Pradhan, Gadratagi Basana Gowda*, Totan Adak*, Govindharaj Guru-Pirasanna-Pandi, Naveenkumar B. Patil, Mahendiran Annamalai and Prakash Chandra Rath
Division of Crop Protection, ICAR-National Rice Research Institute, Cuttack, Odisha, India

*Address all correspondence to: basanagowda.g@icar.gov.in; totan.adak@icar.gov.in

## References

[1] Godfray HC, Beddington JR, Crute IR, Haddad L, Lawrence D, Muir JF, et al. Food security: the challenge of feeding 9 billion people. Science. 2010;**327**:812-818. DOI: 10.1126/science.1185383

[2] McKenzie FC, Williams J. Sustainable food production: Constraints, challenges and choices by 2050. Food Security. 2015;7:221-233. DOI: 10.1007/s12571-015-0441-1

[3] Rosegrant MW, Ringler C, Zhu T. Water for agriculture: maintaining food security under growing scarcity. Annual Review of Environment and Resources. 2009;**34**:205-222. DOI: 10.1146/annurev.environ.030308.090351

[4] Pimentel D, Berger B, Filiberto D, Newton M, Wolfe B, Karabinakis E, et al. Water resources: agricultural and environmental issues. BioScience. 2004;**54**:909-918. DOI: 10.1641/0006-3568(2004)054[0909:WRAAEI]2.0.CO;2

[5] Shortle JS, Abler DG, Ribando M. Agriculture and water quality. In: Environmental Policies for Agricultural Pollution Control. Oxon, NY: CABI Publishing; 2001. pp. 1-8

[6] Akashe MM, Pawade UV, Nikam AV. Classification of pesticides: A review. International Journal of Research in Ayurveda and Pharmacy. 2018;**9**:144-150

[7] Zacharia JT. Identity, physical and chemical properties of pesticides. In: Pesticides in the Modern World-Trends in Pesticides Analysis. Rijeka: InTech; 2011. pp. 3-20

[8] Knowles A. Recent developments of safer formulations of agrochemicals. The Environmentalist. 2008;**28**:35-44

[9] De Oliveira-Filho EC, Lopes RM, Paumgartten FJ. Comparative study on the susceptibility of freshwater species to copper-based pesticides. Chemosphere. 2004;**56**:369-374. DOI: 10.1016/j.chemosphere.2004.04.026

[10] Zhang W, Jiang F, Ou J. Global pesticide consumption and pollution: with China as a focus. Proceedings of the international academy of ecology and environmental sciences. 2011;**1**:125-144

[11] Sharma A, Shukla A, Attri K, Kumar M, Kumar P, Suttee A, et al. Global trends in pesticides: A looming threat and viable alternatives. Ecotoxicology and Environmental Safety. 2020;**201**:110812. DOI: 10.1016/j.ecoenv.2020.110812

[12] Bhardwaj T, Sharma JP. Impact of pesticides application in agricultural industry: An Indian scenario. International Journal of Agriculture and Food Science Technology. 2013;**4**:817-822

[13] FAO WHO. Manual on Development and Use of FAO and WHO Specifications for Pesticides, Third Revision. FAO Plant Production and Protection Paper. 2016;228

[14] Suarez-Lopez JR, Nazeeh N, Kayser G, Suárez-Torres J, Checkoway H, López-Paredes D, et al. Residential proximity to greenhouse crops and pesticide exposure (via acetylcholinesterase activity) assessed from childhood through adolescence. Environmental Research. 2020;**188**:1-8. DOI: 10.1016/j.envres.2020.109728

[15] Lorenz S, Rasmussen JJ, Süß A, Kalettka T, Golla B, Horney P, et al. Specifics and challenges of assessing exposure and effects of pesticides in

small water bodies. Hydrobiologia. 2017;**793**:213-224

[16] Cohen SZ, Wauchope RD, Klein AW, Eadsforth CV, Graney RL. Pesticides report 35. Offsite transport of pesticides in water: Mathematical models of pesticide leaching and runoff (Technical Report). Pure and Applied Chemistry. 1995;**67**:2109-2148

[17] Lushchak VI, Matviishyn TM, Husak VV, Storey JM, Storey KB. Pesticide toxicity: A mechanistic approach. EXCLI Journal. 2018;**17**:1101-1136. DOI: 10.17179%2Fexcli2018-1710

[18] Steffens K, Larsbo M, Moeys J, Jarvis N, Lewan E. Predicting pesticide leaching under climate change: Importance of model structure and parameter uncertainty. Agriculture, Ecosystems and Environment. 2013;**172**:24-34. DOI: 10.1016/j.agee.2013.03.018

[19] El-Nahhal Y, Nir S, Margulies L, Rubin B. Reduction of photodegradation and volatilization of herbicides in organo-clay formulations. Applied Clay Science. 1999;**14**:105-119. DOI: 10.1016/S0169-1317(98)00053-2

[20] Pimentel D, Levitan L. Pesticides: Amounts applied and amounts reaching pests. BioScience. 1986;**36**:86-91. DOI: 10.2307/1310108

[21] McKnight US, Rasmussen JJ, Kronvang B, Binning PJ, Bjerg PL. Sources, occurrence and predicted aquatic impact of legacy and contemporary pesticides in streams. Environmental Pollution. 2015;**200**:64-76. DOI: 10.1016/j.envpol.2015.02.015

[22] Tanabe A, Kawata K. Daily variation of pesticides in surface water of a small river flowing through paddy field area. Bulletin of Environmental Contamination and Toxicology. 2009;**82**:705-710. DOI: 10.1007/s00128-009-9695-7

[23] Kaushik CP, Sharma HR, Kaushik A. Organochlorine pesticide residues in drinking water in the rural areas of Haryana, India. Environmental Monitoring and Assessment. 2012;**184**:103-112. DOI: 10.1007/s10661-011-1950-9

[24] Adeyemi D, Anyakora C, Ukpo G, Adedayo A, Darko G. Evaluation of the levels of organochlorine pesticide residues in water samples of Lagos Lagoon using solid phase extraction method. Journal of Environmental Chemistry and Ecotoxicology. 2011;**3**:160-166. DOI: 10.5897/JECE.9000023

[25] Chowdhury AZ, Jahan SA, Islam MN, Moniruzzaman M, Alam MK, Zaman MA, et al. Occurrence of organophosphorus and carbamate pesticide residues in surface water samples from the Rangpur district of Bangladesh. Bulletin of Environmental Contamination and Toxicology. 2012;**89**:202-207. DOI: 10.1007/s00128-012-0641-8

[26] Kafilzadeh F. Assessment of organochlorine pesticide residues in water, sediments and fish from Lake Tashk, Iran. Achievements in the Life Sciences. 2015;**9**:107-111. DOI: 10.1016/j.als.2015.12.003

[27] Kafle BK, Pokhrel B, Shrestha S, Raut R, Dahal BM. Determination of pesticide residues in water and soil samples from Ansikhola watershed, Kavre, Nepal. International Journal of Geology, Earth & Environmental Sciences. 2015;**5**:119-127

[28] Hasanuzzaman M, Rahman MA, Salam MA. Identification and

quantification of pesticide residues in water samples of Dhamrai Upazila, Bangladesh. Applied Water Science. 2017;7:2681-2688. DOI: 10.1007/s13201-016-0485-1

[29] Deknock A, De Troyer N, Houbraken M, Dominguez-Granda L, Nolivos I, Van Echelpoel W, et al. Distribution of agricultural pesticides in the freshwater environment of the Guayas river basin (Ecuador). The Science of the Total Environment. 2019;**646**:996-1008. DOI: 10.1016/j.scitotenv.2018.07.185

[30] Nag SK, Saha K, Bandopadhyay S, Ghosh A, Mukherjee M, Raut A, et al. Status of pesticide residues in water, sediment, and fishes of Chilika Lake. India. Environmental Monitoring and Assessment. 2020;**192**:1

[31] Zhou Y, Wu J, Wang B, Duan L, Zhang Y, Zhao W, et al. Occurrence, source and ecotoxicological risk assessment of pesticides in surface water of Wujin District (northwest of Taihu Lake), China. Environmental Pollution. 2020;**265**:114953. DOI: 10.1016/j.envpol.2020.114953

[32] Elfikrie N, Ho YB, Zaidon SZ, Juahir H, Tan ES. Occurrence of pesticides in surface water, pesticides removal efficiency in drinking water treatment plant and potential health risk to consumers in Tengi River Basin, Malaysia. Science of the Total Environment. 2020;**712**:136540. DOI: 10.1016/j.scitotenv.2020.136540

[33] Meffe R, de Bustamante I. Emerging organic contaminants in surface water and groundwater: A first overview of the situation in Italy. The Science of the Total Environment. 2014;**481**:280-295. DOI: 10.1016/j.scitotenv.2014.02.053

[34] Aktar MW, Sengupta D, Chowdhury A. Impact of pesticides use in agriculture: their benefits and hazards. Interdisciplinary Toxicology. 2009;**2**:1-12. DOI: 10.2478/v10102-009-0001-7

[35] Liu J, Qi S, Yao J, Yang D, Xing X, Liu H, et al. Contamination characteristics of organochlorine pesticides in multimatrix sampling of the Hanjiang River Basin, Southeast China. Chemosphere. 2016;**163**:35-43. DOI: 10.1016/j. chemosphere.2016.07.040

[36] Li B, Qu C, Bi J. Identification of trace organic pollutants in drinking water and the associated human health risks in Jiangsu Province, China. Bulletin of Environmental Contamination and Toxicology. 2012;**88**:880-884. DOI: 10.1007/s00128-012-0619-6

[37] Zhou R, Zhu L, Chen Y. Levels and source of organochlorine pesticides in surface waters of Qiantang River, China. Environmental Monitoring and Assessment. 2008;**136**:277-287. DOI: 10.1007/s10661-007-9683-5

[38] Bulut S, Erdogmus SF, Konuk M, Cemek M. The organochlorine pesticide residues in the drinking waters of Afyonkarahisar, Turkey. Ekoloji. 2010;**19**:24-31

[39] Fatoki OS, Awofolu OR. Levels of organochlorine pesticide residues in marine-, surface-, ground-and drinking waters from the Eastern Cape Province of South Africa. Journal of Environmental Science and Health, Part B. 2004;**39**:101-114. DOI: 10.1081/PFC-120027442

[40] Díaz G, Ortiz R, Schettino B, Vega S, Gutiérrez R. Organochlorine pesticides residues in bottled drinking water from Mexico City. Bulletin of Environmental Contamination and Toxicology. 2009;**82**:701-704. DOI: 10.1007/s00128-009-9687-7

[41] Navarrete IA, Tee KA, Unson JR, Hallare AV. Organochlorine pesticide

residues in surface water and groundwater along Pampanga River, Philippines. Environmental Monitoring and Assessment. 2018;**190**:1-8. DOI: 10.1007/s10661-018-6680-9

[42] Eitzer BD, Chevalier A. Landscape care pesticide residues in residential drinking water wells. Bulletin of Environmental Contamination and Toxicology. 1999;**62**:420-427

[43] McManus SL, Richards KG, Grant J, Mannix A, Coxon CE. Pesticide occurrence in groundwater and the physical characteristics in association with these detections in Ireland. Environmental Monitoring and Assessment. 2014;**186**:7819-7836. DOI: 10.1007/s10661-014-3970-8

[44] Yang X, Wang S, Bian Y, Chen F, Yu G, Gu C, et al. Dicofol application resulted in high DDTs residue in cotton fields from northern Jiangsu province, China. Journal of Hazardous Materials. 2008;**150**:92-98. DOI: 10.1016/j.jhazmat.2007.04.076

[45] Hoh E, Hites RA. Sources of toxaphene and other organochlorine pesticides in North America as determined by air measurements and potential source contribution function analyses. Environmental Science & Technology. 2004;**38**:4187-4194. DOI: 10.1021/es0499290

[46] Khuman SN, Vinod PG, Bharat G, Kumar YM, Chakraborty P. Spatial distribution and compositional profiles of organochlorine pesticides in the surface soil from the agricultural, coastal and backwater transects along the south-west coast of India. Chemosphere. 2020;**254**:126699 DOI: 10.1016/j.chemosphere.2020.126699

[47] Bakore N, John PJ, Bhatnagar P. Organochlorine pesticide residues in wheat and drinking water samples from Jaipur, Rajasthan, India. Environmental Monitoring and Assessment. 2004;**98**:381-389. DOI: 10.1023/B:EMAS.0000038197.76047.83

[48] Agarwal A, Prajapati R, Singh OP, Raza SK, Thakur LK. Pesticide residue in water—A challenging task in India. Environmental Monitoring and Assessment. 2015;**187**:1-21. DOI: 10.1007/s10661-015-4287-y

[49] Yadav IC, Devi NL, Syed JH, Cheng Z, Li J, Zhang G, et al. Current status of persistent organic pesticides residues in air, water, and soil, and their possible effect on neighboring countries: A comprehensive review of India. The Science of the Total Environment. 2015;**511**:123-137. DOI: 10.1016/j.scitotenv.2014.12.041

[50] Umulisa V, Kalisa D, Skutlarek D, Reichert B. First evaluation of DDT (dichlorodiphenyltrichloroethane) residues and other Persistence Organic Pollutants in soils of Rwanda: Nyabarongo urban versus rural wetlands. Ecotoxicology and Environmental Safety. 2020;**197**:110574. DOI: 10.1016/j.ecoenv.2020.110574

[51] Ukalska-Jaruga A, Bejger R, Ćwieląg-Piasecka I, Weber J, Jamroz E, Debicka M, et al. Interaction of soil humin fraction with pesticides—A review. EGU2020. 2020;**9**. DOI: 10.5194/egusphere-egu2020-4609

[52] Bhandari G, Atreya K, Scheepers PT, Geissen V. Concentration and distribution of pesticide residues in soil: Non-dietary human health risk assessment. Chemosphere. 2020;**253**:126594

[53] Hearon SE, Wang M, Phillips TD. Strong adsorption of dieldrin by parent and processed montmorillonite clays. Environmental Toxicology and

Chemistry. 2020;**39**:517-525. DOI: 10.1002/etc.4642

[54] Jeon JW, Kim CS, Kim L, Lee SE, Kim HJ, Lee CH, et al. Distribution and diastereoisomeric profiles of hexabromocyclododecanes in air, water, soil, and sediment samples in South Korea: Application of an optimized analytical method. Ecotoxicology and Environmental Safety. 2019;**181**:321-329. DOI: 10.1016/j.ecoenv.2019.06.015

[55] Snow DD, Chakraborty P, Uralbekov B, Satybaldiev B, Sallach JB, Hampton LT, et al. Legacy and current pesticide residues in Syr Darya, Kazakhstan: Contamination status, seasonal variation and preliminary ecological risk assessment. Water Research. 2020;**184**:116141. DOI: 10.1016/j.watres.2020.116141

[56] Zhang H, Dorr GJ, Hewitt AJ. Retention and efficacy of ultra-low volume pesticide applications on Culex quinquefasciatus (Diptera: Culicidae). Environmental Science and Pollution Research. 2015;**22**:16492-16501. DOI: 10.1007/s11356-015-5480-9

[57] Ccanccapa A, Masiá A, Andreu V, Picó Y. Spatio-temporal patterns of pesticide residues in the Turia and Júcar Rivers (Spain). The Science of the Total Environment. 2016;**540**:200-210. DOI: 10.1016/j.scitotenv.2015.06.063

[58] Albuquerque AF, Ribeiro JS, Kummrow F, Nogueira AJ, Montagner CC, Umbuzeiro GA. Pesticides in Brazilian freshwaters: A critical review. Environmental Science: Processes & Impacts. 2016;**18**:779-787. DOI: 10.1039/c6em00268d

[59] Woudneh MB, Ou Z, Sekela M, Tuominen T, Gledhill M. Pesticide multiresidues in waters of the lower Fraser valley, British Columbia, Canada. Part I. Surface Water. Journal of Environmental Quality. 2009;**38**:940-947. DOI: 10.2134/jeq2007.0524

[60] Lockridge O, Verdier L, Schopfer LM. Half-life of chlorpyrifos oxon and other organophosphorus esters in aqueous solution. Chemico-Biological Interactions. 2019;**311**:108788. DOI: 10.1016/j.cbi.2019.108788

[61] Pehkonen SO, Zhang Q. The degradation of organophosphorus pesticides in natural waters: A critical review. Critical Reviews in Environmental Science and Technology. 2002;**32**:17-72. DOI: 10.1080/10643380290813444

[62] Alvarenga N, Birolli WG, Seleghim MH, Porto AL. Biodegradation of methyl parathion by whole cells of marine-derived fungi Aspergillus sydowii and Penicillium decaturense. Chemosphere. 2014;**117**:47-52. DOI: 10.1016/j.chemosphere.2014.05.069

[63] Kunstadter P, Prapamontol T, Sirirojn BO, Sontirat A, Tansuhaj A, Khamboonruang C. Pesticide exposures among Hmong farmers in Thailand. International Journal of Occupational and Environmental Health. 2001;7:313-325. DOI: 10.1179/107735201800339227

[64] Ruggieri F, D'Archivio AA, Fanelli M, Mazzeo P, Paoletti E. A multi-lysimeter investigation on the mobility and persistence of pesticides in the loam soil of the Fucino Plain (Italy). Journal of Environmental Monitoring. 2008;**10**:747-752. DOI: 10.1039/b800559a

[65] Hulbert D, Jamil RZ, Isaacs R, Vandervoort C, Erhardt S, Wise J. Leaching of insecticides used in blueberry production and their toxicity to red worm. Chemosphere. 2020;**241**:125091. DOI: 10.1016/j.chemosphere.2019.125091

[66] Willis GH, Smith S, McDowell LL, Southwick LM. Carbaryl washoff from soybean plants. Archives of Environmental Contamination and Toxicology. 1996;**31**:239-243. DOI: 10.1007/BF00212372

[67] Wolfe NL, Zepp RG, Paris DF. Carbaryl, propham and chlorpropham: a comparison of the rates of hydrolysis and photolysis with the rate of biolysis. Water Research. 1978;**12**:565-571. DOI: 10.1016/0043-1354(78)90134-3

[68] Derbalah A, Sunday M, Kato R, Takeda K, Sakugawa H. Photoformation of reactive oxygen species and their potential to degrade highly toxic carbaryl and methomyl in river water. Chemosphere. 2020;**244**:125464. DOI: 10.1016/j.chemosphere.2019.125464

[69] Birolli WG, Alvarenga N, Seleghim MH, Porto AL. Biodegradation of the pyrethroid pesticide esfenvalerate by marine-derived fungi. Marine Biotechnology. 2016;**18**:511-520. DOI: 10.1007/s10126-016-9710-z

[70] Affum AO, Acquaah SO, Osae SD, Kwaansa-Ansah EE. Distribution and risk assessment of banned and other current-use pesticides in surface and groundwaters consumed in an agricultural catchment dominated by cocoa crops in the Ankobra Basin, Ghana. Science of The Total Environment. 2018;**633**:630-640. DOI: 10.1016/j.scitotenv.2018.03.129

[71] Lehmann E, Turrero N, Kolia M, Konaté Y, De Alencastro LF. Dietary risk assessment of pesticides from vegetables and drinking water in gardening areas in Burkina Faso. The Science of the Total Environment. 2017;**601**:1208-1216. DOI: 10.1016/j.scitotenv.2017.05.285

[72] Mahai G, Wan Y, Xia W, Wang A, Shi L, Qian X, et al. A nationwide study of occurrence and exposure assessment of neonicotinoid insecticides and their metabolites in drinking water of China. Water Research. 2021;**189**:116630. DOI: 10.1016/j.watres.2020.116630

[73] Rafique N, Tariq SR. Photodegradation of α-cypermethrin in soil in the presence of trace metals ($Cu^{2+}$, $Cd^{2+}$, $Fe^{2+}$ and $Zn^{2+}$). Environmental Science: Processes & Impacts. 2015;**17**:166-176. DOI: 10.1039/C4EM00439F

[74] Taylor M, Lyons SM, Davie-Martin CL, Geoghegan TS, Hageman KJ. Understanding trends in pesticide volatilization from agricultural fields using the pesticide loss via volatilization model. Environmental Science & Technology. 2019;**54**:2202-2209. DOI: 10.1021/acs.est.9b04762

[75] Palma P, Kuster M, Alvarenga P, Palma VL, Fernandes RM, Soares AM, et al. Risk assessment of representative and priority pesticides, in surface water of the Alqueva reservoir (South of Portugal) using on-line solid phase extraction-liquid chromatography-tandem mass spectrometry. Environment International. 2009;**35**:545-551. DOI: 10.1016/j. envint.2008.09.015

[76] Dores EF, Carbo L, Ribeiro ML, De-Lamonica-Freire EM. Pesticide levels in ground and surface waters of primavera do leste region, mato grosso, Brazil. Journal of Chromatographic Science. 2008;**46**:585-590. DOI: 10.1093/chromsci/46.7.585

[77] Van Toan P, Sebesvari Z, Bläsing M, Rosendahl I, Renaud FG. Pesticide management and their residues in sediments and surface and drinking water in the Mekong Delta. Vietnam. Science of the Total Environment. 2013;**452**:28-39. DOI: 10.1093/chromsci/46.7.585

[78] Wan Y, Tran TM, Nguyen VT, Wang A, Wang J, Kannan K. Neonicotinoids, fipronil, chlorpyrifos, carbendazim, chlorotriazines, chlorophenoxy herbicides, bentazon, and selected pesticide transformation products in surface water and drinking water from northern Vietnam. Science of the Total Environment. 2021;**750**:141507. DOI: 10.1016/j.scitotenv.2020.141507

[79] Postle JK, Rheineck BD, Allen PE, Baldock JO, Cook CJ, Zogbaum R, et al. Chloroacetanilide herbicide metabolites in Wisconsin groundwater: 2001 survey results. Environmental Science & Technology. 2004;**38**:5339-5343. DOI: 10.1021/es040399h

[80] Shi L, Jiang Y, Wan Y, Huang J, Meng Q, He Z, et al. Occurrence of the insecticide fipronil and its degradates in indoor dust from South, Central, and North China. The Science of the Total Environment. 2020;**741**:140110. DOI: 10.1016/j. scitotenv.2020.140110

[81] Stehle S, Bline A, Bub S, Petschick LL, Wolfram J, Schulz R. Aquatic pesticide exposure in the U.S. as a result of non-agricultural uses. Environment International. 2019;**133**:105234. DOI: 10.1016/j.envint. 2019.105234

[82] Rendón-von Osten J, Dzul-Caamal R. Glyphosate residues in groundwater, drinking water and urine of subsistence farmers from intensive agriculture localities: a survey in Hopelchén, Campeche, Mexico. International Journal of Environmental Research and Public Health. 2017;**14**:595. DOI: 10.3390/ ijerph14060595

[83] Li X, Tian T, Shang X, Zhang R, Xie H, Wang X, et al. Occurrence and health risks of organic micro-pollutants and metals in groundwater of Chinese rural areas. Environmental Health Perspectives. 2020b;**128**:107010. DOI: 10.1289/EHP6483

[84] El-Nahhal I, El-Nahhal Y. Pesticide residues in drinking water, their potential risk to human health and removal options. Journal of Environmental Management. 2021;**299**:113611. DOI: 10.1016/j.jenvman.2021.113611

[85] Yang KJ, Lee J, Park HL. Organophosphate pesticide exposure and breast cancer risk: a rapid review of human, animal, and cell-based studies. International Journal of Environmental Research and Public Health. 2020;**17**:5030. DOI: 10.3390/ijerph17145030

[86] Pardo LA, Beane Freeman LE, Lerro CC, Andreotti G, Hofmann JN, Parks CG, et al. Pesticide exposure and risk of aggressive prostate cancer among private pesticide applicators. Environ. Health: a global access science source. 2020;**19**:30. DOI: 10.1186/ s12940-020-00583-0

[87] Ewence A, Brescia S, Johnson I, Rumsby PC. An approach to the identification and regulation of endocrine disrupting pesticides. Food and Chemical Toxicology. 2015;**78**:214-220. DOI: 10.1016/j.fct.2015.01.011

[88] Pouchieu C, Piel C, Carles C, Gruber A, Helmer C, Tual S, et al. Pesticide use in agriculture and Parkinson's disease in the AGRICAN cohort study. International Journal of Epidemiology. 2018;**47**:299-310. DOI: 10.1093/ije/dyx225

[89] Frasco MF, Fournier D, Carvalho F, Guilhermino L. Cholinesterase from the common prawn (Palaemon serratus) eyes: Catalytic properties and sensitivity to organophosphate and carbamate compounds. Aquatic Toxicology. 2006;**77**:412-421. DOI: 10.1016/j. aquatox.2006.01.011

[90] Eddleston M, Phillips MR. Self poisoning with pesticides. BMJ. 2004;**328**:42-44. DOI: 10.1136/bmj.328.7430.42

[91] Gunnell D, Eddleston M, Phillips MR, Konradsen F. The global distribution of fatal pesticide self-poisoning: systematic review. BMC Public Health. 2007;**7**:1-5

[92] Lapertot M, Pulgarín C, Fernández-Ibáñez P, Maldonado MI, Pérez-Estrada L, Oller I, et al. Enhancing biodegradability of priority substances (pesticides) by solar photo-Fenton. Water Research. 2006;**40**:1086-1094. DOI: 10.1016/j.watres.2006.01.002

[93] Lagadec AJ, Miller DJ, Lilke AV, Hawthorne SB. Pilot-scale subcritical water remediation of polycyclic aromatic hydrocarbon-and pesticide-contaminated soil. Environmental Science & Technology. 2000;**34**:1542-1548. DOI: 10.1021/es990722u

[94] Masselon C, Krier G, Muller JF, Nélieu S, Einhorn J. Laser desorption Fourier transform ion cyclotron resonance mass spectrometry of selected pesticides extracted on C 18 silica solid-phase extraction membranes. The Analyst. 1996;**121**:1429-1433

[95] Aungpradit T, Sutthivaiyakit P, Martens D, Sutthivaiyakit S, Kettrup AA. Photocatalytic degradation of triazophos in aqueous titanium dioxide suspension: identification of intermediates and degradation pathways. Journal of Hazardous Materials. 2007;**146**: 204-213

[96] Mahalakshmi M, Arabindoo B, Palanichamy M, Murugesan V. Photocatalytic degradation of carbofuran using semiconductor oxides. Journal of Hazardous Materials. 2007;**143**:240-245. DOI: 10.1016/j.jhazmat.2006.09.008

[97] Ali I, Gupta VK. Advances in water treatment by adsorption technology. Nature Protocols. 2006;**1**:2661-2667

[98] Ahmad AL, Tan LS, Shukor SA. Dimethoate and atrazine retention from aqueous solution by nanofiltration membranes. Journal of Hazardous Materials. 2008;**151**:71-77. DOI: 10.1016/j.jhazmat.2007.05.047

[99] Badawy MI, Ghaly MY, Gad-Allah TA. Advanced oxidation processes for the removal of organophosphorus pesticides from wastewater. Desalination. 2006;**194**:166-175. DOI: 10.1016/j.desal.2005.09.027

[100] Ribeiro AR, Nunes OC, Pereira MF, Silva AM. An overview on the advanced oxidation processes applied for the treatment of water pollutants defined in the recently launched Directive 2013/39/EU. Environment International. 2015;**75**:33-51. DOI: 10.1016/j.envint.2014.10.027

[101] Kitous O, Cheikh A, Lounici H, Grib H, Pauss A, Mameri N. Application of the electrosorption technique to remove Metribuzin pesticide. Journal of Hazardous Materials. 2009;**161**:1035-1039. DOI: 10.1016/j.jhazmat.2008.04.091

[102] Ohnoa K, Minamia T, Matsuia Y, Magara Y. Effects of chlorine on organophosphorus pesticides adsorbed on activated carbon: Desorption and oxon formation. Water Research. 2008;**42**:1753-1759. DOI: 10.1016/j.watres.2007.10.040

[103] López-Ramón MV, Fontecha-Cámara MA, Alvarez-Merino MA, Moreno-Castilla C. Removal of diuron and amitrole from water under static and dynamic conditions using activated carbons in form of fibers, cloth, and grains. Water Research. 2007;**41**:2865-2870. DOI: 10.1016/j.watres.2007.02.059

[104] Qiu Y, Xiao X, Cheng H, Zhou Z, Sheng GD. Influence of environmental factors on pesticide adsorption by black carbon: pH and model dissolved organic matter. Environmental Science & Technology. 2009;**43**:4973-4978. DOI: 10.1021/es900573d

[105] Castro CS, Guerreiro MC, Goncalves M, Oliveir LCA, Anastacio AS. Activated carbon/iron oxide composites for the removal of atrazine from aqueous medium. Journal of Hazardous Materials. 2009;**164**:609-614. DOI: 10.1016/j.jhazmat.2008.08.066

[106] Shao L, Cheng XQ, Liu Y, Quan S, Ma J, Zhao SZ, et al. Newly developed nanofiltration (NF) composite membranes by interfacial polymerization for Safranin O and Aniline blue removal. Journal of Membrane Science. 2013;**430**:96-105. DOI: 10.1016/j.memsci.2012.12.005

[107] Chian ES, Bruce WN, Fang HH. Removal of pesticides by reverse osmosis. Environmental Science & Technology. 1975;**9**:52-59. DOI: 10.1021/es60099a009

[108] Liu Z, Mu Q, Sun Y, Gao P, Yu Y, Gao J, et al. Effective adsorption of chloroanilines from aqueous solution by m-phenylenediamine modified hyper-cross-linked resin: Kinetic, equilibrium, and thermodynamic studies. Colloids and Surfaces A: Physicochemical and Engineering Aspects. 2020;**601**:124996. DOI: 10.1016/j.colsurfa.2020.124996

[109] Oliveira BR, Penetra A, Cardoso VV, Benoliel MJ, Barreto Crespo MT, Samson RA, et al. Biodegradation of pesticides using fungi species found in the aquatic environment. Environmental Science and Pollution Research. 2015;**22**: 11781-11191. DOI: 10.1007/s11356-015-4472-0

[110] Hai FI, Modin O, Yamamoto K, Fukushi K, Nakajima F, Nghiem LD. Pesticide removal by a mixed culture of bacteria and white-rot fungi. Journal of the Taiwan Institute of Chemical Engineers. 2012;**43**:459-462. DOI: 10.1016/j.jtice.2011.11.002

[111] Castro JV Jr, Peralba MC, Ayub MA. Biodegradation of the herbicide glyphosate by filamentous fungi in platform shaker and batch bioreactor. Journal of Environmental Science and Health, Part B. 2007;**42**:883-886. DOI: 10.1080/03601230701623290

[112] Abraham J, Silambarasan S. Biodegradation of chlorpyrifos and its hydrolysis product 3, 5, 6-trichloro-2-pyridinol using a novel bacterium Ochrobactrum sp. JAS2: a proposal of its metabolic pathway. Pesticide Biochemistry and Physiology. 2016;**126**:13-21. DOI: 10.1016/j.pestbp.2015.07.001

[113] Gangola S, Sharma A, Bhatt P, Khati P, Chaudhary P. Presence of esterase and laccase in Bacillus subtilis facilitates biodegradation and detoxification of cypermethrin. Scientific Reports. 2018;**8**:1-11. DOI: 10.1038/s41598-018-31082-5

[114] Khajezadeh M, Abbaszadeh-Goudarzi K, Pourghadamyari H, Kafilzadeh F. A newly isolated Streptomyces rimosus strain capable of degrading deltamethrin as a pesticide in agricultural soil. Journal of Basic Microbiology. 2020;**60**:435-443. DOI: 10.1002/jobm.201900263

[115] Luo X, Zhang D, Zhou X, Du J, Zhang S, Liu Y. Cloning and characterization of a pyrethroid pesticide decomposing esterase gene, Est3385, from Rhodopseudomonas palustris PSB-S. Scientific Reports. 2018;**8**:1-8. DOI: 10.1038/s41598-018-19373-3

[116] Jariyal M, Jindal V, Mandal K, Gupta VK, Singh B. Bioremediation of organophosphorus pesticide phorate in soil by microbial consortia. Ecotoxicology and Environmental Safety. 2018;**159**:310-316. DOI: 10.1016/j.ecoenv.2018.04.063

[117] Erguven GO. Comparison of some soil fungi in bioremediation of herbicide acetochlor under agitated culture media. Bulletin of Environmental Contamination and Toxicology. 2018;**100**:570-575. DOI: 10.1007/s00128-018-2280-1

[118] Wang X, Hou X, Liang S, Lu Z, Hou Z, Zhao X, et al. Biodegradation of fungicide Tebuconazole by Serratia marcescens strain B1 and its application in bioremediation of contaminated soil. International Biodeterioration and Biodegradation. 2018;**127**:185-191. DOI: 10.1016/j.ibiod.2017.12.001

[119] Purnomo AS, Sariwati A, Kamei I. Synergistic interaction of a consortium of the brown-rot fungus Fomitopsis pinicola and the bacterium Ralstonia pickettii for DDT biodegradation. Heliyon. 2020;**6**:e04027. DOI: 10.1016/j.heliyon.2020.e04027